Lecture Notes in Artificial Intelligence   1849

Subseries of Lecture Notes in Computer Science
Edited by J. G. Carbonell and J. Siekmann

Lecture Notes in Computer Science

Edited by G. Goos, J. Hartmanis and J. van Leeuwen

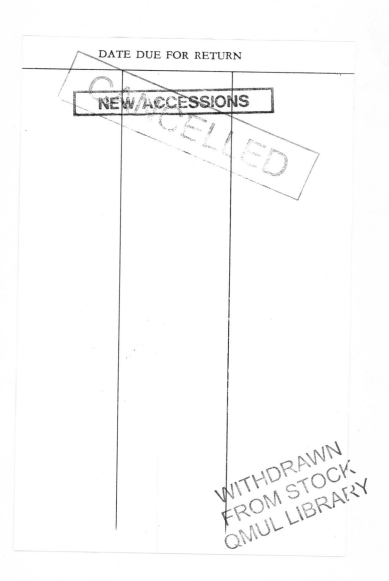

Springer

Berlin
Heidelberg
New York
Barcelona
Hong Kong
London
Milan
Paris
Singapore
Tokyo

Christian Freksa   Wilfried Brauer
Christopher Habel   Karl F. Wender (Eds.)

Spatial Cognition II

Integrating Abstract Theories,
Empirical Studies, Formal Methods,
and Practical Applications

Springer

Series Editors

Jaime G. Carbonell,Carnegie Mellon University, Pittsburgh, PA, USA
Jörg Siekmann, University of Saarland, Saarbrücken, Germany

Volume Editors

Christian Freksa
Christopher Habel
Universität Hamburg, Fachbereich Informatik
Vogt-Kölln-Str. 30, 22527 Hamburg, Germany
E-mail:{freksa/habel}@informatik.uni-hamburg.de

Wilfried Brauer
Technische Universität München, Fakultät für Informatik
80290 München, Germany
E-mail: brauer@informatik.tu-muenchen.de

Karl F. Wender
Universität Trier, FB 1 - Psychologie,
54286 Trier, Germany
E-mail: wender@cogpsy.Uni-Trier.de

Cataloging-in-Publication Data applied for

Die Deutsche Bibliothek - CIP-Einheitsaufnahme

Spatial cognition / Christian Freksa ... (ed.). - Berlin ; Heidelberg ;
New York ; Barcelona ; Hong Kong ; London ; Milan ; Paris ;
Singapore ; Tokyo : Springer
2. Integrating abstract theories, empirical studies, formal methods,
and practical applications. - 2000
  (Lecture notes in computer science ; 1849 : Lecture notes in
  artificial intelligence)
  ISBN 3-540-67584-1

CR Subject Classification (1998): I.2.4, I.2, J.2, J.4, E.1, I.3, I.7

ISBN 3-540-67584-1 Springer-Verlag Berlin Heidelberg New York

Springer-Verlag is a company in the BertelsmannSpringer publishing group.
© Springer-Verlag Berlin Heidelberg 2000
Printed in Germany

Typesetting: Camera-ready by author, data conversion by DA-TeX Gerd Blumenstein
Printed on acid-free paper   SPIN 10722109   06/3142   5 4 3 2 1 0

Preface

Spatial cognition is concerned with the ways humans, animals, or machines think about real or abstract space and also with the ways spatial structures can be used for reasoning. Thus, space is considered both, as an *object* of cognition and as a *means* of cognition. Spatial cognition is an interdisciplinary research area involving approaches from artificial intelligence, cognitive psychology, geography, mathematics, biology, design, theoretical computer science, architecture, and philosophy. Research on spatial cognition has progressed rapidly during the past few years. The disciplines contributing to the field have moved closer together and begin to speak a common language. They have found ways of merging the research results obtained through different approaches. This allows for developing more sophisticated hybrid approaches that overcome intrinsic limitations of the individual disciplines.

Research on spatial cognition has drawn increased attention in recent years for at least three different reasons: (1) basic research dimension: there is a growing awareness of the importance of spatial cognitive abilities in biological systems, specifically with respect to perception and action, to the organization of memory, and to understanding and producing natural language; (2) computational dimension: spatial representations and spatial inference may provide suitable limitations to enhance the computational efficiency for a large and relevant class of problems; (3) application dimension: a good understanding of spatial processes is essential for a wide variety of challenging application areas including Geographic Information Systems (GIS), pedestrian and vehicle navigation aids, autonomous robots, smart graphics, medical surgery, information retrieval, virtual reality, Internet navigation, and human-computer interfaces.

This is the second volume published in the framework of the Spatial Cognition Priority Program. It augments the results presented in Freksa et al. 1998. The interdisciplinary research program (www.spatial-cognition.de) was established by the Deutsche Forschungsgemeinschaft in 1996. It consists of 16 research projects at 13 research institutions throughout Germany. Besides carrying out research in individual projects and joint research between projects, the Spatial Cognition Priority Program organizes 'topical colloquia', partly with international participation. A colloquium on *Types of spatial knowledge* was held in Göttingen in May 1997; a colloquium on *Spatial cognition and soft computing* was held in Hamburg in June 1997; a colloquium on *Qualitative and metric approaches to spatial inference and motion analysis* was held in Berlin in June 1997; a colloquium on *Space and action* was held in Ohlstadt in December 1997; a colloquium on *Route and survey knowledge* was held in Bremen in February 1998; a colloquium on the *Representation of motion* was held in Munich in October 1998; a colloquium on *Spatial inference* was held in Freiburg in February 1999; a colloquium on *Systems of reference for spatial knowledge* was held in Hamburg in April 1999; a colloquium on *Spatial cognition in real and virtual environments* was held in Tübingen in April 1999; and a colloquium on *Maps and diagrammatic representations of the environment* was held in Hamburg in August 1999.

The volume contains 28 articles and is structured into five sections: The section *Maps and diagrams* consists to a large extent of contributions to the international colloquium on *Maps and diagrammatic representations of the environment*. The work presented in the section *Motion and spatial reference* was discussed at the colloquium on the *Representation of motion* or at the colloquium on *Systems of reference for spatial knowledge*. The section *Spatial relations and spatial inference* draws largely from contributions to the colloquium on *Spatial inference*. Most of the work published in the section *Navigation in real and virtual systems* was first presented at the international colloquium on *Spatial cognition in real and virtual environments*. Some of the work in the section *Spatial memory* had been discussed at the colloquium on *Route and survey knowledge*. All contributions underwent a thorough reviewing procedure. The articles reflect the increased cooperation among the researchers in the area of spatial cognition.

We thank all authors for their careful work and for observing our tight deadlines and formatting conventions. We thank the reviewers of the contributions for their insightful and thorough comments and suggestions for improvement. We thank Thora Tenbrink, Karin Colsman, and Thitsady Kamphavong for their superb editorial support. We thank Alfred Hofmann, Antje Endemann and Anna Kramer of Springer-Verlag for the pleasant cooperation. We gratefully acknowledge the guidance by Andreas Engelke and the support of the Deutsche Forschungsgemeinschaft. We thank the reviewers of the DFG spatial cognition priority program for their valuable advice.

April 2000   Christian Freksa
                                                         Wilfried Brauer
                                                      Christopher Habel
                                                         Karl F. Wender

Commentators and Reviewers

Related Book Publications

Freksa, C. and Habel, C., Hrsg., *Repräsentation und Verarbeitung räumlichen Wissens,* Informatik-Fachberichte 245, Springer, Berlin 1990.

Mark, D.M., Frank, A.U., eds., *Cognitive and Linguistic Aspects of Geographic Space*, 361-372, Kluwer, Dordrecht 1991.

Frank, A.U., Campari, I., and Formentini, U., eds., *Theories and Methods of Spatio-Temporal Reasoning in Geographic Space*, Lecture Notes in Computer Science 639, 162-178, Springer, Berlin 1992.

Frank, A.U. and Campari, I., eds. *Spatial Information Theory: A Theoretical Basis for GIS*, Lecture Notes in Computer Science 716, Springer, Berlin 1993.

Frank, A.U. and Kuhn, W., eds. *Spatial Information Theory: A Theoretical Basis for GIS*, Lecture Notes in Computer Science 988, Springer, Berlin 1995.

Burrough, P., Frank, A. *Geographic Objects with Indeterminate Boundaries,* Taylor and Francis, London 1996.

Hirtle, S.C. and Frank, A.U., eds. *Spatial Information Theory: A Theoretical Basis for GIS*, Lecture Notes in Computer Science 1329, Springer, Berlin 1997.

Egenhofer, M.J. and Golledge, R.G., eds. *Spatial and Temporal Reasoning in Geographic Information Systems.* Oxford University Press, Oxford 1997.

Freksa, C., Habel, C., and Wender, K.F., eds. *Spatial Cognition.* Lecture Notes in Artificial Intelligence 1404, Springer, Berlin 1998.

Freksa, C., Mark, D.M., eds., *Spatial Information Theory. Cognitive and Computational Foundations of Geographic Information Science.* Lecture Notes in Computer Science 1661, Springer, Berlin 1999.

Table of Contents

Maps and Diagrams

Cognitive Zoom: From Object to Path and Back Again 1
Carol Strohecker

Monitoring Change: Characteristics of Dynamic Geo-spatial Phenomena
for Visual Exploration 16
Connie Blok

The Use of Maps, Images and "Gestures" for Navigation 31
Stephen C. Hirtle

Schematizing Maps: Simplification of Geographic Shape
by Discrete Curve Evolution 41
Thomas Barkowsky, Longin Jan Latecki and Kai-Florian Richter

Schematic Maps as Wayfinding Aids 54
*Hernan Casakin, Thomas Barkowsky, Alexander Klippel
and Christian Freksa*

Some Ways that Maps and Diagrams Communicate 72
Barbara Tversky

Spatial Communication with Maps: Defining the Correctness
of Maps Using a Multi-Agent Simulation 80
Andrew U. Frank

Schematic Maps for Robot Navigation 100
Christian Freksa, Reinhard Moratz and Thomas Barkowsky

Motion and Spatial Reference

From Motion Observation to Qualitative Motion Representation 115
*Alexandra Musto, Klaus Stein, Andreas Eisenkolb, Thomas Röfer,
Wilfried Brauer and Kerstin Schill*

Lexical Specifications of Paths 127
Carola Eschenbach, Ladina Tschander, Christopher Habel and Lars Kulik

Visual Processing and Representation of Spatio-temporal Patterns 145
*Andreas Eisenkolb, Kerstin Schill, Florian Röhrbein, Volker Baier,
Alexandra Musto and Wilfried Brauer*

Orienting and Reorienting in Egocentric Mental Models 157
Robin Hörnig, Klaus Eyferth and Holger Gärtner

Investigating Spatial Reference Systems through Distortions
in Visual Memory 169
Steffen Werner and Thomas Schmidt

Spatial Relations and Spatial Inference

Towards Cognitive Adequacy of Topological Spatial Relations 184
Jochen Renz, Reinhold Rauh and Markus Knauff

Interactive Layout Generation with a Diagrammatic Constraint Language . 198
Christoph Schlieder and Cornelius Hagen

Inference and Visualization of Spatial Relations 212
Sylvia Wiebrock, Lars Wittenburg, Ute Schmid and Fritz Wysotzki

A Topological Calculus for Cartographic Entities 225
Amar Isli, Lledó Museros Cabedo, Thomas Barkowsky and Reinhard Moratz

The Influence of Linear Shapes on Solving Interval-Based
Configuration Problems 239
Reinhold Rauh and Lars Kulik

Navigation in Real and Virtual Spaces

Transfer of Spatial Knowledge from Virtual to Real Environments 253
Patrick Péruch, Loïc Belingard and Catherine Thinus-Blanc

Coarse Qualitative Descriptions in Robot Navigation 265
*Rolf Müller, Thomas Röfer, Axel Lankenau, Alexandra Musto,
Klaus Stein and Andreas Eisenkolb*

Oblique Angled Intersections and Barriers: Navigating through
a Virtual Maze 277
Gabriele Janzen, Theo Herrmann, Steffi Katz and Karin Schweizer

Modelling Navigational Knowledge by Route Graphs 295
Steffen Werner, Bernd Krieg-Brückner and Theo Herrmann

Using Realistic Virtual Environments in the Study of Spatial Encoding ... 317
Chris Christou and Heinrich H.Bülthoff

Navigating Overlapping Virtual Worlds: Arriving in One Place
and Finding that You're Somewhere Else 333
Roy A. Ruddle

Spatial Memory

Influences of Context on Memory for Routes 348
Sabine Schumacher, Karl Friedrich Wender and Rainer Rothkegel

Preparing a Cup of Tea and Writing a Letter: Do Script-Based Actions
Influence the Representation of a Real Environment? 363
Monika Wagener, Silvia Mecklenbräuker, Werner Wippich,
Jörg E. Saathoff and André Melzer

Action Related Determinants of Spatial Coding in Perception
and Memory 387
Bernhard Hommel and Lothar Knuf

Investigation of Age and Sex Effects in Spatial Cognitions as Assessed
in a Locomotor Maze and in a 2-D Computer Maze 399
Bernd Leplow, Doris Höll, Lingju Zeng and Maximilian Mehdorn

Author Index 419

Cognitive Zoom: From Object to Path and Back Again

Carol Strohecker

MERL - Mitsubishi Electric Research Laboratory,
201 Broadway, Cambridge, MA 02139, USA
stro@merl.com

Abstract. This paper posits the usefulness of mental shifts of scale and perspective in thinking and communicating about spatial relations, and describes two experimental techniques for researching such cognitive activities. The first example involves mentally expanding a hand-sized piece of entangled string, a knot, so that following a portion of the string into a crossing resembles the act of walking along a path and over a bridge. The second example involves transforming experience and conceptions of the large-scale environment to small-scale representations through the act of mapmaking, and then translating the map to depictions of street-level views. When used in the context of clinical research methodologies, these techniques can help to elicit multimodal expressions of conceived topological relationships and geographical detail, with particular attention to individual differences.

1 Introduction

As advancements in computer graphics and communication technologies enable realization of geographic information systems (GIS), researchers are identifying needs to better understand how people perceive, represent, and communicate geographical information in their processes of constructing geographical knowledge. Researchers take varying approaches when applying results of cognitive studies to technology development and when using technologies as supports for cognitive studies. Additionally, some researchers seek to identify and formalize processes of human reasoning for embodiment within a computational system, while other researchers envision computational systems as information-supply and decision-support tools. In the latter views the human intelligence resides in the person using the machine rather than the machine itself, and the emphasis for system development is on the nature and range of functionality and its presentation in the human interface. The varying purposes lead to varying modes of inquiry, analysis, and interpretation.

Spatial cognition researchers employ different methods for different settings (such as the environment or the laboratory), and different methods for studying different aspects of spatial phenomena (Spencer et al. 1989). Generally, however, researchers aim to elicit and describe human perception and cognition of space. The descriptions may be verbal, pictorial, numerical, or logical. Nevertheless, patterns that emerge through the descriptions lead to theorizing and to further questions for research.

Ch. Freksa et al. (Eds.): Spatial Cognition II, LNAI 1849, pp. 1-15, 2000.
© Springer-Verlag Berlin Heidelberg 2000

Among spatial cognition researchers interested in GIS, high-priority research agendas include conceptualizations of geographical detail, understandings of dynamic phenomena and representations, and roles of multiple modalities for thinking about space and spatial relations (Mark et al. 1999).

The terms "geographic information" and "geographic knowledge" imply cognition of large-scale spaces. At this point our understanding of this sort of cognition is insufficient to say whether the same or similar processes may pertain to cognition of space at other scales:

> It is important to recognize the distinction between geographical space and space at other scales or sizes. Palm-top and table-top spaces are small enough to be seen from a single point, and typically are populated with manipulative objects, many of which are made by humans. In contrast, geographical or large-scale spaces are generally too large to be perceived all at once, but can best be thought of as being transperceptual…, experienced only by integration of perceptual experiences over space and time through memory and reasoning, or through the use of small-scale models such as maps. Some of our discussions of geographical cognition might not apply to spatial cognition at other scales (ibid., p. 748).

However, members of an earlier survey characterized research in spatial cognition in terms of a developmental connection between "fundamental concepts of space," implicitly at the scale of the human body, and conceptions of environments at geographic scales:

> Researchers in spatial cognition focus on two kinds of conceptual growth: …development of fundamental concepts of space, and the further differentiation and elaboration of these concepts into the development and representation of large-scale environments (Hart and Moore 1973, p. 248).

More work is needed to ascertain the presence and nature of such varying concepts and to understand whatever relations may exist between them.

There is a wide range of detail at the geographic scale. Just as it includes differences in degree as profound as those between cities and countries, the human scale ranges from palm-size and table-size to distances reachable through the course of a comfortable walk.[1] How common-sense notions of space may relate to faculties for spatial cognition at other scales remains a question. In fact, researchers interested in such phenomena must address a host of questions, such as whether certain spatial processes emerge at specific scales, and what sensory forms most appropriately represent conceptions at different scales (Mark et al. 1999, p. 761). A question of particular concern here is how changes of scale may influence granularity or clarity of data.

This paper posits the usefulness of mental shifts of scale and perspective in thinking and communicating about spatial relations, and describes two examples of experimental approaches for researching such cognitive activities. The first example involves mentally expanding a hand-sized piece of entangled string, a knot, so that following a portion of the string into a crossing resembles the act of walking along a path and over a bridge. These degrees of detail are arguably within the same scale,

---

[1] (Montello 1993) reviews and elaborates classification schemes for such variances.

that of the human body. The second example involves transforming conceptions of the large-scale environment to small-scale representations through the act of mapmaking, and then translating the map to depictions of street-level views. These degrees of detail concern different spatial scales, the geographic and that of the human body. The example also involves detailed translations within the pictorial scale.

An important distinction pertains to both of these examples, having to do with correspondences and variances between experiential and conceptual activity. Researchers usually discuss spatial scale with respect to the external world: depending on the classification scheme, the geographic scale may include the planet, countries, or cities; the environmental scale may include cities, town squares, or rooms in buildings; the human scale may include town squares, rooms, or table-tops; and so on. Conceptions and representations of the world may or may not maintain correspondences to these particulars. Here we explore shifts of scale that pertain in different ways to the external world and to internal worlds. Much of the conceptual activity involves changing perspectives and frames of reference.

The knot and mapmaking examples described in this paper stem from questions about how spatial thinking develops, and how understandings of basic spatial relations may be elaborated or otherwise changed through developmental processes. For now, the purpose of the two techniques is simply to elicit expressions of changing conceptions, for description and further study. The ant-crawling and mapmaking techniques are compatible with anthropological and clinical approaches to gathering data and a microgenetic approach to analyzing it (Berg and Smith 1985, Turkle 1984). The concern is with documenting richness and variety in people's ways of thinking and learning about space and spatial relationships (Turkle and Papert 1990). Thus the approach is to employ qualitative research methods, conduct longitudinal studies, and employ or develop experimental situations that will tend to evoke expressions of spatial conceptions.

These expressions take many forms: spoken description, including gestures and specific terminologies; written or typed descriptions and communications; hand-drawn sketches and diagrams; and constructed representations in various media (such as string and computer-based maps). Such representations become forms of data, supplementing the researcher's notes and audio and/or video recordings of discussions, work sessions, and open-ended interviews. All of these data are subject to protocol and microgenetic analysis (Ericsson and Simon 1984, Jacob 1987). The methods are appropriate to elucidation of higher-level conceptual structures rather than lower-level neurological structures.

2   Conceptual Elements

Researchers across the range of scales in spatial cognition are concerned with the notion of conceptual elements. Their vocabularies differ somewhat; the elements may have to do with representations of objects in the environment and/or to relations between them.

> Mental representations of geographical information seem to be constructed from elements, such as roads, landmarks, cities, land masses, the spatial relations among them, and the spatial relations of them to the larger units encompassing them (Mark et al. 1999, p. 757).

Kevin Lynch, an urban designer interested in "the possible connection between psychology and the urban environment" (Lynch 1984), articulated a version of this view by describing five elements of the "city image," a form of mental imagery with which people remember aspects of the places they inhabit (Lynch 1960).[2] City images are highly individual but consistently include the general features of districts, edges, paths, nodes, and landmarks. These elements pertain to the structure of large-scale, built environments. Other aspects of urban experience, such as senses of sociability and aesthetics, are present in the data that Lynch and his colleagues collected but omitted from the focus on general structures. Lynch described the structures as existing both in the external environment and in the subjects' conceptions of the environment. Inevitably, the internal world is more complicated than such a pristine characterization allows, however, and dilemmas arose as urban designers attempted to invert Lynch's analytic process by using the sparse set of conceptual structures as a starting point in design. He later addressed this practical problem, reminding readers that the structural elements represent only part of urban experience and emphasizing the importance of including urban dwellers in design of the places they inhabit (Lynch 1984).

At the human scale, developmental psychologists have identified elementary spatial relationships that form the basis of growing conceptions (Piaget and Inhelder 1967).[3] These elements include relations of surrounding, proximity, separation, and order. Studies of knot-tying and understandings of knots have focused on articulating notions of surrounding and order (ibid., Strohecker 1991, Strohecker 1996a). These relationships take the forms of intertwining and of relative positions that demonstrate the property of between-ness. While simple knots concern relations of surrounding, more complex knots also involve combinations of forms, sequencing of production steps, and grouping types of productions. Thus thinking about complex knots becomes more broadly mathematical in this qualitative sense (Beth and Piaget 1966, Piaget and Inhelder 1967, Strohecker 1991).

---

[2] The studies took place "at a time when most psychologists – at least, those in the field of perception – preferred controlled experiments in the laboratory to the wandering variables of the complicated, real environment. We hoped to tempt some of them out into the light of day" (Lynch 1984). The research established new ground leading to formation of the field of cognitive anthropology.

[3] It is important to clarify which aspects of Piaget's voluminous work apply and which do not. His early notion of stage theory, which he all but abandoned in later writings, is not relevant to this discussion, nor are his notions of assimilation and accommodation. Rather, of interest are his techniques for activity-based experimental design and the clinical method with which Piaget and his colleagues were able to respond promptly and deeply to subjects' productions and reports, forming new hypotheses within the course of an interview and ultimately collecting data beyond the relatively surface levels of attitudes and beliefs (Berg and Smith 1985, Piaget 1951, Turkle 1984). These methods continue to be useful to post-Piagetians who focus on individual differences and the importance of social context in learning (Turkle and Papert 1990, Harel and Papert 1991).

Lynch's elements represent static environmental features, yet he and his colleagues focused on wayfinding in eliciting subjects' verbal, pictorial, and action-based descriptions of mental images. Had they not been so focused on notions of external structure, their data may have yielded formulations of more dynamic and/or relational properties (Lynch 1984, Strohecker 1999).[4] Just as Lynch focused on activity within the environment, and on people's translations of actions to conceptions, Piagetians considered conceptual elements as internalized actions (Gruber and Vonèche 1977). The developmental psychologists' elements are inherently process-based and pertain primarily to relations between spatial elements. In particular, relationships of surrounding are seen as mental analogs of the action of surrounding one thing by another.

3   Knots as Pathways

Even simple knots can seem complicated, and learning to tie them can be challenging. The actions that generate a knot can be as fundamentally different as entwining separate pieces of string in modular fashion, convoluting two ends of a single piece of string, or leading one end of a string around, over, and under relatively fixed portions of the same strand. These methods typically require the use of two hands, but some talented tyers can single-handedly maneuver a string. Some can even toss a string into the air so that it descends into the form of a knot.

There are thousands of different knots, but each one demonstrates some general properties, understanding of which can help in learning to recognize and form new knots. In a study involving twenty-two children, aged ten to fourteen, subjects demonstrated a variety of ways of perceiving, describing, remembering, and effecting the relations of surrounding that characterize particular knots and "families" of knots (Strohecker 1996b). The children came from diverse cultural and socioeconomic backgrounds. They met in a workshop-style setting that grew in response to their interests and creations during the course of three months. Four work groups, consisting of about five children each, met once a week for hour-long sessions in which they worked with string as well as books and videos about knots.[5]

Three important elements of the research occurred through these free-form working sessions: the sessions formed a period of culture-building and of immersion in thinking about knots, so that the culminating interviews fit within a context that all the children shared; the working sessions, in their own right, generated data on thinking about knots; and in the course of the working sessions, many of the children built up a relationship with the researcher that came to involve comfort and trust, which facilitated the "think aloud" nature of the relatively structured final interviews that generated the most comprehensive set of data. By the time of these interviews, the children had become familiar with ways of thinking and talking about knots. I asked the children to differentiate between two similar knots and to arrange a

---

[4] Indeed, the researchers were unable to completely extract this important aspect of urban experience: nodes are not merely static; as decision points, they pertain to the flow of human activity (Lynch 1960).

[5] For particulars, see (Strohecker 1991).

collection of about ten knots into groups according to whatever organizational scheme they thought appropriate. Here I focus on an aspect of these discussions which proved useful as children attempted to differentiate between two similar knots (Strohecker 1996a).

The knots known as the square knot and the thief knot seem identical when regarded locally. When seen in their entirety, however, it becomes clear that the knots are different:

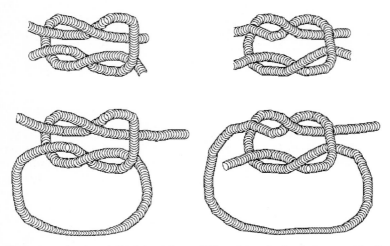

Fig. 1. The square knot and thief knot have different topologies but seem identical when viewed locally

Children employed various ways of differentiating between these two knots. Some considered the configurations of the ends of the strings. However, they focused on the ends in different ways and used different terms even when focusing in the same ways: some children considered the positions of the ends relative to the entangled portion of the knot, while other children considered the end positions relative to the non-entangled loop. The entanglement was usually at the top of the odd-looking object we called a "knot," but several children experimented with inverting and/or flipping the knots through the course of discussion. These moves augmented the vocabularies for describing differences between the configurations. Other children supplemented their vocabularies by comparing the two knots with other similar though different knots, such as the granny knot and the sheet bend.

Many of the children went beyond comparisons of the static objects, tying the square knot and the thief knot while describing the tying processes and then summarizing with a description of the finished knot. These descriptions tended to be richer in expressing details of the knots' topologies: rather than relying on simple terms and gestures to capture differences in positions of the ends ("in," "out," "up," "down," "across," "straight," "above," "inside"), describing a tying process and its outcome requires specifying the "overs" and "unders" of the crossings that form the knot, and understanding their sequential relationships. Tying the square and thief knots also calls attention to a fundamental distinction from the outset: although both knots can be produced by entangling either one or two ends of a string, as the tyer may prefer, it is much easier to produce the thief knot by winding one end around itself, in snake-like fashion.

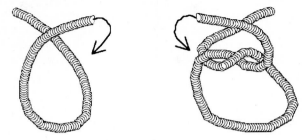

Fig. 2. A version of the usual method of producing a square knot

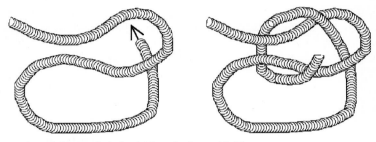

Fig. 3. A version of the usual method of producing a thief knot

Another effective technique in eliciting detailed expressions of conceptions of topological relationships in complex intertwinements is to prompt a shift of scale. Regarding a knot at its usual palm-top scale, children (and adults) often see a hopelessly entangled piece of string. However, they respond readily to the idea of imagining oneself as a tiny ant crawling along the string (Piaget and Inhelder 1967, Strohecker 1991). This mental shift of scale involves both space and time: the knot effectively enlarges as the observer assumes the new perspective, and subjects tend to dwell on each crossing for study and description.

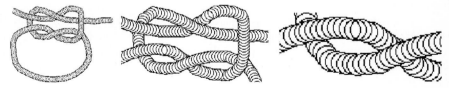

Fig. 4. These illustrations are suggestive of the scale shift that subjects manage as they imagine themselves to be a "tiny ant" walking along a knot. However, it is important to remember that these shifts are mental, not external or pictorial. The thinking is also an active process, involving change over time as the "ant" moves along the string

Thus the shift helps people to focus on details of the knot's topology, to make sense of the sequence of over/under relationships that form the knot, and to notice other relevant relations of surrounding and proximity. The shift has the effect of transforming the experience of regarding the knot to that of a leisurely walk, invoking conceptions, language, and multimodal representations appropriate to that scale. For example, people engaging in this thought experiment sometimes refer to over-crossings as "bridges" and under-crossings as "underpasses."

Wayfinding is usually defined as finding a route between two points. The crawling-ant technique is a simpler exercise involving path-following by maintaining focus on a single, though moving, point. In this way it is more comparable to "route planning":

> Route planning is modelled as a process of finding the overall direction and topology of the path, then filling in the details by deciding how to go around barriers (Mark et al. 1999, p. 758).[6]

Nevertheless, ant-crawling, route planning, and wayfinding are relevant to the same scale, that of a comfortable human walk, and the ant technique is effective in eliciting multimodal representations that describe the basic spatial relations. Subjects express these conceptions through speech, including word choice and varying intonations; hand gestures; coarser body language and movements; drawings; and written descriptions, including metaphors and narratives (Strohecker 1991).

4   Pathways as Transformers of Scale and Perspective

"WayMaker" is a graphical software tool with which researchers can prompt similar shifts of scale, along with changes of perspective and frame of reference (Strohecker and Barros 2000). Representations of Lynch's elements become facilities with which subjects construct map-like images of the structure of urban environments. In so doing they invoke conceptions at the geographic scale and translate them for diagrammatic representation. The resulting constructions are overviews of the large-scale domain, consisting of representations that may be abstract symbols (triangles for landmarks, lines for paths, etc.) or depictions of details of particular places (substituting towers for triangles, textured paths for lines, etc.).

Not surprisingly, subjects vary in their strategies for placing these elements relative to one another. Some people form a "big picture" by creating a collection of districts and then placing smaller elements within them, while other people concentrate on a network of pathways; some people pay particular attention to landmarks, and so on. Thus the subject creates an individualized frame of reference that can act as a conceptual anchor when comparing the map to additional, shifted-scale representations of the environment. These shifts become apparent in WayMaker's second mode.

A miniature version of the map remains visible while the larger version disappears, as it is replaced by displays of street-level scenes along a chosen pathway. These scenes appear one after the other, like a flip-card animation, illustrating rather than simulating the visual experience of a virtual world.

---

[6] Indeed, "barrier" and "obstacle" are among the terms subjects used to describe under-crossings.

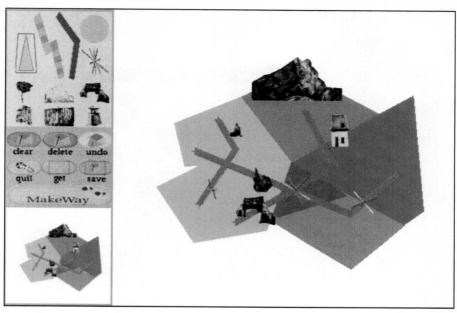

Fig. 5. The WayMaker construction screen presents representations of Lynch's "elements of the city image" (upper left). Users position chosen representations with respect to one another in order to form a map of a large-scale environment (at right, echoed in the miniature at lower left)

The represented scale becomes that of a stroll through various districts. Details are consistent with the subject's placement and characterization of the structural elements. They appear along the pathways and in the distance according to their relative placements on the map, moving forward and appearing around bends in the path to approximate perspective in the street-level views. Thus the second mode of the software maintains spatial relationships among the elements but effects shifts of representation, view, and scale.

The manner of representation for the scenes is impressionistic and painterly in the current version of the software. The program composes scenes with images segmented from paintings by Paul Cézanne. We interpreted a similarity of purpose between our effort and Cézanne's: many of his paintings represent studies of environmental structure and of relationships between built and natural environments (Machotka 1996). Additionally, we hoped that the appeal of this distinctive imagery would survive the scene-generation technique, which is fairly crude in this early prototype of the mapping tool. Though our approach to scene structure is different from Cézanne's, WayMaker users often express enjoyment of the visual results, and this helps to sustain their continued experimentation with the prototype. In subsequent versions we hope to replace the Cézanne database with images drawn, photographed, or otherwise generated by the users themselves.

Fig. 6. WayMaker transforms the map to a series of street-level views along pathways through the environment. A moving dot on the miniature version of the map locates each view within the user's construction. The scenes are impressionistic but maintain spatial relationships among the districts, edges, paths, nodes, and landmarks

Fig. 7. The program composites scenes automatically, drawing from a database of images that consists of ground, back, and sky planes, as well as building bases, facades, and rooftops

Fig. 8. The program displays scenes along a pathway in the manner of a frame-by-frame animation

The subject can go easily back and forth between the construction and view modes. Thus there is an interesting combination of represented scales: the represented walk-scale of the views accompanies the diagrammatic scale of the map, which stems from the geographic scale (both external and represented) of the mapped environment. The geographic scale is both external and represented because the subject creates the map, an activity that involves drawing on direct experience of large-scale environments, conceptualizing features of such environments, and representing them diagrammatically. Although the subject composes the map as a computer-screen sized city layout, it is echoed in a smaller version that provides a basis for comparison with the street-level views. Thus there are arguably two diagrammatic scales for the map.

All of these represented levels are within the pictorial (palm-top/table-top) scale, yet the subject's act of creating the map differentiates use of WayMaker from mere map use.

> Maps represent environmental and geographic spaces, but are themselves instances of pictorial space! ...the psychology of map use ... draw[s] directly on the psychology of pictorial space rather than on the psychology of environmental space (Montello 1993).

The psychology of map creation, however, draws on the psychology of environmental space. As map producers, WayMaker users must consider space at the environmental scale in order to compose a map for dynamic transformation. This exercise is comparable to that of other mapmakers who work with elemental representations, as for production of typical subway maps. We anticipate that WayMaker users' narrations and other expressions of expected and sensed experiences of the shifts of scale, view, and representation will be deepened by the high degree of control afforded by the software: beyond constructing their own maps, subjects can interact in a variety of ways.

A red dot on the miniature map pinpoints the location of a given street view. The subject can click somewhere else on the miniature to define a route automatically for frame-by-frame animation. The program then calculates a route to this second point along the network of pathways, choosing turns when necessary based on which would constitute the shortest route. In the resulting dynamic display, the dot illustrates these programmatic decisions by moving along the paths, locating each new view within the environment. This manner of interaction is similar to path following. The subject may decide instead to click somewhere directly on a chosen path, so that the program does not need to make navigational decisions. This manner of interaction is more like the two-point technique of wayfinding. Alternatively, the subject can reposition the dot manually somewhere else along a pathway, to maximize control over the movement and effect a kind of fast-forward of the corresponding views.

Most importantly, the range, nature, and degree of interactivity encourage the subject to control shifts of his or her own mental states, invoking conceptualizations of environments, environmental features, and relations between environmental features formed through experience at environmental scales but actively represented and manipulated at smaller scales.

WayMaker extends Lynch's formulations by setting the environmental elements within a dynamic context, more consistent with the experience of a city and perhaps with the nature of its mental representations (Lynch 1984, Strohecker 1999). The program supports multiple perspectives and brings map construction and map reading together with cognition of direct environmental experience. Nevertheless, as a research tool, the program should be supplemented by other tools and techniques, just as Lynch's cohort combined multimodal representations and walk-and-talk personal interviews in their explorations of the city image.[7]

5   Further Work

These studies do not answer the question of whether the same conceptual structures are present in thinking about small-scale spaces and large-scale spaces. However, the ant-crawling technique highlights the facility with which people can shift from one

---

[7] See (Strohecker 1999) for a discussion of the importance of participatory design contexts (Schuler and Namioka 1993, Henderson et al. 1998) as amelioration for WayMaker users' inversion of Lynch's analytic process. While it seemed that this inversion contributed to problems in urban design, Lynch contended that the real problem was the designers' omission of city dwellers from the design process. Understanding the dwellers' conceptions of their environments motivated his effort, and city dwellers' input in urban design was meant to return the richness of human experience to the relatively sparse results of the analytic process. Lynch contended that inhabitants of a city should be involved in its design. Similarly, extending WayMaker use beyond the prototype should involve participants in contexts shaped by the environmental structure design tool. Such modes might include neighborhood planning groups, multiuser virtual environments, and other domains that situate the tool use within appropriate social contexts.

conceptual level to the other. In so doing they may use structures at one level to illuminate or boost understandings at another. Further work is needed in order to document such strategies and examine relationships between spatial structures at different scales. Combined with qualitative research techniques, WayMaker may help in elucidating various forms of articulation of such conceptual elements.

Much of this sort of research is descriptive, and it is important not to generalize too quickly from responses to specific representations and depictions. A subject's difficulty of understanding may stem from a poorly interpreted image rather than inadequacy of some cognitive faculty, and poor interpretation of an image may result from bad design or an incomplete design process rather than some incapacity on the part of the viewer. Good design results from and depends on a cyclical, empirical process in which the designer generates ideas for various solutions, tests them with various people, and uses the results to inform subsequent solutions (Schön 1983). Interactive software applications pose particular, complex design problems pertaining to representation, sequencing of actions, and so on (DiBiase et al. 1992, Winograd 1996). Design of graphical displays and diagrammatic reasoning currently have low research priority (Mark et al. 1999), but may prove increasingly important as more geographic information systems emerge and as researchers seek to ensure that data guiding their design are sufficiently robust for this sort of generalization.

WayMaker needs further development in order to best support research in spatial cognition, though even in its current state it could help in generating data suggestive of research directions relevant to scale and other issues. The most important improvement may be development of new databases for the composited views, consisting of images created by the subjects themselves. Such personalization could help subjects get more deeply involved with the interactions, supporting the research potential to understand individual differences in the associated thinking (Papert 1980, Harel and Papert 1991, Kafai and Resnick 1996). Researchers focusing on varying levels and conditions of cognition recognize the significance of individual differences (Hutchins 1983, Kosslyn et al. 1984, Levinson 1996, Mark et al. 1999, Montello 1993, Turkle and Papert 1990). The population of subjects should vary across age, gender, and culture.

Acknowledgments

For collaboration in creating and developing the WayMaker software I thank Barbara Barros, Adrienne Slaughter, Daniel Gilman, and Maribeth Back; for usage trials, as well as discussions of the tool's design and its potential usefulness as a design tool, I thank students at the Harvard University Graduate School of Design; and for recent discussions of the tool's use in research on spatial cognition I thank Christian Freksa, Gabriela Goldschmidt, and other participants in the Visual and Spatial Reasoning in Design conference (MIT 1999) and the COSIT workshop (Hamburg 1999). I am grateful to the reviewers whose comments helped to improve this paper. MERL supports the work.

References

Berg DN, Smith KK (1985) Exploring Clinical Methods for Social Research. Sage, Beverly Hills

Beth EW, Piaget J (1966) Mays W (trans) Mathematical Epistemology and Psychology. Reidel, Dordrecht (1966)

DiBiase D, MacEachren A, Krygier J, Reeves C (1992) Animation and the role of map design in scientific visualization. Cartography and Geographic Information Systems 19, pp 201-214.

Ericsson KA, Simon HA (1984) Protocol Analysis: Verbal Reports as Data. MIT Press: Cambridge, MA

Gruber HE, Vonèche JJ (eds, 1977) The Essential Piaget. Basic Books, New York

Harel I, Papert S (eds, 1991) Constructionism. Ablex, Norwood NJ

Hart RA, Moore GT (1973) The development of spatial cognition: A review. In Stea B, Downs R (eds) Image and Environment, pp 226-234. University of Chicago Press, Chicago

Henderson RH, Kuhn S, Muller M (1998) Proceedings of the Participatory Design Conference: Broadening Participation. ACM Press, New York

Hutchins E (1983) Understanding Micronesian navigation. In Gentner D, Stevens AL (eds) Mental Models, pp 191-226. Lawrence Erlbaum, Hillsdale NJ

Jacob E (1987) Qualitative Research Traditions: A Review. Review of Educational Research 57:1, pp 1-50

Kafai Y, Resnick M (eds) 1996 Constructionism in Practice: Designing, Thinking, and Learning in a Digital World. Lawrence Erlbaum, Mahwah NJ

Kosslyn SM, Brunn J, Cave KR, Wallach RW (1984) Individual differences in mental imagery: A computational analysis. In Pinker S (ed) Visual Cognition, pp 195-243. MIT Press, Cambridge MA

Levinson SC (1996) Frames of reference and Molyneaux's question. In Bloom P, Peterson MA, Nadel L, Garrett MF (eds) Language and Space, pp 109-170. MIT Press, Cambridge MA

Lynch K (1960) The Image of the City. MIT Press, Cambridge MA

Lynch K (1984) Reconsidering the image of the city. In Rodwin, L. and Hollister, R, M. (eds), Cities of the Mind: Images and Themes of the City in the Social Sciences, pp 151-161. Plenum Press, New York

Machotka P (1996) Cézanne: Landscape into Art. Yale University Press, New Haven

Mark DM, Freksa C, Hirtle SC, Lloyd R, Tversky B (1999) Cognitive models of geographical space. Int. J. Geographical Information Science 13:8, pp 747-774

Montello DR (1993) Scale and multiple psychologies of space. Proceedings of COSIT'93, Spatial Information Theory: A Theoretical Basis for GIS, pp 312-321. Springer-Verlag, Berlin

Papert S (1980) Mindstorms: Children, Computers, and Powerful Ideas. Basic Books, New York

Piaget J (1951 [1929]). Tomlinson J and Tomlinson A (trans) The Child's Conception of the World. Humanities Press, New York

Piaget J, Inhelder B (1967 [1956, 1948]) Langdon FJ, Lunzer JL (trans) The Child's Conception of Space. Norton, New York

Schön D (1983) The Reflective Practitioner. Basic Books, New York

Schuler D, Namioka A (eds, 1993) Participatory Design: Principles and Practices. Lawrence Erlbaum, Hillsdale, NJ

Spencer C, Blades M, Morsley K (1989) The Child in the Physical Environment: The Development of Spatial Knowledge and Cognition. John Wiley & Sons, Chichester

Strohecker C (1991) Why knot? Ph.D. diss., Massachusetts Institute of Technology, Media Laboratory, Epistemology and Learning Group

Strohecker C (1996a) Understanding topological relationships through comparisons of similar knots. AI&Society: Learning with Artifacts 10, pp 58-69

Strohecker C (1996b) Learning about topology and learning about learning. Proceedings of the Second International Conference on the Learning Sciences, Association for the Advancement of Computing in Education

Strohecker C (1999) Toward a developmental image of the city: Design through visual, spatial, and mathematical reasoning. Proceedings of the International Conference on Visual and Spatial Reasoning in Design, pp 33-50. Key Centre of Design Computing and Cognition, University of Sydney

Strohecker C, Barros B (2000 [1997]) Make way for WayMaker. Presence: Teleoperators and Virtual Environments 9:1

Turkle S (1984) The Second Self: Computers and the Human Spirit. Simon and Schuster, New York

Turkle S, Papert S (1990) Epistemological pluralism: Styles and voices within the computer culture. Signs 16(1)

Winograd T (1996) Bringing Design to Software. ACM Press, New York

Monitoring Change:
Characteristics of Dynamic Geo-spatial Phenomena
for Visual Exploration[1]

Connie Blok

ITC, Geoinformatics, Cartography and Visualisation Division, P.O.Box 6, 7500 AA
Enschede, the Netherlands
blok@itc.nl
http://www.itc.nl/~carto

Abstract. In the context of a research about application of dynamic visualisation parameters in animations for monitoring purposes, this paper reports on characteristics of dynamic geo-spatial phenomena for visual exploration. A framework of concepts is proposed to describe a variety of phenomena in the physical environment. The nature of monitoring tasks is taken into account. Explicit attention is paid to context-sensitivity and to the description of the temporal behaviour of dynamic phenomena. If it proves to be a suitable framework, visualisation variables will be linked to the general concepts to form a theoretical basis for animated representation of dynamic geo-spatial phenomena. Empirical testing of animations will follow to evaluate whether dynamic visualisation variables can be used to prompt thinking about the processes at work and about spatio-temporal relationships by domain experts who perform monitoring tasks. The ultimate goal is to contribute to the development of representation methods and interaction tools.

1   Introduction

Monitoring of geographic phenomena starts with the acquisition of data about spatial dynamics, followed by exploration and analysis of the spatio-temporal data in order to gain insights in processes and relationships. Insights may enable scientists to warn or act otherwise in case of undesired developments, like hazards. It may also lead to the generation of models of reality which are, for example, used to estimate the effects of interference versus no interference in a development, or to extrapolation of trends in order to predict future developments. Almost all geographic phenomena are dynamic

---

[1] This is the revised and updated version of a paper previously published as: Blok, C.A.: Monitoring of Spatio-temporal Changes: Characteristics of Dynamics for Visual Exploration. In: Keller, C.P. (ed.): Proceedings 19th International Cartographic Conference, Ottawa, Canada, Vol. 1. Canadian Institute of Geomatics, Ottawa (1999) 699–709.

Ch. Freksa et al. (Eds.): Spatial Cognition II, LNAI 1849, pp. 16-30, 2000.

and many geoscientists are, for various monitoring applications, interested in tracking changes that are directly or indirectly visible in the landscape (see e.g. publications in journals such as *Environmental Monitoring and Assessment* and *International Journal of Remote Sensing*). The research described here refers mainly to dynamic geo-spatial phenomena in the physical environment, such as vegetation, erosion, atmospheric processes, etc., because it is directed to environmental applications. Dynamics related to social and socio-economic phenomena, like the movement of people, are not explicitly taken into account.

Domain specialists who are involved in the monitoring of geographic phenomena have *questions* that are related to the geometric, thematic and temporal characteristics of data in order to learn more about [8], [10], [24]:

- the occurrence of anomalies: deviations from 'normal' values may require immediate action;
- ongoing processes;
- relationships in space, between themes and/or in time;
- causes;
- trends.

In a GIS environment, *answers* to questions that are relevant for monitoring are commonly sought by using analytical and computational functions. It means that a query has to be defined, which leads to some kind of processing of the data [15], [24]. The computed results can be graphically represented, for example in a map. Alternatively, it is possible to use graphic representations as a *starting point* for analysis and exploration of the data. In that case, geographic visualisation is applied, a process which is commonly considered as making geographic data visible in a particular use context: visual exploration. Visual exploration is characterised by highly interactive and private use of graphic representations of predominantly unknown data [16]. It is being assumed that visualisation (a process which generates not only graphic, but also cognitive representations of reality) stimulates thought, and may reveal patterns and relationships that might otherwise remain hidden, e.g. see [6], [19], [18]. Offering different perspectives to the data, dynamics and interaction possibilities play an important role in this context [17].

In the current GIS practice, answers to questions are sought through data analysis in which either the spatial aspects of dynamic phenomena are emphasised by comparison of a (limited) number of snap shots, or the temporal aspects of objects or pixels are for further analysis represented outside the spatial context (e.g. in a graph). These approaches lead to respectively spatial or temporal reasoning, but *spatio-temporal reasoning*, taking the behaviour of a phenomenon in space *and* time into account, is not easy in these ways (figure 1). Combined analysis will be easier in a temporal GIS environment (TGIS), but apart from a few prototypes, e.g. [21], a fully operational TGIS does not yet exist [5]. Animated visualisation of dynamic phenomena might also facilitate spatio-temporal reasoning.

In order to investigate the behaviour of dynamic geo-spatial phenomena, it is not enough to analyse or visually explore the 'footprints' of dynamics in the spatial context. Information on temporal characteristics is required as well, such as the order and duration of the various stages in a development, the rate of change, etc. Availability of this information (hence, the possibility to reason with spatio-temporal data) depends, amongst others, on factors related to data acquisition, like the moments in time, the frequency, scale and resolution.

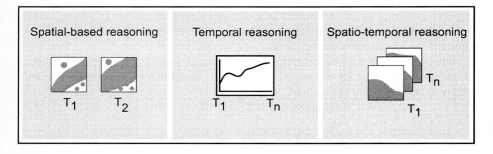

Fig. 1. Types of reasoning with dynamic geo-spatial data

Acquisition decisions are often not based on the occurrence of change, but on other (e.g. economic) grounds (figure 2). However, change can be considered as indicating ongoing developments. In the current context *change* is defined as: variations in time in the geometric, thematic or temporal characteristics of geographic phenomena, or in a combination of these characteristics.

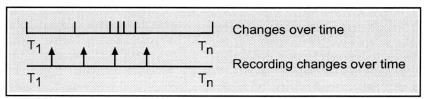

Fig. 2. Data acquisition time is usually not guided by the occurrence of change

Change can be graphically represented for visual exploration. Vasiliev [22] proposed a framework for the graphic representation of spatio-temporal data in static maps. However, in dynamic maps (animations) the temporal characteristics of change in 'world time' can be viewed in the time during which the animation is running, or in 'display time' [13]. Dynamic (and interactive) graphic representation of spatio-temporal data, preferably fully integrated in a TGIS environment, can therefore perhaps add a dimension to the interpretation of geographic data.

A domain expert who uses animations to monitor geographic phenomena can extract information about dynamics from changes occurring in the conventional (graphic) representation variables (such as position, form, colour, size etc.). The question is what role the so-called *dynamic visualisation variables* play in this context. Such variables are, for example, identified by DiBiase et al. [6] and

MacEachren [16]. Examples are the *moment* at which change becomes visible in the display, the *order* and *duration* of the various stages, the *frequency* at which a stage is repeated and *the rate of change,* see also [14].

MacEachren [17] and Köbben &Yaman [12] attempted to link application of the dynamic variables to the measurement levels of the data that are to be represented. This approach is established by Bertin [2] for the graphic variables. Another option is to link the dynamic visualisation variables to dynamic characteristics of the data. In the context of a research about the role of dynamic visualisation variables in animated representations for the monitoring of phenomena in the physical environment, this paper deals with characteristics of dynamic phenomena for representation purposes. The paper reports on a part of the research, of which the ultimate goal is to contribute to the development of visualisation methods and interaction tools that can be used by domain specialists for visual data exploration during the execution of monitoring tasks.

In the next sections, questions and tasks that are relevant for the monitoring of dynamic geo-spatial phenomena in the physical environment are first discussed. Then a framework of concepts to characterise those phenomena is proposed. Finally, some aspects related to the next phase in the research are discussed.

2   Geo-spatial Dynamics and Monitoring

Characterising the dynamic phenomena that are relevant for monitoring by a framework of concepts which describe those phenomena in general terms can be useful, especially if the concepts can be linked to representation variables. Such a framework needs to fulfill a number a prerequisites. The concepts should:

- be applicable to a broad range of phenomena in the physical environment (unique links between individual phenomena and representation variables cannot be expected);
- be stated in common linguistic expressions that can act to trigger domain-specific knowledge about dynamic phenomena;
- be geared to changes that can be *visually* explored in dynamic representations;
- describe changes in the spatial domain; explicitly incorporate the temporal characteristics of changes exhibited in the spatial domain to facilitate the conceptualisation of dynamic phenomena.

It has been described above that domain experts who are involved in monitoring tasks are interested in anomalies, processes, relationships, causes and trends. Insight in recent developments and in longer term dynamics are both important. In the context of visual exploration of data representations, it means that questions can partly be addressed to relatively short sequences of images containing (usually recent) changes, and partly to longer time series in which also less recent images are involved (for convenience further indicated as respectively 'short series' and 'longer series'). A number of basic questions can be inferred from literature. Questions that play a role in visual exploration of short series are: "Is there any change?" And if a known process is going on: "What are the developments?" Discovering new processes and trends, however, will usually require examination of longer series. These basic questions enable *identification* of the geometric, thematic and temporal characteristics of change (figure 3). In addition to identification, *comparison* of characteristics of geographic phenomena is a task which is supported by visual exploration [17]. Comparison can be applied to different phenomena at the same location and time (e.g. vegetation and rainfall), or to one phenomenon at different locations or in different times (e.g. vegetation in two areas, or in two growing seasons). "Are there changes?" and "Are there anomalies?" are basic questions that can be addressed to short series. The last question can also be addressed to longer series. In addition, questions with respect to possible relationships and causes are relevant, see e.g. [8], [10], [24].

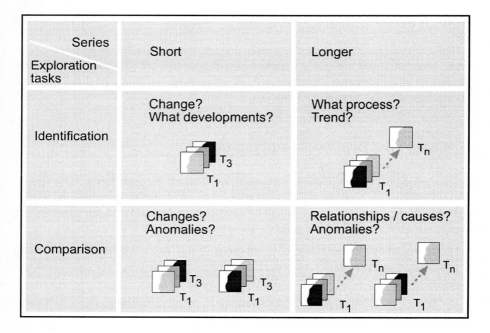

Fig. 3. Monitoring questions in relation to visual exploration tasks

The tasks and the inferred basic questions that play a role in monitoring can be used as a starting point for the selection of descriptive concepts for the framework. Although relevance and completeness of the proposed framework have to be validated in the next research phase (see also the section 'Discussion and Outlook' below), it is postulated that domain experts are interested in the following characteristics of dynamic phenomena:

- the appearance of new phenomena;
- the disappearance of existing phenomena;
- changes in the geometric, thematic and temporal characteristics;

- spatio-temporal behaviour over short series (to track changes and act in case of undesired developments);
- spatio-temporal behaviour over longer series (to gain insights in processes and relationships and to be able to distinguish trends and extrapolate developments into the future).

3 Spatio-temporal Characteristics

Various existing categorizations of spatio-temporal phenomena have been investigated to assess their usefulness for the research described here. However, none of the classification found meets all the prerequisites that have been defined in the previous section. Some relevant examples are described below.

In some classifications categories are distinguished that emphasize changes in the spatial domain; they do not explicitly take the temporal aspects into account. Typical examples of concepts used are appearance/disappearance, movement, expansion/shrinkage, increase/decrease or change in internal/external morphology, e.g. [7].

Hornsby and Egenhofer [11] propose a *change description language* to characterise sequences of object changes. It consists of a detailed description of transition types that either preserve or change the identity of single and composite objects (e.g. create, generate, reincarnate, divide, merge, mix). Domain-specific knowledge will often be required to distinguish between transition types. The language is meant to build formal data models of change, it does not fully describe the changes in *graphic* models (e.g. animations) to be used for monitoring applications. For instance, movement, geometric changes such as boundary shifts and temporal characteristics are not (explicitly) incorporated.

Eschenbach [9] attempts to classify movement based on the spatial structure of objects. The two main classes are movement along trajectories (complete shifts in position) and internal motion (changes in the position of parts of an object). The last category is further subdivided into growth/shrinkage, internal rotation, movement of parts, and a category referring to movement of large bodies that might be along trajectories, but are too short to result in a complete shift of position. The distinction between the last two subcategories is a not very clear, and probably not relevant for monitoring. The author indicates that the categories are not exclusive, some movements are, for instance, trajectory-based with internal motion. This classification also emphasizes changes in the spatial domain.

Yattaw [23] classifies movement by taking the spatial characteristics point, line, area, volume and the temporal characteristics continuous, cyclical, intermittent into account. A matrix of these characteristics results in twelve classes of movement. She also mentions the influence of spatial and temporal scale and of context for the assignment of a particular movement to a class. An advantage is that temporal aspects are explicitly included, although the concepts mainly describe patterns over longer periods. Characteristics like duration and rate of change, relevant for shorter *and* longer periods of time, are not accommodated. Another drawback is that the spatial characteristics distinguished do not adequately describe changes in the spatial domain.

Building on the work of these authors, however, and on parameters of display time in animations [6], [16], a framework of general linguistic expressions is proposed to characterise changes that are relevant for monitoring and that can (at least in theory) be discovered by visual exploration. Four main categories are distinguished in the framework: concepts to describe change in the spatial domain, change in the temporal domain, overall spatio-temporal patterns over longer series, and relative similarity in comparisons (figure 4). The main categories are related to visual exploration tasks and to the focus of attention of the viewer of the representations: recent developments only (short series) or dynamics over longer periods of time (longer series).

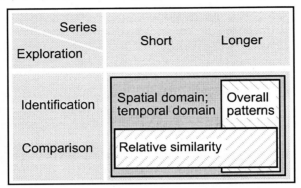

Fig. 4. Main categories of the framework of concepts in relation to visual exploration tasks and the focus of attention: short or longer series

3.1 Change in the Spatial Domain

The basic concepts proposed to describe change in the spatial domain are (figure 5):

- *Appearance/disappearance:* refers to the emergence (the 'birth') of a new phenomenon or the vanishing (the 'death') of an existing one (e.g. a tornado, a volcano, a wild fire, pollution). Changes in the nature of an already existing phenomenon (such as an inactive volcano that becomes active, or the change from forest stand to arable land) are not characterised by appearance/disappearance, but by concepts from the next category.
- *Mutation:* refers to a transformation that affects the thematic attribute component of an existing phenomenon; it does not refer to changes in geometric characteristics. Two subtypes are distinguished.

  - *Mutation at nominal level of measurement:* refers to a change in the nature or character of a phenomenon (e.g. change from rain to snow, from gully to sheet erosion, from forest to burned area, from a dry to a water containing intermittent river).

- *Increase/decrease:* refers to a change at ordinal, interval or ratio level of measurement (e.g. changes in the force of a tornado, the thickness of the cloud cover, the amount of precipitation, the vegetation index).

• *Movement:* refers to a change in the spatial position and/or the geometry of a phenomenon. Again, two subtypes are distinguished.

- *Movement along a trajectory:* refers to a movement by which the whole phenomenon changes its position. A kind of path is followed, hence it can be assumed that movement takes more than a single instant of time, some continuity is involved. Along the path, the geometric characteristics of the phenomenon may change (e.g. tornadoes and pollution carried by running water exhibit movement along a trajectory).
- *Boundary shift*: refers to movement where at least part of the phenomenon maintains its location (e.g. the jet stream; expansion of the area occupied by an existing phenomena, such as a cleared area, an eroded area, an area with a high vegetation index, or a polluted area). These types of movement may either happen at a single instant of time, or take a longer period.

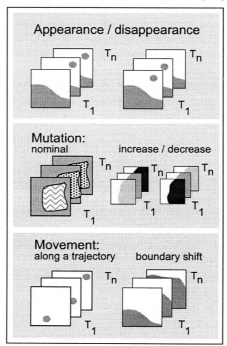

Fig. 5. Basic characteristics of change in the spatial domain

A number of remarks can be made here. Firstly, most dynamic phenomena can be characterised by more than one of the proposed basic concepts for change in the spatial domain. For example, a tornado appears, increases in force, moves along a trajectory, decreases again, and finally disappears. In other words: to be able to describe the behaviour of a phenomenon in space, the concepts often have to be combined.

Secondly, the concepts used to characterise geo-spatial phenomena often vary: *"...the problem of putting a label on a geographic feature is much a matter of context."* [1, pp.73]. That also applies to dynamic phenomena. A number of factors play a role, such as the perspective of the expert or the aspect of a phenomenon being studied [25]. E.g. if the *behaviour* of a wild fire is examined, relevant concepts to characterise it are appearance/disappearance, mutations (changes in intensity) and, most likely, movement along a trajectory. If, however, the wild fire *effect* is examined and the expert is interested in *burned* areas, new areas appear and boundary shifts occur in existing areas. But if the effect is studied in terms of changes in *land use*, it is more like mutations and boundary shifts. Characterising is also influenced by the phenomenon itself. Incidental pollution (e.g. carried by running water or moving air) is characterised by movement along a trajectory, but in case of more or less continuous pollution of water, soil or the atmosphere from a fixed source, boundary shifts occur. Display scale or size of the area represented also influence characterization. At a global scale the changes from cloudy to cloud free skies can be conceived as mutations, but if a small area is represented, clouds appear and disappear. A large scale display of the eruption of a volcano shows boundary shifts in lava streams. In a small scale representation of the volcanic activity in a region, however, movement cannot be expected. The time frame considered is another factor influencing the way in which a phenomenon is characterised. If a short series of changes is considered, clouds exhibit boundary shifts, in longer series movement along trajectories are more likely.

Hence, some basic concepts to characterise change in the spatial domain have been proposed, but use and display contexts seem to influence which (combination of) concepts (is) are most appropriate to describe dynamic phenomena. The relative importance of context sensitivity will be further considered in the next research phase, where various monitoring applications will be investigated to validate the relevance and the completeness of the defined concepts for these applications. Some refinements might be necessary. For instance, it is not clear at this stage whether changes such as splitting and merging need to be distinguished separately, or whether they can be considered as combinations of appearance and disappearance. Also, many dynamic phenomena exhibit almost continuous change of the contours of patterns, in different directions (e.g. cloud cover, vegetation values). For such movement patterns it is not easy to determine whether a phenomenon only changes in form, or in form and size, and perhaps even direction. Therefore, the category 'boundary shift' is not further subdivided, but if useful, another level can be added to the hierarchical categorization of basic changes (e.g. expansion/shrinkage and other geometric changes). Representation variables will only be linked to the framework of concepts after this further investigation into applications.

3.2 Change in the Temporal Domain

To further characterise the behaviour of dynamic geo-spatial phenomena, concepts that describe change in the temporal domain are also required. Proposed are (figure 6):

- *Moment in time*: refers to the instant of time or the time interval at which a change is initiated, it allows the location in time of a change in the spatial domain.
- *Pace*: refers to the rate of change over time. It can be expressed in terms such as 'slow/fast'; or 'at an increasing/decreasing/constant rate of change' [17].
- *Duration*: refers to the length of time involved in a change and/or the time between changes. It can be expressed in absolute or in relative terms (respectively number of time units, and notions such as 'short/long').
- *Sequence*: refers to the order of phases in a series of changes in the spatial domain.
- *Frequency*: refers to the number of times that a phase is repeated in a series of changes in the spatial domain.

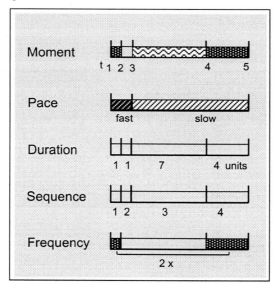

Fig. 6. Characteristics of change in the temporal domain

3.3 Overall Spatio-temporal Patterns over Longer Series

Characterising change over longer series of images requires integration of individual changes into an overall spatio-temporal pattern of change. Hence, some additional concepts are introduced to enable the description of spatio-temporal patterns over longer series (although recognition of those patterns depends on the selected time frame and resolution). The patterns can, if required, be further characterised by concepts proposed for the spatial and temporal domains (see also figure 4). The new concepts may refer to the geometric, thematic or temporal data component, or to any combination of these components. Proposed for monitoring applications are (figure 7):

- *Cycle:* refers to a periodical return to a previous state/condition [20]. Cycles are quite common in the physical environment, e.g. atmospheric processes and changes in vegetation may exhibit cyclic patterns. If a cycle is discovered, developments can perhaps be predicted, although disturbance of usual patterns may always occur.

- *Trend:* refers to a structured, but non-cyclical pattern [20]. It is the general direction, or tendency, of a development over a period of time. Examples are developing spatial clustering (geometric data component), increase in value (thematic data component) and higher frequencies (temporal data component). If a trend can be observed, extrapolation in time may be possible. However, changes in the direction of a development may always occur.

If no cycle or trend can be discovered, the pattern is unstructured. Characterising unstructured patterns does not seem useful, because they are hard to interpret.

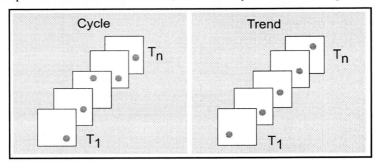

Fig. 7. Overall spatio-temporal patterns over longer series (including recent and less recent change)

3.4 Relative Similarity in Comparisons

Comparison of changes (in space, between themes and in time) is an exploration task at a higher conceptual level than identification. It can, however, only be performed in a meaningful way if the phenomena are also identified. Therefore, the concepts proposed for identification in short and long series above are considered relevant for comparison as well. In addition, characterizing the relative similarity is desired.

For comparison in *short series* (with usually recent images) the following concepts will probably suffice (figure 8):

- *Same/different:* refers to changes that are comparable/incomparable, particularly in the geometric and/or the thematic data component. Same/different observations may point to the occurrence of anomalies and lead to immediate action of a domain expert.

Concepts proposed for the comparison of change over *longer series* (containing recent and less recent images) are:

- *Same/opposite/different:* refers to patterns that show respectively comparable (proportional), inversely proportional and incomparable changes. Same patterns may point to a positive (cor)relation, opposite patterns to a negative one, and different patterns again to anomalies.
- *In phase (synchronous)/phase difference:* refers to the simultaneousness of pattern developments, which might be particularly relevant for same and opposite patterns. If pattern developments start and end at the same time, patterns are in phase, and perhaps somehow correlated. If same or opposite patterns are observed with a time lag, there is a phase difference. Exploration of those pattern developments can still be interesting, because that may point to a causal relationship. For instance, vegetation might develop similar patterns as precipitation, but somewhat later in time, see also [17], [4].

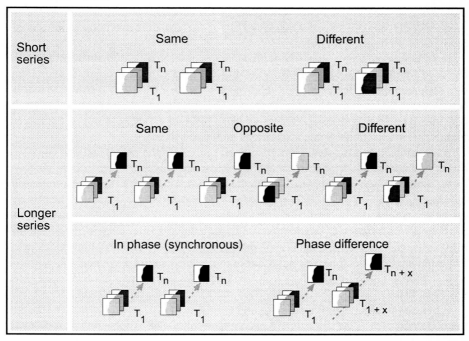

Fig. 8. Comparing spatio-temporal patterns: relative similarity

4 Discussion and Outlook

Retrieval of information from geo-spatial data in a GIS environment is usually accomplished by querying the data, followed by some kind of geocomputation. Intermediate and final results of the queries are often graphically represented. Nowadays, research is also directed to another way to gain information from geo-spatial data and that is by visual exploration, where graphic representations are the starting point for further querying and analysis of the underlying data. For visual exploration of dynamic geo-spatial phenomena, dynamic representations (animations) seem to be attractive media. In dynamic representations, use is being made of dynamic visualisation parameters. Currently there is no empirically tested theoretical framework for application of the dynamic visualisation parameters to geo-spatial data representation. The research described here hopes to contribute to the development of a theoretical framework. The approach is to link characteristics of dynamic geo-spatial phenomena to representation variables.

As a first step, the characteristics of phenomena of interest to a particular type of application, the monitoring of change in the physical environment, have been investigated. By taking monitoring tasks into account, a theoretical framework of concepts to describe general characteristics of dynamic phenomena that can be visually explored is proposed in this paper. The general characteristics should be able to trigger more specific domain expert knowledge about the dynamic phenomena by people involved in monitoring tasks.

In particular, concepts used to describe change in the spatial domain seem to be influenced by the use and display contexts, but within a particular context, unambiguous assignments are required. The relative importance of context sensitivity will be further examined in the next research phase. The objective of the next phase is to validate the relevance and completeness of the proposed framework for a number of monitoring applications. Some adaptations might be necessary.

Adaptations may also stem from literature review into the perceptual and cognitive aspects related to geo-spatial dynamics. Looking into ways in which people perceive and reason with geo-spatial dynamics, in the real world as well as in graphic representations, is considered as an important step, to be taken before characteristics of dynamic phenomena are related to characteristics of dynamic representations. Studying the cognitive aspects of dynamic representations of geo-spatial data is one of the priority research items on the agenda of the Commission on Visualisation and Virtual Environments of the International Cartographic Association [18] and the research described here hopes to contribute to it. For questions that remain to be answered, see e.g. Blok [3].

If perceptual and cognitive aspects can be satisfactory incorporated in visualisation methodologies (or if, to some extent, user experiences and expectations can be met), dynamic representations of geo-spatial data might be powerful instruments for the monitoring of dynamic phenomena. This is based on the assumption underlying exploratory visualisation that vision and cognition together are powerful in pattern seeking [18]. Visual representations can be explored without the necessity to predefine a query in which at least one of the data components is specified (location, thematic attributes or time), as is the case in current GIS's, see e.g. [15], [24]. However, analytical and computational functions will remain important, for example to calculate the strength of a possible spatio-temporal relationship between two phenomena, discovered by visual exploration. Visual and computational methods each have strengths and weaknesses, but full integration of a dynamic visualisation and a (T)GIS environment seems, potentially at least, beneficial for tasks in which identification and comparison of spatio-temporal patterns and relationships are important.

The link between the framework of concepts proposed here and the visualisation variables will form the theoretical basis for animated representation of the dynamic phenomena that are relevant for monitoring. Animations in which the theory is implemented will be empirically tested to evaluate which (combination of) dynamic visualisation variables can be used to prompt thinking about the processes at work and about spatio-temporal relationships by domain experts who perform monitoring tasks. The ultimate goal is to contribute to the development of representation methods and interaction tools that can be helpful in the execution of tasks described above.

References

1. Ahlqvist, O., Arnberg, W.: A Theory of Geographic Context, the Fifth Dimension in Geographical Information. In: Hauska, H. (ed.): Proceedings of the 6th Scandinavian Research Conference on Geographical Information Systems. ScanGIS'97, Stockholm, Sweden (1997) 67–84
2. Bertin, J.: Semiology Graphique. Mouton, Den Haag (1967)
3. Blok, C.A.: Cognitive Models of Dynamic Phenomena and their Representation: Perspective on Aspects of the Research Topic. Presented at NCGIA's Varenius Project Meeting on Cognitive Models of Dynamic Phenomena and their Representations, University of Pittsburgh, PA, U.S.A. (1998) <http://www2.sis.pitt.edu/~cogmap/ncgia/blok.html>
4. Blok, C.A., Köbben, B., Cheng, T., Kuterema, A.A.: Visualization of Relationships between Spatial Patterns in Time by Cartographic Animation. Cartography and Geographic Information Science 26.2 (1999) 139–151
5. Castagneri, J.: Temporal GIS Explores New Dimensions in Time. GIS World 11.9 (1998) 48–51
6. DiBiase, D., MacEachren, A.M., Krygier, J.B., Reeves, C.: Animation and the Role of Map Design in Scientific Visualization. Cartography and Geographic Information Systems 19.4 (1992) 201–214, 265–266
7. Dransch, D.: Temporale und Nontemporale Computer-Animation in der Kartographie. Reihe C, Kartographie, Band 15. Freie Universität, Technische Universität, Technische Fachhochschule, Berlin (1995)
8. Eastman, R.J., McKendry, J.E., Fulk, A.: Change and Time Series Analysis. 2nd edn. Explorations in Geographic Information Systems Technology, Vol. 1. Unitar, Geneva (1995)
9. Eschenbach, C.: Research Abstract on Dynamic Phenomena in Space and their Representation. Presented at NCGIA's Varenius Project Meeting on Cognitive Models of Dynamic Phenomena and their Representations, University of Pittsburgh, PA, U.S.A. (1998) <http://www2.sis.pitt.edu/~cogmap/ncgia/eschenbach.html>

10. Groten, S.M.E., Ilboudo, J.: Food Security Monitoring Burkina Faso. NRSP Report 95–32. BCRS (Netherlands Remote Sensing Board), Delft (1996)
11. Hornsby, K., Egenhofer, M.J.: Qualitative Representation of Change. In: Hirtle, S.C., Frank, A.U. (eds.): Spatial Information Theory; a Theoretical Basis for GIS. Lecture Notes in Computer Science, Vol. 1329. Springer-Verlag, Berlin Heidelberg New York (1997) 15–33
12. Köbben, B., Yaman, M.: Evaluating Dynamic Visual Variables. In: Ormeling, F.J., Köbben, B., Perez Gomez, R. (eds.): Proceedings of the Seminar on Teaching Animated Cartography. International Cartographic Association, Utrecht (1996) 45–51
13. Kraak, M.J., MacEachren, A.M.: Visualization of the Temporal Component of Spatial Data. In: Waugh, T.C., Healey, R.C. (eds.): Advances in GIS Research Proceedings, Vol. 1. Sixth International Symposium on Spatial Data Handling, Edinburgh, Scotland, UK (1994) 391–409
14. Kraak, M.J., Ormeling, F.J.: Cartography, Visualization of Spatial Data. Addison Wesley Longman Limited, Harlow (1996)
15. Langran, G.: Time in Geographic Information Systems. Taylor & Francis, London (1992)
16. MacEachren, A. M.: Visualization in Modern Cartography: Setting the Agenda. In: MacEachren, A.M., Taylor, D.R.F. (eds.): Visualization in Modern Cartography. Pergamon/Elsevier Science Ltd., Oxford (1994) 1–12
17. MacEachren, A.M.: How Maps Work; Representation, Visualization and Design. The Guilford Press, New York (1995)
18. MacEachren, A.M., the ICA Commission on Visualization: Visualization, Cartography for the 21st Century. Presented at the Polish Spatial Information Association Conference, Warsaw, Poland (1998) <http://www.geovista.psu.edu/ica/draftagenda.html>
19. Monmonier, M., Gluck, M.: Focus Groups for Design Improvements in Dynamic Cartography. Cartography and Geographic Information Systems 21.1 (1994) 37–47
20. Muehrcke, P.C., Muehrcke, J.O.: Map Use, Reading, Analysis, and Interpretation. 3rd Edn. JP Publications, Maddison, Wisconsin (1992)
21. Peuquet, D., Wentz, E: An Approach for Time-Based Spatial Analysis of Spatio-temporal Data. Advances in GIS Research, Proceedings 1 (1994) 489–504
22. Vasiliev, I.R.: Mapping Time. Monograph 49, Cartographica 34.2 (1997)
23. Yattaw, N.J.: Conceptualizing Space and Time: a Classification of Geographic Movement. Cartography and Geographic Information Science 26.2 (1999) 85–98
24. Yuan, M.: GIS Data Schemata for Spatio-temporal Information. In: Goodchild, M. (ed.): Proceedings of the Third International Conference on Integrating GIS and Environmental Modelling. NCGIA, Santa Fe, New Mexico (1996)
25. Yuan, M.: Use of Knowledge Acquisition to Build Wildfire Representation in Geographical Information Systems. International Journal of Geographical Information Science 11.8 (1997) 723–745

The Use of Maps, Images and "Gestures" for Navigation

Stephen C. Hirtle

University of Pittsburgh
School of Information Sciences
Pittsburgh, PA 15260
hirtle+@pitt.edu

Abstract. The problem of designing an information system for locating objects in space is discussed. Using the framework of a cognitive collage, as developed by Tversky (1993), it is argued that redundant information from a variety of media is most useful in constructing a spatial information broker. This approach is supported by examining four case studies in detail, you-are-here maps, information kiosks, an information browser, and library locator system. Implications for design of future navigation systems are discussed.

1   Introduction

Interest in wayfinding continues to be strong, even after many decades of empirical research (Golledge, 1999). Current interest on this topic is the direct result of the availability of low-cost high-speed computing. Computational advances have dramatically altered the ability to provide information about spatial locations on a real-time basis. In addition, the World Wide Web and other non-mobile information sources are providing the first, and often the only, source of information to travelers.

This raises the issue of how to best design an information system for providing spatial information. Such systems are not designed to be used in a mobile environment, such as in a car, but rather in a stationary environment, such as information kiosk in a museum. By combining the foundations of cognitive mapping with the principles of good design, this paper sets forth principles for building an effective, off-line, navigation system.

2   Cartographic Evolution

Modern, paper-based cartographic maps present singular views of an environment using a rich set of accepted standards for representing objects, boundaries, and other cartographic details. However, it is interesting to consider the history of cartography and how early maps differed from current standards. MacEachren (1995) notes that

Ch. Freksa et al. (Eds.): Spatial Cognition II, LNAI 1849, pp. 31-40, 2000.

early maps were often drawn from a planar perspective and included three-dimensional images. It was not until recent times that cartographic representations started to take on bird's eye view of the environment. As maps moved towards adoption of an aerial perspective, many symbols on the map kept a three-dimensional orientation. The representation of a church by a box with a cross "on top" or representing mountains with wavy lines are indications of how a three-dimensional perspective is maintained at the level of cartographic symbols, even today.

In contrast to static images, computer generated maps have the flexibility to present three dimensional images from a user centered perspective and provide zoom, pan, and indexing capabilities (e.g., Masui, Minakuchi, Borden and Kashiwagi, 1995). While not without problems, such systems expand the capabilities of information displays and remove the boundary between strict 2-D and what has been termed 2.1 D representations by Chalmers (1992).[1] In the next section, theories of the cognition of space are reviewed.

3   Cognitive Collages

In discussing how spatial knowledge is represented in memory, Barbara Tversky (1993) has argued that the term 'cognitive collage,' rather than cognitive map (Tolman, 1948), does a better job of capturing the richness of spatial representations. That is, Tversky views our spatial knowledge to be a collection of partial bits of multimedia knowledge that is combined into a collage or mosaic. For example, one's knowledge of New York City is a combination of small knowledge chunks, such as images of Central Park, the Empire State Building and Times Square on New Year's Eve, knowledge that streets in midtown Manhattan form a grid, knowledge that Manhattan is an island and that it is connected to land by tunnels and bridges, knowledge of a subway system, which can be represented by the color lines on a subway map, and so on. None of these pieces are necessarily connected, but together they form our atlas of spatial knowledge (Hirtle, 1998).

In many ways, early maps reflect a kind of multimedia knowledge by combining pure spatial information with visual information in the shapes of buildings or texture of the terrain. Hirtle and Sorrows (1998) discussed a variant on this principle that can be seen in an 1891 map of Molde, Norway, as shown in Figure 1. The map indicates many features of landscape, including circles for mountain peaks, which are labeled with the mountain name and the elevation of the peak. However, in addition to the cartographic information, a panoramic drawing of the actual mountains, as view from a lookout point above the town center, is presented around the perimeter of the map.

---

[1] A 2.1 D representation is based on the analogy of flying just above the surface of the space, so that the elevation of objects can easily be noted. At the same, navigation is constrained by the landscape, so there is a smaller likelihood of becoming disoriented (Chalmers, 1992).

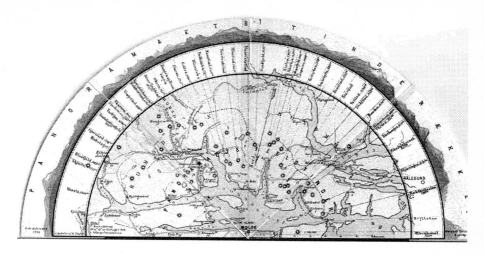

Fig. 1. 1891 map of Molde, Norway

Thus, spatial, verbal, and visual information are provided in a single, compact representation.

One corollary of the cognitive collage principle is that redundant information is a necessary part of spatial knowledge. We encode multiple representations, often in different formats, whenever we can. Redundancy is needed to align these bits of knowledge across formats. The notion of redundancy is more likely to be considered, when a design perspective is adopted, as opposed to an experimental perspective that often pits one factor against another in a factorial experimental design. In the next section, the role of redundancy is discussed in detail for four different domains.

4   Examples

Two basic principles begin to emerge from the analysis. First, the inclusion of multiple modes of data formats is important to the design of navigational systems. Second, the use of redundancy is not only positive, but also critical to the mapping of component parts onto each other and on to the environment.

4.1   You-Are-Here Maps

Levine (1982) has carried out extensive analyses of the 'you-are-here' maps, as you might find in a shopping mall or office building. The problem inherent in interpreting a 'you-are-here' map is to make the correct translation between a paper map, which is typically posted in a vertical manner, with an environment. Four different principles were outlined by Levine as being critical for assisting in this transformation. First, 'up' on the map should refer to 'straight-ahead' in the space. Second, the map should be placed in non-symmetric (off-centered) location with unique identifying features to minimize reversals in the mapping. Third, signs and labels should be provided on both the map and the environment. Fourth, and most important for this discussion, a map with all of these features presents the easiest translation from the map to the environment. That is to say, the redundancy is a positive characteristic, in contrast to other forms of communication where concise, non-redundant communication might be preferred (e.g., Grice, 1975).

4.2   Explanation by Navigation

In his book entitled *Visual Explanation*, Tufte (1997) examines how images can be used to assist in the generation of explanation. One relevant example to the discussion here concerns a guide for directing patrons at the National Gallery in Washington, DC. By giving patrons a three-dimensional map, a sparse set of critical written instructions, and an image of the patron's current location, the information kiosk is able to direct the patrons on a very complex path from one building in the museum to another with a minimum of confusion. Each piece of the spatial collage provides a different role. The image of the immediate environment directs the patron using almost gesture like annotations to begin the route down the correct set of stairs. The map presents the overall plan and gives the patron an immediate feel for the complexity of the task. Finally, the written instructions provide the details that can be carried with the patron on the journey to the exhibit.

Each piece of information is redundant with the other and yet provides a fundamentally different perspective to the problem. Tufte (1997) adds that for this particular interface, information content remains high, representing approximately 90% of the image data. He states that when content becomes secondary to the design, the system begins to lose its usability. In an effective system, "the information becomes the interface" (Tufte, 1997, p. 146).

4.3   Navigation through Cyberworlds

Navigational systems are not only used for navigation through physical space, but also through cyberworlds of various kinds. Rather than survey all of the navigational systems for cyberworlds, in this section, one system is highlighted in detail. Beaudoin, Parent, and Vroomen (1996) introduced Cheops, for exploring complex hierarchies. In this system, a hierarchical structure of thousands of nodes is presented on compact, triangular display as shown in Figure 2.

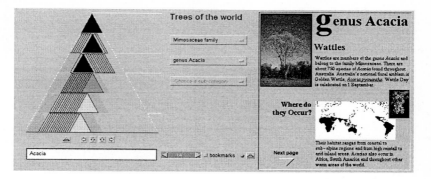

Fig. 2. Example of Cheops display

Rather than use a type of fish-eye view, such as a hyperbolic browser (Lamping, Rao, and Pirolli, 1995), the Cheops approach uses overlapping triangles which can be sorted and highlighted at will by the user. The approach is not unlike overlapping cards in a game of solitaire to use less space. Furthermore, as various nodes are chosen the corresponding silhouettes of parents and children nodes are highlighted and textual and visual information about the node is displayed in the far right panel. By actively selecting nodes and exploring the space, it is possible to navigate through large hierarchies in a very efficient manner.

This system also maintains the two properties of successful navigational tools. Images are central to the system, both in terms of target information, but also in terms of the image silhouettes that emerge from the exploration. Secondly, the location in the hierarchy is represented in a redundant fashion, that is, hierarchically, visually, and through textual labels.

4.4   Navigation through a Campus

The fourth and final example is the Library Locator (LibLoc) system of Hirtle and Sorrows (1998). This system is a web-based browser that was designed to locate satellite libraries on the University of Pittsburgh campus. There are 17 small libraries on the campus and while most students would be familiar with a few of library locations, many of library locations are not well known. Furthermore, many satellite libraries are located in isolated locations deep inside an academic building.

To assist students in finding libraries, a prototype of the LibLoc system was constructed. It consists of four frames, as shown in Figure 3. The upper-left quadrant is the spatial frame that contains either a map of the campus or a floor plan of the building. The upper-right quadrant shows a key image along the navigational path, such as the target building, the front door of the target building or the inside of the library. The lower-right quadrant gives verbal instructions as to the location of the library and the lower-left window provides navigational instructions for the system.

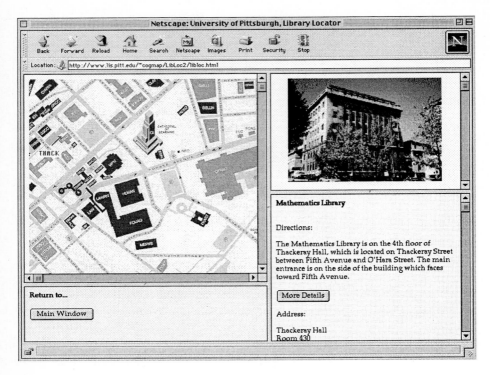

Fig. 3. A sample screen of the Library Locator system showing the Mathematics Library

At any level of the system, more detail can be requested. That is, once the navigational instructions for finding the building on campus have been given, the user can request instructions and floor plans for finding the library inside the building, as shown in Figure 4. Throughout all levels of the system, the three content frames provide three different kinds of overlapping and redundant information about the physical environment. The system is particularly useful for students on campus, who have some knowledge of the buildings, but may not have yet associated a label with every building.

The building of the LibLoc system suggested five principles about the design of navigation systems (Hirtle and Sorrows, 1998). First, it was impossible to standardize the design in terms of the number of levels of information. That is, some locations are just harder to locate and require additional steps. Second, what determined the most informative medium (or frame, in this case) was dependent on the characteristics of the library location. Some locations are easier to pinpoint in words, others in maps, and still others by image. An image has the ability to spark the strongest 'ah-ha' response for distinctive looking buildings, yet it can also be the most uninformative, particularly when a building is non-distinctive in terms of its visual characteristics.

The third observation of Hirtle and Sorrows (1998) was the ability of the LibLoc system to support individual preferences. Some users prefer maps, while others want written instructions. All frames are designed to be clickable when providing more detailed information. Thus, the system can follow the user's focus of attention, rather than requiring the processing of a separate menu bar. Fourth, the information across the frames is overlapping and redundant to maximize the possible connections between the representations. Finally, the system encourages the incorporation of the new knowledge into the existing spatial knowledge of the campus, as opposed to providing one-time instructions about a single route.

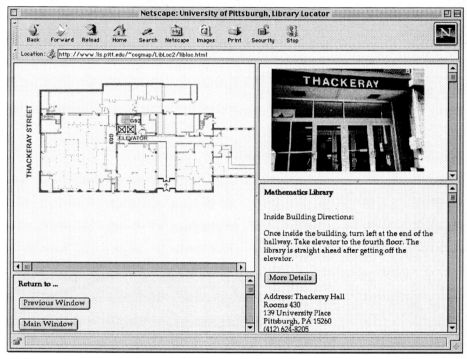

Fig. 4. A sample screen of the Library Locator system showing the building floor plan

5   Summary

In this paper, it was argued for the inclusion of redundant, multi-media information for building navigational systems. Four different examples of these principles were illustrated using you-are-here maps (Levine, 1982), information kiosks (Tufte, 1997), an information browser (Beaudoin, Parent, and Vroomen, 1996) and the library locator system (Hirtle and Sorrows, 1998). The design perspective imbedded in each of these projects highlights specific ways to incorporate spatial knowledge.

It is also interesting to note that many of these systems include the use of annotations on images or maps. The annotations that are used in the LibLoc system provide two benefits to the user. First, they direct the user's attention to a specific location and to a specific path on the map. Second, they provide the equivalent of a visual gesture, e.g, 'walk through the main door and head over this way.' Like a gesture, this is not a detailed articulation with complete information, as might be provided by the map, but rather a generalized abstraction of one key idea. Thus, by its very nature such annotations represent schematized spatial information, not unlike schematized maps or other schematic diagrams. Diagrams can be distinguished from maps by the absence of fidelity. Thus, the London Underground map presents a well-documented example that abstracts general principles of the underground network, without being faithful to the details. Yet, at the same time the London Underground Map presents more information than the network nodes, in the same way that complex gesture can express more than just 'go to the right', but instead can indicate the complexity of the turn.

Fig. 5. A prototypical example of a map gesture, indicating the start and finish locations of a route

A map given to the participants of the 1999 COSIT meeting gave an excellent example of the use of a map gesture. As shown in Figure 5, the large arrow from the subway stop to Dock #4 on the quay is not the route taken by participants. In fact, the route, in one sense, is a complicated sequence of lefts, rights, ups and downs, with several route options available. However, in another sense, the problem is trivial, in part because of the many options and in part because of the physical constraints on the navigation, as result of the river. Therefore, the arrow acts a gesture for gathering attention and for indicating the general direction that the participant needs to traverse.

The analysis that is presented here is in no way complete. There are many other principles, such as the identification of landmarks and neighborhoods, the inclusion of appropriate navigation tools, and the application of a theory of symbols and sign, that will add to the usability of navigation systems. Navigational aids tend to be idiosyncratic and many spatial problems require specific solutions, such as painting footprints on the floor of a hospital ward. However, there are benefits at examining the general principles. The benefits of using the metaphor of a cognitive collage were supported through the examples presented.

In summary, it has been argued that a navigation system for learning about a spatial layout requires the use of redundant, multi-media information. Furthermore, the inclusion of a theoretical framework based on Tversky's (1993) cognitive collage will constrain and direct some of the possible design considerations for such a system.

Acknowledgements

The author thanks Molly Sorrows, Misook Heo, Guoray Cai, and the other members of the Spatial Information Research Group (SIRG) at the University of Pittsburgh for their assistance in cultivating the ideas presented in this paper. The ideas also benefited from discussions through the Varenius Project of the National Center for Geographic Information and Analysis. I am particularly grateful to Max Egenhofer, John Gero, David Mark, and Alan MacEachren for their insights and contributions to the ideas presented here. A shorter version of this paper was presented at the *International Conference on Visual and Spatial Reasoning in Design*, June 15-17, MIT.

References

Beaudoin, L., Parent, M.-A., and Vroomen, L. C. 1996, Cheops: A compact explorer for complex hierarchies. In the Proceedings of *Visualization'96*, San Francisco.

Chalmers, M. 1993., Using a landscape metaphor to represent a corpus of documents. In A. U. Frank and Campari, I. (eds.) *Spatial Information Theory: A Theoretical Basis for GIS: COSIT '93*. Springer-Verlag. Heidelberg.

Golledge, R. G. (ed.), 1999, *Wayfinding behavior: Cognitive mapping and other spatial processes,* Johns Hopkins Press, Baltimore, MD.

Grice, H. P., 1975, Logic and conversation. In P. Cole and J. Morgan (eds.), *Syntax and semantics*, Academic Press, New York.

Hirtle, S. C. 1998, The cognitive atlas: Using a GIS as a metaphor for memory. In Egenhofer, M. J., and Golledge, R. G. (eds.)., *Spatial and temporal reasoning in geographic information systems.* (pp. 263-271). Oxford, New York.

Hirtle, S. C., and Sorrows, M. E. 1998, Designing a multi-modal tool for locating buildings on a college campus. *Journal of Environmental Psychology, 18*, 265-276.

Lamping, J., Rao, R., and Pirolli, P. 1995, A focus+context technique based geometry for visualizing large hierarchies, In *Proceedings of CHI'95 Human Factors in Computing Systems,* ACM Press.

Levine, M. 1982, You-are-here maps: Psychological considerations. *Environment and Behavior,* **14**, 221-237.

MacEachren, A. 1995, *How maps work: Representation, visualization, and design.* Guilford Press, New York.

Masui, T., Minakuchi, M., Borden, G. R. and Kashiwagi, K., 1995. Multiple-view approach for smooth information retrieval. *ACM Symposium on User Interface Software and Technology,* Pittsburgh, pp. 199-206.

Tolman, E. C. 1948, Cognitive maps in rats and men. *Psychological Review,* **55**, 189-208.

Tufte, E. R. 1997, *Visual Explanation*, Graphics Press, Cheshire, CT.

Tversky, B. 1993, Cognitive maps, cognitive collages, and spatial mental model. In A. U. Frank and I. Campari (Eds.), *Spatial information theory: Theoretical basis for GIS*, Springer-Verlag, Heidelberg-Berlin.

Schematizing Maps: Simplification of Geographic Shape by Discrete Curve Evolution[1]

Thomas Barkowsky[1], Longin Jan Latecki[2], and Kai-Florian Richter[1]

[1] University of Hamburg, Department for Informatics
and Cognitive Science Program
Vogt-Kölln-Str. 30, 22527 Hamburg, Germany
{barkowsky, 4krichte}@informatik.uni-hamburg.de

[2] TU Munich, Zentrum Mathematik and
University of Hamburg, Department of Applied Mathematics
Bundesstr. 55, 20146 Hamburg, Germany
latecki@math.uni-hamburg.de

Abstract. Shape simplification in map-like representations is used for two reasons: either to abstract from irrelevant detail to reduce a map user's cognitive load, or to simplify information when a map of a smaller scale is derived from a detailed reference map. We present a method for abstracting simplified cartographic representations from more accurate spatial data. First, the employed method of *discrete curve evolution* developed for simplifying perceptual shape characteristics is explained. Specific problems of applying the method to cartographic data are elaborated. An algorithm is presented, which on the one hand simplifies spatial data up to a degree of abstraction intended by the user; and which on the other hand does not violate local spatial ordering between (elements of) cartographic entities, since local arrangement of entities is assumed to be an important spatial knowledge characteristic. The operation of the implemented method is demonstrated using two different examples of cartographic data.

1 Map Schematization

Maps and map-like representations are a common means for conveying knowledge about spatial environments that usually cannot be surveyed as a whole, like, for example, local built areas, cities, or entire states or continents. Besides the large spatial extent of these *geographic spaces* (Montello, 1993) an important characteristic is their complexity what regards possible aspects that can be depicted in a map. Dependent on its scale, a general topographic map is intended to depict as much spatial information as possible, since the specific purpose the map will be used for is not known in advance.

---

[1] This work is supported by the Deutsche Forschungsgemeinschaft (DFG) under grants Fr 806-8 ('Spatial Structures in Aspect Maps', Spatial Cognition Priority Program) and Kr 1186-1 ('Shape in Discrete Structures').

Ch. Freksa et al. (Eds.): Spatial Cognition II, LNAI 1849, pp. 41-53, 2000.
© Springer-Verlag Berlin Heidelberg 2000

Performing a given task, however, usually only requires a rather small subset of spatial knowledge aspects extractable from a general purpose geographic map. Therefore special purpose *schematic maps* are generated which are only suitable for restricted purposes, but which, on the other hand, ease their interpretation by concentrating on relevant aspects of information by abstracting from others.

Schematic public transportation network maps are a common example of schematic maps (e.g. Morrison, 1996). In this type of map-like representations most entities not directly relevant for using busses, underground trains, etc. are omitted (see Fig. 1 as an example). So, these kinds of maps concentrate on stations, the lines connecting them, and some typical features helpful for the overall orientation within the city at hand. Especially, they usually abstract from detailed shape information concerning the course of the lines and other spatial features, like waters. As a consequence, schematic maps often convey qualitative spatial concepts thus adapting to common characteristics of mental knowledge representation (Freksa et al., 1999).

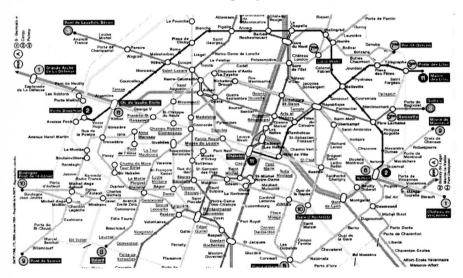

Fig. 1. Example of a schematic public transportation network map (Paris)

Intending to derive a schematic map with reduced shape information from detailed spatial data, we need techniques for reducing spatial accuracy of shape aspects to abstract from the exact course of linear entities or the detailed shape of areal geographic objects.

Shape simplification processes are needed not only for the generation of schematic map-like representations but also in general cartographic contexts. When a map of a smaller scale (e.g. an overview map of a larger area) is intended to be constructed using reference map data of larger scale, it is usually necessary to perform a reduction in spatial detail. Therefore, apart from the use of symbolic replacement, displacement, and resizing of cartographic entities, an important step in cartographic generalization is the simplification of details of spatial features (Hake & Grünreich, 1994). Automated cartographic generalization is a major research issue in the area of *geographic information systems* (GISs) (for an overview, see Müller et al., 1995; Jones, 1997).

We employ discrete curve evolution (Latecki & Lakämper, 1999a, 1999b, 1999c) as a technique for simplifying shape characteristics in cartographic information, which will be presented in the following section.

2   Discrete Curve Evolution

The main accomplishment of the discrete curve evolution process described in this section is automatic simplification of polygonal curves that allows to neglect minor distortions while preserving the perceptual appearance. The main idea of discrete curve evolution is a stepwise elimination of kinks that are least relevant to the shape of the polygonal curve. The relevance of kinks is intended to reflect their contribution to the overall shape of the polygonal curve. This can be intuitively motivated by the example objects in Fig. 2. While the bold kink in (a) can be interpreted as an irrelevant shape distortion, the bold kinks in (b) and (c) are more likely to represent relevant shape properties of the whole object. Clearly, the kink in (d) has the most significant contribution to the overall shape of the depicted object.

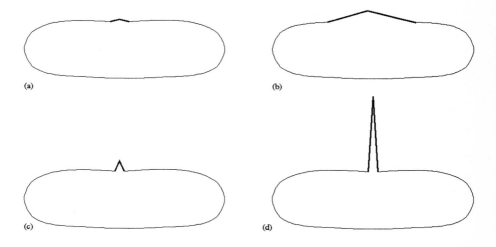

Fig. 2. The relevance measure *K* of the bold arcs is in accord with our visual perception

There exist simple geometric concepts that can explain these differences in the shape contribution. The bold kink in Fig. 2 (b) has the same turn angle as the bold kink in (a) but is longer. The bold kink in (c) has the same length as the one in (a) but its turn angle is greater. The contribution of the bold kink in Fig. 2 (d) to the shape of the displayed object is the most significant, since it has the largest turn angle and its line segments are the longest.

It follows from this example that the shape relevance of every kink can be defined by the turn angle and the lengths of the neighboring line segments. We have seen that the larger both the relative lengths and the turn angle of a kink, the greater is its contribution to the shape of a curve. Thus, a cost function K that measures the shape relevance should be monotone increasing with respect to the turn angle and the lengths of the neighboring line segments. This assumption can also be justified by the rules on salience of a limb in (Siddiqi & Kimia, 1995).

Based on this motivation we give a more formal description now. Let s_1, s_2 be two consecutive line segments of a given polygonal curve. More precisely, it seems that an adequate measure of the relevance of kink $s_1 \cup s_2$ for the shape of the polygonal curve can be based on turn angle $\beta(s_1, s_2)$ at the common vertex of segments s_1, s_2 and on the lengths of the segments s_1, s_2. Following (Latecki & Lakämper, 1999a), we use the relevance measure K given by

$$K(s_1, s_2) = \frac{\beta(s_1, s_2) l(s_1) l(s_2)}{l(s_1) + l(s_2)} \tag{1}$$

where l is the length function. We use this relevance measure, since its performance has been verified by numerous experiments (e.g. Latecki & Lakämper, 2000a, 2000b). The main property of this relevance measure is the following:

- The higher the value of $K(s_1, s_2)$, the larger is the contribution of kink $s_1 \cup s_2$ to the shape of the polygonal curve.

Now we describe the process of *discrete curve evolution*. The minimum of the cost function K determines the pair of line segments that is substituted by a single line segment joining their endpoints. The substitution determines a single step of the discrete curve evolution. We repeat this process for the new curve, i.e., we determine again the pair of line segments that minimizes the cost function, and so on. The key property of this evolution is the order of the substitution determined by K. Thus, the basic idea is the following:

- In every step of the evolution, a pair of consecutive line segments s_1, s_2 with a smallest value of $K(s_1, s_2)$ is substituted by a single line segment joining the endpoints of $s_1 \cup s_2$.

Observe that although the relevance measure K is computed locally for every stage of the evolution, it is not a local property with respect to the original input polygonal curve, since some of the line segments have been deleted. As can be seen in Fig. 3, the discrete curve evolution allows to neglect minor distortions while preserving the perceptual appearance.

A detailed algorithmic definition of this process is given in (Latecki & Lakämper, 1999a). A recursive set theoretic definition can be found in (Latecki & Lakämper, 1999c). Online examples are given on the web page www.math.uni-hamburg.de/home/lakaemper/shape. This algorithm is guaranteed to terminate, since in every evolution step, the number of line segments in the curve decomposition decreases by one (one line segment replaces two adjacent segments). It is also obvious that this

evolution converges to a convex polygon for closed polygonal curves, since the evolution will reach a state where there are exactly three line segments in the curve decomposition, which clearly form a triangle. Of course, for many closed polygonal
curves, a convex polygon with more then three sides can be obtained in an earlier
stage of the evolution.

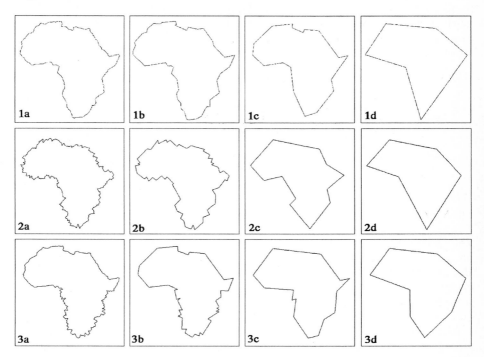

Fig. 3. Examples of discrete curve evolution. Each row shows only a few stages

3   Application to Map Schematization

In the previous section we have seen how discrete curve evolution is used to simplify
the shape of closed curves. Entities depicted in geographic maps are point-like, linear,
or areal. Shape simplification can be applied both to linear and areal entities.
However, when dealing with geographic information, the discrete curve evolution
method has to be extended in several respects.

We have seen that the relevance measure of a kink is computed from the two adjacent line segments. Therefore, for geographic objects represented by a single isolated
point and for the endpoints of linear objects in spatial data sets no relevance measure
can be computed as there is none or only one adjacent line segment (cf. Fig. 4a). As a
consequence, these points cannot be included in the discrete curve evolution process
(they should not be intended to be eliminated, either).

On the other hand, if points belong to more than one object, we may not want to
eliminate them, as this might seriously change spatial information. Consider as an

example a linear object (e.g. a state border) being in part connected to the boundary of an areal object (e.g. a lake) as illustrated in Fig. 4b. When such points are eliminated or displaced in either of the two objects by the discrete curve evolution the spatial feature of being connected to each other may be (at least in part) modified. We will call such points *fix points*.

These two cases make it plausible that not every point in an object may be eliminated or displaced by the discrete curve evolution. Therefore, *fix points* are introduced as points not to be considered by the simplification procedure, be it that it is not possible to assign a relevance measure to them, be it that they must not be eliminated.

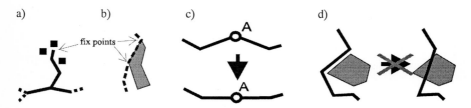

Fig. 4. Examples for cases in cartographic data which require the extension of the discrete curve evolution method

Thus, point-like cartographic entities are considered as fix points as the simplification process does not apply to them; consequently, their positions are not changed during shape simplification. However, in geographic data, point-like objects are often located on linear objects (for example cities lying on roads they are connected by, or train stations in the above public transportation network example). In this case the position of the point-like objects will have to be moved when the linear object they belong to is modified. So, when a line segment is straightened in the discrete curve evolution process a point-like object located on this line has to be projected back onto the line afterwards (Fig. 4c). We will call such point-like entities *movable points*.

According to the above discussion, we obtain the following classification of points in a given map:

1. *Fix points* – points that represent geographic features that cannot be removed and whose position cannot be changed.
2. *Movable points* – points that represent geographic entities that cannot be removed but whose position can be changed.
3. *Removable points* – points that can be deleted; they usually occur as parts of polygonal curves and do not represent individual geographic entities.

However, changing a point's position or deleting a point is only possible if the local arrangement of points is not changed. Before we state this restriction more precisely, we give an intuitive motivation for it. There are many cases in which we will not want to straighten kinks between line segments. Consider, for example, the case of a point-like or an areal entity (e.g. a city, a lake) being located on the concave side of a kink belonging to another object (e.g. a street). If the kink is to be straightened in a discrete curve evolution step it may be the case that the obtained straight line segment intersects the other object (Fig. 4d). In the case of a point-like object located next to the line segment the simplification step may result in the point being located on the other side of the line. Clearly, we usually should not modify the spatial relationships between geographic objects in such a severe way. The restriction that local arrangements of spatial entities must not be changed is a fundamental spatial knowledge aspect used in reasoning with geographic maps (cf. Berendt et al., 1998a, 1998b).

Therefore, we will check whether local ordering relations (cf. Schlieder, 1996) between objects are violated before a simplification is performed (i.e., the schematization process is performed in a context-sensitive way)[2]. Notice that this check only concerns (parts of) entities in immediate vicinity to the line segments in question. Violations of spatial ordering between objects in farther distances - although they may occur during the schematization process as a whole - can be assumed not to affect the extraction of appropriate information from the map by a map user.

In the next section we will show how these requirements are implemented by the simplification process for cartographic representations.

4   The Algorithm

As elaborated in the previous section, the algorithm for performing the simplification task based on the discrete curve evolution requires some additional considerations. However, it is still fairly easy.

Point-like entities are considered as *fix points* as stated in the previous section. What regards linear and areal objects we start by computing the relevance measures for all (non-end) points and sort them by their respective relevance measures. Notice that points that have a relevance measure of zero do not need to be dealt with, since they are already on a straight line.

Then we consider the first point in the sorted list (i.e. the point which has the lowest relevance measure). Two cases have to be distinguished: (A) The coordinates of the point at hand are unique or (B) there is more than one object at the given point's coordinates.

Case A: Unique coordinates. Since the point's coordinates are unique there is no interaction with any other object in this point. Endpoints of linear objects are treated as *fix points* in this case as no relevance measure can be computed for them. For non-endpoints two cases have to be distinguished:

1. If the point at hand is a *removable point* we may delete it unless this does not violate the local arrangement of spatial entities. Therefore we check whether there is any other point inside the triangle formed by the original and the new straightened line. If so, it is not possible to straighten the line since this would change the local spatial arrangement of the two objects involved. The point at hand is treated as *fix point* in this case.

---

[2] Clearly, it would also be possible to modify both, the object to be simplified and the conflicting object (e.g. by moving it aside). However, due to the degrees of freedom (as well as further possible conflicts) this does not seem to be a feasible approach.

If there is no point in the triangle the line is straightened by removing the point at hand. Finally, the relevance measures for the adjacent points are recalculated. When relevance measures are not normalized with respect to the total length of entities (see previous section) it is sufficient to recalculate just the relevance measures for the immediately neighboring points since in this case the relevance measure is only based on local shape characteristics. Afterwards we resort the list of relevance measures and start all over again.

2. If the point at hand is a *movable point* (i.e. it has to be preserved) it is treated like described before. Instead of removing it, however, its new coordinates on the straightened line are calculated and the point is projected to this position. After moving the point its relevance measure is zero for it is now on a straight line; therefore, it will not receive further consideration in discrete curve evolution.

Case B: Multiple objects at the given coordinate. If the point's coordinates belong to more that one object, there are interactions between at least two objects in the point at hand. Three cases have to be distinguished:

1. If the coordinates of the immediately neighboring points of all objects involved are pairwise identical, then the objects the points belong to have common coordinates for at least three subsequent vertices. In this case the middle vertex (the point position at hand) is a *removable point* as long as simplification is performed simultaneously for all objects involved. Therefore we can proceed in analogy to removable points as described for case A (see above). *Movable points* (which shall not be eliminated from either object) can also be treated in analogy to movable points in case A. In this case all points at the middle vertex's coordinates are simultaneously transferred to their new common position on the straightened line.

2. The current point's position coincides with an endpoint of one object and for the other objects holds that the coordinates of the immediately neighboring points are pairwise identical. In this case the point at hand is treated as a *movable point* (which is - regarding the endpoints - in contrast to case A). The respective endpoint is projected onto the straightened line. So, although the endpoints still have no relevance measure, they are moved in accordance with the points they coincide with.

3. In any other case the point at hand has to be treated as a *fix point* and cannot be modified.

If the point at hand has been identified as a fix point by the test under cases A or B we continue with the next point in the sorted list of relevance measures (i.e. the point with the next lowest relevance measure). If there is no non-fix point left in the list or the abort criterion (see below) is reached we are done and the algorithm terminates.

Usually a certain value of the relevance measure is chosen as abort criterion but it is also possible to specify a value of turn angle as a stop condition of the simplification. These abort criteria specify the degree of schematization and have to be chosen in accordance with the schematic map's intended purpose of use or other design requirements. So, if the point at hand chosen for elimination has a higher relevance measure (or a larger turn angle), then the abort criterion is fulfilled and the algorithm stops. It is important to realize that this check is made against the point that is about to be changed and not against the point with the lowest relevance measure. Since points that do not fulfill the above preconditions may not be considered any more, it is possible that a point is tested whose relevance measure is substantially higher than the point with the lowest relevance measure.

The algorithm has been implemented and tested with different types of geographic data. In the following section we demonstrate the operation of the algorithm showing two examples.

5   Examples

The first example employs a simple schematic public transportation network map as basic data for the algorithm. Schematic transportation maps usually abstract both from the exact positions of stations and the detailed course of lines connecting the stations.

Figure 5a shows the original data set containing linear and point-like entities. The schematic map exhibits the exact positions of subway and city train stations of a part of the Hamburg public transportation network. Obviously, some stations belong to more than one subway line. The stations at the ends of the lines have to be considered as fix points as they do not allow for computing a relevance measure for them. The line courses connecting the stations have already been abstracted to straight lines. In the course of the simplification process minor kinks in the line courses are eliminated (Fig. 5b and c) while the overall appearance of the network map is preserved.

This simple example already exhibits the most important features of the simplification process developed in the previous sections:

- As the point-like stations (depicted as squares) must be preserved (i.e. must not be deleted during the simplification process) they are always projected back onto the lines during the evolution process. So they are treated as *movable points*. As only straight line connections are used in this example, there are no *removable points*.
- Endpoints of lines (usually to be treated as *fix points*) which are simultaneously part of another line are moved according to the simplification of the other line.
- Simplification is only performed as far as the local spatial arrangement between lines and stations is not violated. In other words, in the simplification process no station changes sides with respect to any line (seen from a given direction).

The employed map simplification method, however, does not deal with the local concentration of entities on the map. To improve the readability of schematic maps, the map's scale may be locally distorted (for example in the downtown region where concentration of relevant entities is higher than in the outskirts). For this purpose the presented method may be combined with constraint-based methods for graph layout (e.g. Kamps et al., 1995; He & Marriott, 1996).

For the second and more complex example we used an original ARCINFO data set of the Cologne (Germany) conurbation (Fig. 6a). The original data set showing cities, streets and waters of the region (Fig. 6b) comprises approx. 2700 points. Figure 6d shows the maximal simplification that can be performed without violating local spatial arrangement between spatial entities. This maximally simplified data set consists of approx. 600 points. An intermediate simplification level, i.e. a simplification up to a specified relevance measure is shown in Fig. 6c (approx. 750 points).

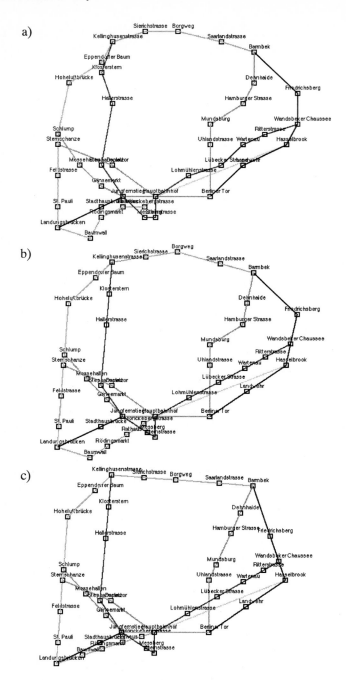

Fig. 5. Simplification of a public transportation network map (Hamburg). Original plan (a) and two schematizations up to different simplification levels

In this example the point-like entities representing cities have to be treated as *fix points*, i.e. they must neither be deleted nor moved. The points in the linear entities (streets and waters) can generally be considered as removable; i.e., they can be eliminated during the evolution process as far as spatial ordering relations between point-like and linear features are preserved. Observe that no further point in the data set shown in Fig. 6d can be removed without violating local spatial arrangements.

As the different types of linear entities (i.e., streets and waters) play different roles in the interpretation of the resulting map, it may be sensible to use different degrees of schematizations for either of them. For example, streets may be simplified up to the highest possible degree, whereas waters may remain curved to a certain extent to convey their characteristics of being natural entities. For this purpose, different types of geographic objects may be assigned different threshold values for schematization.

6   Conclusion and Outlook

We presented a method for simplifying cartographic shape information based on discrete curve evolution. The basic method of the discrete curve evolution has been extended to meet specific requirements of geographic data depicted in map-like representations.

These qualitative restrictions suitable for geographic data sets can be further extended to make the method usable for more constrained spatial tasks, which may be necessary in some applications. For example, it may be particularly important to preserve the order of objects with respect to their relative distances or with respect to their cardinal directions. Also, concepts for spatial proximity (for example a city being located near the coast) may be intended to be preserved by automatic map simplification processes.

The modified discrete curve evolution presented in this contribution can also be used as a pre-filter for similarity measures of geographic representations. Similar to the shape similarity measure of object contours in (Latecki & Lakämper, 1999b), it seems to be cognitively and computationally desirable to simplify representations before determining their similarity. The fact that the modified discrete curve evolution preserves relevant qualitative properties is of primary importance for this task, since similarity of geographic representations not only depends on metric features but also on qualitative properties (e.g., is a point-like object A to the right or to the left of a line object B?). Similarity measures of geographic representations have many potential applications, e.g., in *geographic information systems* (GISs) or in the automatic matching of satellite images with existing geographic data.

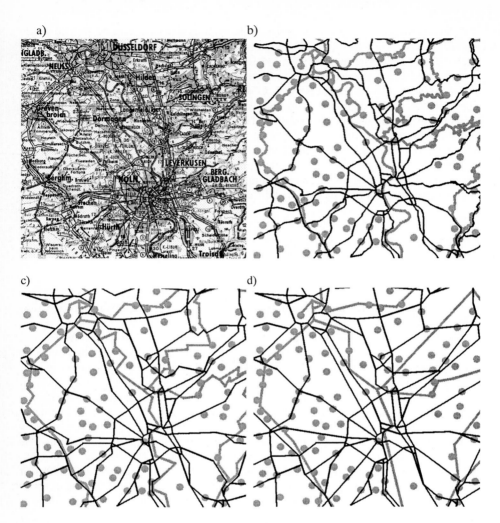

Fig. 6. Example of simplification of an ARCINFO data set: (a) conventional map of the Cologne (Germany) conurbation; (b) the original ARCINFO data set containing cities, streets, and waters; (c) intermediate simplification level; (d) maximal simplification preserving local spatial arrangement

Acknowledgments

We would like to thank Sabine Timpf and an anonymous reviewer for helpful comments on this contribution.

References

Berendt, B., Barkowsky, T., Freksa, C., & Kelter, S. (1998a). Spatial representation with aspect maps. In C. Freksa, C. Habel, & K. F. Wender (Eds.), *Spatial cognition - An interdisciplinary approach to representing and processing spatial knowledge* (pp. 313-336). Berlin: Springer.

Berendt, B., Rauh, R., & Barkowsky, T. (1998b). Spatial thinking with geographic maps: an empirical study. In H. Czap, H. P. Ohly, & S. Pribbenow (Eds.), *Herausforderungen an die Wissensorganisation: Visualisierung, multimediale Dokumente, Internetstrukturen* (pp. 63-73). Würzburg: ERGON Verlag.

Freksa, C., Barkowsky, T., & Klippel, A. (1999). Spatial symbol systems and spatial cognition: A computer science perspective on perception-based symbol processing. *Behavioral and Brain Sciences, 22*(4), 616-617.

Hake, G., & Grünreich, D. (1994). *Kartographie*. Berlin, New York: de Gruyter.

He, W., & Marriott, K. (1996). Constrained graph layout. In S. North (Ed.), *Graph drawing (Proceedings of the Symposium on Graph Drawing, GD'96, Berkeley, CA, USA)* (pp. 217-232). Berlin: Springer.

Jones, C. B. (1997). *Geographical information systems and computer cartography*. Essex: Longman.

Kamps, T., Kleinz, J., & Read, J. (1995). Constraint-based spring-model algorithm for graph layout. In R. Tamassia & I. G. Tollis (Eds.), *Graph drawing* (pp. 349-360). Berlin: Springer.

Latecki, L. J., & Lakämper, R. (1999a). Convexity rule for shape decomposition based on discrete contour evolution. *Computer Vision and Image Understanding, 73*(3), 441-454.

Latecki, L. J., & Lakämper, R. (1999b). Contour-based shape similarity. *Proc. of Int. Conf. on Visual Information Systems*, Amsterdam, June 1999. Berlin: Springer.

Latecki, L. J., & Lakämper, R. (1999c). Polygon Evolution by Vertex Deletion. *Proc. of Int. Conf. on Scale-Space Theories in Computer Vision*, Corfu, Greece, September 1999.

Latecki, L. J., & Lakämper, R. (2000a). Application of planar shape comparison to object retrieval in image databases. *Pattern Recognition (PR)*, to appear.

Latecki, L. J., & Lakämper, R. (2000b). Shape similarity measure based on correspondence of visual parts. *IEEE Trans. Pattern Analysis and Machine Intelligence (PAMI)*, to appear.

Montello, D. R. (1993). Scale and multiple psychologies of space. In A. Frank & I. Campari (Eds.), *Spatial information theory: A theoretical basis for GIS. LNCS No. 716* (pp. 312-321). Berlin: Springer.

Morrison, A. (1996). Public transport maps in western European cities. *The Cartographic Journal, 33*(2), 93-110.

Müller, J. C., Lagrange, J. P., & Weibel, R. (Eds.). (1995). *GIS and generalization: Methodology and practice*. London: Taylor & Francis.

Schlieder, C. (1996). Qualitative shape representation. In P. Burrough & A. Frank (Eds.), *Geographic objects with indeterminate boundaries* (pp. 123-140). London: Taylor & Francis.

Siddiqi, K., Kimia, B. B. (1995). Parts of visual form: Computational aspects. *IEEE Trans. PAMI*, 17, 239–251.

Schematic Maps as Wayfinding Aids [1]

Hernan Casakin[1,2], Thomas Barkowsky[2], Alexander Klippel [2],
and Christian Freksa[2]

[1] The Yehuda and Shomron Academic College,
Department of Architecture, P.O.Box 3, 44837-Ariel, Israel
hernan@techunix.technion.ac.il

[2] University of Hamburg,
Department for Informatics and Cognitive Science Program,
Vogt-Kölln-Str. 30, D - 22527 Hamburg, Germany
{barkowsky,klippel,freksa}@informatik.uni-hamburg.de

Abstract. Schematic maps are effective tools for representing information about the physical environment; they depict specific information in an abstract way. This study concentrates on spatial aspects of the physical environment such as branching points and connecting roads, which play a paramount role in the schematization of wayfinding maps. Representative classes of branching-points are identified and organized in a taxonomy. The use of prototypical branching points and connecting road types is empirically evaluated in the schematization of maps. The role played by the different functions according to which the map is classified is assessed, and main strategies applied during the schematization process are identified. Implications for navigational tasks are presented.

1 Introduction

Due to their abstracting power, schematic maps are ideal means for representing specific information about a physical environment. They play a helpful role in spatial problem solving tasks such as wayfinding. An important class of features for these tasks are route intersections (cf. Janzen et al., 2000, this volume). One of the challenges in constructing schematic maps consists in establishing clear relationships between detailed information found in the environment, and abstract / conceptual structures contained in the map. An important aim of any schematic wayfinding map is to efficiently support information to find a destination. Therefore, it is crucial to determine in which way and to what extent spatial information such as branching points and their connecting roads can be regarded essential or irrelevant, and therefore must be preserved or discarded during the schematization of a map.

---

[1] This work has been supported by the Deutsche Forschungsgemeinschaft (DFG) in the Doctoral Program in Cognitive Science at the University of Hamburg and in the framework of the Spatial Cognition Priority Program.

Ch. Freksa et al. (Eds.): Spatial Cognition II, LNAI 1849, pp. 54-71, 2000.

Aspects related to schematic maps as visual tools for communicating spatial concepts for wayfinding tasks must also be taken into account. Accordingly, the focus of this contribution is directed towards exploring spatial characteristics of branching points and their connecting roads in the schematization of wayfinding maps. The idea is to analyze how these features are affected during the process of schematization. After a brief review of selected literature in architecture, cognitive psychology, geography, and environmental design, we present a description of an empirical investigation, interpret results, and offer conclusions for the schematization of maps as wayfinding aids.

1.1   Spatial Orientation

A person's ability to establish his or her location in an environment is termed *spatial orientation* (e.g., Arthur & Passini, 1992; Correa de Jesus, 1994). From a cognitive point of view spatial orientation is considered as the capability to form a cognitive map (Tolman, 1948; Golledge et al., 1996; Downs & Stea, 1973). Successful spatial orientation involves the representation of a suitable cognitive map of the environment, within which the subject is able to establish his or her position. The concept of spatial orientation has been demonstrated to be helpful in exploring some of the spatial characteristics that facilitate cognitive mapping. Lynch (1960) was one of the pioneers who established direct relationships between the spatial orientation of people and their physical environments. Since then, extensive research has been done in this domain. As the concepts of spatial orientation and cognitive mapping mainly refer to the static relationship between a subject and a specific spatial layout, they often neglect the dynamic effect of moving through a spatial environment (Passini, 1984). In the light of this situation, a new perspective was needed to define spatial orientation whilst moving in space.

1.2   Wayfinding as Spatial Problem Solving

In the recent years, the notion of spatial orientation has been replaced by the concept of wayfinding, which refers to the process of moving through space and encompasses the goal of reaching a spatial destination (e.g. Garling et al., 1986; Downs & Stea, 1973; Kaplan, 1976; Passini, 1998). Since finding a way is concerned with perceiving, understanding, and manipulating space, this concept is understood as a process of spatial problem solving. Wayfinding involves cognitive and behavioral abilities which are performed to reach a destination (Arthur & Passini, 1992; Lawton et al., 1996). These abilities, which play a critical role in achieving a wayfinding goal (e.g. reaching a desired destination), can be classified into: i) decision making, ii) decision executing, and iii) information processing (Passini, 1984; 1998).

The availability of relevant information about the environment is an important factor in the process of decision making. In wayfinding tasks, successful decisions are generally based on suitable information about physical characteristics of the environment. These account for a description of a net of connecting roads to be followed, as well as for different types of branching points that may serve to connect roads to reach a specific destination.

In order to reach a destination, decisions have to be transformed into actions. It is in this process that decision execution takes place in a specific position within a certain environment. According to Arthur and Passini (1992) and Passini (1984) the execution of a decision involves the matching of a representation (mental or external) of the environment with the real environment itself.

Both decision making and decision execution are supported by information processing. Spatial perception (related to the process of acquiring knowledge from the environment) and spatial reasoning (related to the manipulation of spatial information) constitutes the main interrelated components of information processing (Passini, 1984; 1998).

In spatial problem solving, a map is seen as a base of knowledge instrumental for supporting information processing. By focusing on the visual representation of physical entities, such as connecting roads and branching points, our work will attempt to address main aspects derived from schematic maps.

1.3   Maps and Representation

While behavioral abilities are performed in the physical environment, cognitive abilities are carried out in the domain of cognitive representation. The relationships between the cognitive domain and the environment are essential for providing meaning to the structures and operations in the cognitive representation (Freksa, 1999). Accordingly, structures and operations in the representation of elements of the environment are responsible for the success or failure of wayfinding procedures (Downs & Stea, 1973).

Theoretical frameworks dealing with representation focus on the correspondence between elements and existing relationships in a represented world, and elements and existing relationships in a representing world (Palmer, 1978; Furbach et al., 1985). Whereas physical elements of an environment constitute the represented world, a map-like structure is considered a suitable medium for the representing world (MacEachren, 1995). Maps can be seen as derived from an aerial view, where meaningful entities and spatial relationships between entities are partially replaced by symbols (Berendt et al., 1998). Among different types of map representations, there are sketch maps and schematic maps. Sketch maps are related to verbal descriptions about spatial features (e.g. Suwa et al., 1998; Tversky & Lee, 1999; Freksa et al., 2000, this volume), schematic maps, on the other hand, are obtained by relaxing spatial and other constraints from more detailed maps (e.g. Barkowsky & Freksa, 1997).

1.4   Schematic Maps and Wayfinding

While topographic maps are intended to represent the real world as faithfully as possible, schematic maps are seen as conceptual representations of the environment. "For many purposes, it is desirable to distort maps beyond the distortions required for representational reasons to omit unnecessary details, to simplify shapes and structures, or to make the maps more readable" (Freksa et al., 2000, this volume). Schematic maps provide a suitable medium for representing meaningful entities and spatial relationships between entities of the represented world. Moreover, they relate concrete and detailed spatial information from the physical environment to abstract and conceptual information stored in our brain. However, an instrumental condition is that a suitable relationship must exist between schematic maps and the represented world. The more clearly we relate schematic maps to the critical elements of the represented environment, the more easily users will find a wayfinding solution.

Schematic maps are external pictorial representations. An important feature of these external pictorial representations is that they reflect to a great degree the way we reason through abstract structures (Freksa, 1991). Thus, the way we represent spatial information will strongly depend on how we perceive and conceive of the environment.

Entities and conceptual aspects of the depicted domain that are symbolized in the schematic map vary according to the schematic map's purpose. In wayfinding, the schematization of branching points is important for reducing the cognitive effort when trying to find a destination. Freksa (1999) has proposed that an appropriate representation tool should include the following processes: i) identifying and selecting relevant aspects from the physical environment; ii) choosing an appropriate structure for the inferences to be made between the represented world and the representing world; iii) interpreting the results of the inferences.

A key question for constructing schematic maps is to select aspects that make them suitable for solving wayfinding problems. Identifying these aspects is what turns a schematic map into an ideal tool for supporting navigation and orientation. We claim that identifying relevant branching point situations from reality, building a taxonomy of branching points, and establishing a hierarchy of connecting roads will help produce schematic maps to ease wayfinding.

2   Branching Points and Wayfinding

In wayfinding tasks, spatial problem solving is strongly influenced by the nature of the intersections. By developing a wayfinding model for built environments, Raubal and Egenhofer (1998) pointed out that decision making is affected by the complexity of a wayfinding task. Major characteristics of branching points, where decisions have to be taken when seeking a destination, provide parameters to measure the complexity of a spatial problem solving task (Hillier et al., 1984; O'Neill, 1991a, 1991b; Peponis et al., 1990; Weisman, 1981), as well as for the construction of schematic maps.

Arthur and Passini (1992) noted that the number of branching points has an important influence on the difficulty of performing wayfinding. In solving spatial problems, the process of constructing a schematic map is directly affected by the number of routes that converge in a branching point, and by their respective angles. These are seen as essential elements for the organization of branching points into a general taxonomy.

2.1   A Taxonomy of Branching Points

Branching points are classified by the number of routes, roads, or branches that meet at an intersection and by the angles in which they intersect. Although an infinite number of cases is obtained by combining these two variables, we reduce the number of classes to a small set that can be qualitatively distinguished. Our qualitative approach intends to provide a suitable framework for dealing with major similarities and major differences among branching points.

We classify branching points into a taxonomy of two to eight routes or roads that meet at an intersection, and we distinguish four qualitatively different intersection angles: acute, perpendicular, obtuse, and straight on; moreover, we take into account rotation and mirror symmetries. We represent these categories by prototypes derived from an equiangular eight-sector-model (cf. Hernández, 1994; Frank, 1992). The concept of taxonomy helps to define and classify different types of complex intersections. In wayfinding tasks, this is more advantageous than simply referring to intersecting angles between roads. For example, it contributes to establishing and preserving important and specific relationships in the schematic map that otherwise may be lost. The schematization and classification of representative branching points of the physical environment is illustrated in Fig. 1 through the different categories and sub-categories of the proposed taxonomy. Although the taxonomy of representative branching points proposed in this study includes categories that embraces two to eight branches, the prototypes studied in this work were restricted to the categories of three and four branches.

We propose that the organization of branching point situations into main categories will contribute to perceiving, understanding, and organizing complex information from the environment through a reduced number of cases. Understanding about how representative a particular branching point is for actual problem solving situations will help to ease the schematization of maps.

2.2   Schematizing Maps

One of the problems in schematizing maps is how to deal with the relationships between all the details found in the environment and the abstract structures to be represented in the schematic map. We have posited that in the process of schematization, entities, relationships, and conceptual aspects of the real environment have to be represented according to the schematic map's intended function and purpose. Since branching points and connecting roads are relevant components of the physical environment, we claim that both play an important role in the schematization process.

Understanding about the main aspects and criteria for preserving, including, or eliminating these entities will contribute to the construction of efficient schematic maps. The classification of representative branching points into a general taxonomy is seen as an attempt to provide a structure for mapping relevant relationships between the represented world and the representing world. However, the question of how such a taxonomy could be used for the schematization of maps that are supposed to serve as wayfinding aids is still unclear.

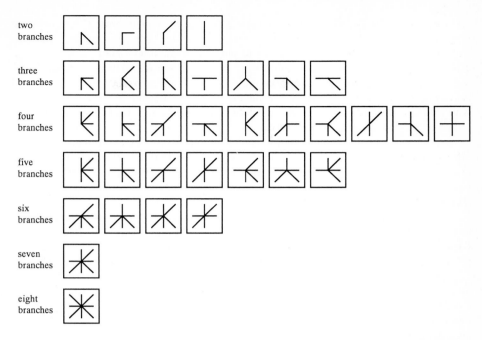

two branches

three branches

four branches

five branches

six branches

seven branches

eight branches

Fig. 1. Taxonomy of branching points for intersections with two to eight branches. Acute, perpendicular, obtuse, and straight on intersections are distinguished. Rotation and mirror symmetries are taken into account

3   Empirical Study

In this section we describe an empirical study which investigates the schematization of a wayfinding map by employing a taxonomy of qualitative branching points. We first describe the general objectives of the study and motivate the selection of a particular case study. Thereafter, the empirical task is described. This exploratory study is intended as an empirical contribution from a theoretical and computer science perspective.

3.1   Goals

Whilst referring to schematic maps, most spatial cognition literature focuses on specific aspects dealing with the issue of representation. However, it does not really concentrate its attention on the process of schematization that leads to the representation of the map itself. For this reason, an important aim of this empirical study is to focus both on aspects related to schematic map representation, as well as on the schematization process itself. Therefore, the empirical study aims at the following goals:

First, we explore in which way and to what extent a taxonomy of prototypical branching points can be successfully utilized for the schematization of maps. Particularly, we would like to confirm whether branching points found in the schematic map are constructed by referring to their corresponding prototypes of the proposed taxonomy. Moreover, we explore if right-left positions between schematized branching points are preserved with respect to those of the original map. Additionally, we analyze whether close distances between neighboring branching points are eliminated by merging with one another.

Second, we analyze in which way and to what extent different road types are affected during the process of schematization. A particular focus is set to examine the extent to which roads of the original map are eliminated. We are also interested in evaluating whether subjects are able to schematize connecting roads according to hierarchical structures (cf. Timpf, 1998).

Third, we study the role played in the schematization process by the different functions (e.g. cemetery, park, etc.) according to which the map is divided (from now on termed as *functional areas*). We are particularly interested in evaluating to what extent each function will affect the inclusion or elimination of the different road types.

Fourth, we explore possible reasoning strategies that could be applied during the map schematization process. A basic study will be carried out to identify cognitive patterns of behavior that may be applied by subjects while schematizing a map. Particularly, we would like to exhibit if any sequence of schematization can be found. Additionally, we explore whether hierarchical structures between roads are considered during the schematization process.

3.2   A Case Study: The Stadium

With the aim of studying map schematization, we carried out a case study in the 'real world', the Hamburg Volksparkstadion area (see Fig. 2). One of the reasons for choosing this environment is that it represents interesting examples of wayfinding problems. The Volksparkstadion area is surrounded by a variety of urban features such as parks, a cemetery, and parking lots that are organized according to a rather complex geometric layout. Furthermore, an intricate circulation system comprising streets, a freeway, and footpaths constitutes the network of ways leading to the stadium.

For the purposes of organizing and analyzing this spatial structure, a hierarchical representation was proposed a priori. Components of the circulation system were classified with respect to their assumed relative importance for the wayfinding task. Accordingly, while *border main roads* and *internal main roads* are part of a higher level hierarchy, *external roads*, *secondary roads* and *dead ends* form the lower levels of that hierarchy.

Another motivation for selecting this environment for a case study was an inadequate 'official' schematic map displayed at many locations in the stadium area (see Fig. 3). Presumably, this map was intended as a wayfinding aid for stadium visitors. In this map, the circulation system of the area can be considered over-simplified. As a result, essential spatial information has been discarded, and some spatial relationships of the physical environment have been severely distorted in the map. For example, critical spatial relationships like types of branching points and shape of roads leading to the stadium were altered, some are missing. The mismatch between characteristics of the physical environment, and their pictorial representation in the map is a critical factor that may impair wayfinding. We can assume that in some cases it might be better not to have a map at all than having a confusing or inaccurate map (cf. Levine, 1982).

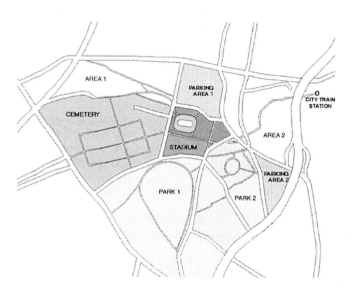

Fig. 2. City map of the Hamburg Volksparkstadion area

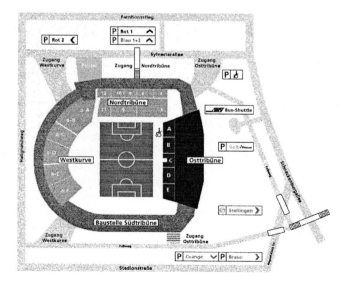

Fig. 3. Inadequate schematic map for wayfinding tasks in the Hamburg Volksparkstadion area

As this example of an inadequate schematic wayfinding map triggered the question for the aspects to be considered in producing more suitable maps, we investigated processes and strategies involved in producing schematic depictions of the illustrated area.

3.3   Description of the Empirical Study

Subjects. 15 subjects participated in the empirical tasks conducted in this study. Most of them were graduate students of the University of Hamburg Department for Informatics and the Doctoral Program in Cognitive Science.

Empirical Task and Materials. Subjects were provided with a task sheet containing procedural instructions, a warm-up task, and a description of the main task. In the warm-up task, subjects were presented a picture of a non-complex representation of a city area and were required to produce a new simplified map. An important requirement to carry out this task was to use the available taxonomy of 3 and 4 branches (see Fig. 1). In the main task, subjects were given visual information consisting of an A4 sheet containing a representation of the Hamburg Volksparkstadion area (see Fig. 2), and a similar picture of the taxonomy of branching points. The map of the area included data about freeways, streets, paths, and some urban features (e.g. stadium, cemetery, parks, and parking lots). Subjects were given the task to produce schematic maps to help wayfinders find their way to the stadium.

Since the schematic map to be produced was intended to help a variety of users find the way to the stadium, the type of transportation (e.g. by car, by bike, or by city train) potentially used by the wayfinders also was to be included. Subjects were asked to use the prototypes provided with the taxonomy of branching points as main constraints to carry out their map schematization task. They were required to evaluate and reconsider which type of information from the detailed map should be essentially included for schematization, and what could be simplified or eliminated when constructing an easily readable map. The task of completing the whole map dealt with the wayfinding aspect, but a main focus of this study was set on the schematization of the map.

Procedure. Sessions were carried out in an empirical research lab room. Individual sessions were restricted to approximately 45 minutes. During the first 5 minutes the subjects were made familiar with the problem and the requirements of the experiment. The next 10 minutes were assigned to carry out a warm-up task. The second part of the session was used to solve the schematization task. During the whole session, subjects were requested to think aloud; both their produced graphic output and their verbalizations were recorded on videotape. Subjects were not restricted with respect to the size of the map they were asked to produce. Nevertheless, they were provided with checkered sheets of A3 paper, which were considered to be large enough to construct the schematic map without facing any problem of lack of space. The experimenter answered procedural questions before starting the drawing procedure, but any further interaction was avoided during the empirical task. A signal was given 5 minutes before the end of each session.

4   Results

In the subsequent evaluation of the data, verbalizations were transcribed and, together with the sketches of the schematic maps produced during the empirical tasks, they were used to prepare a protocol. Both qualitative and quantitative results were evaluated; they are presented below.

4.1   Qualitative Results

Two examples of contrasting cases of schematic maps produced in the empirical study are shown in Fig. 4. Both maps have been schematized using the prototypes of the taxonomy of branching points. By comparing each example with the original map, we see that branching points of both schematic maps are constructed by considering their related prototypes of three and four branches in the given taxonomy. In addition, right / left relationships of branching points found in the new map are largely preserved with respect to those of the original map. However, by comparing one schematic map with the other, different levels of schematization are identified. Whereas in the first case almost no roads are eliminated, in the second one a large number of roads is not included in the schematic map. The group of eliminated roads is mostly constituted by *secondary roads* and *dead ends* that, as we will show in the next sections, often have to be regarded irrelevant for the purposes of wayfinding.

a)   b)

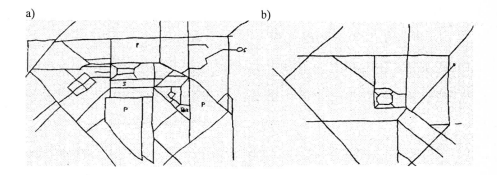

Fig. 4. Examples of two contrasting schematic maps: a) high-resolution depiction b) low-resolution depiction

4.2   Quantitative Results

To achieve the goals conveyed in Section 3.2, the performance of each of the 15 subjects that participated in the empirical task was assessed according to two key elements: i) the schematic map, and ii) the process of schematization. Quantitative results are informally presented as follows:

The Schematic Map. In this study we verified that a vast majority of the different branching point types embraced in the schematic map are preserved taking in consideration their related prototypes of the proposed taxonomy. In the experiments, 89% of branching points were schematized in terms of the prototypical cases. The experiments also showed that right / left relationships between branching points in the schematic map and those of the provided map are largely preserved. In other words, the relative right or left position between branching points has been maintained. 99% of right-left relationships between branching points are preserved. A further analysis was performed to examine if close distances between neighboring branching points are shortened to merge into a single branching point. By focusing on the five most closely located pairs of branching points, it was found that 63% were merged into a single branching point.

Results related to the inclusion or elimination of roads reveal that 34% of the total roads of the original map are eliminated in the new map. Accordingly, we found that within the eliminated roads, 40% correspond to *external main roads*, 52% belong to *secondary roads*, and 76% correspond to *dead ends*. However, only 4% of the eliminated roads belong to *border main roads*, and 5% to *internal main roads*. Additional results about the inclusion or elimination of road types with respect to their relationship with the functional areas upon which the map is organized were found:

Low percentages of connecting road elimination were found in the different functional areas of the map with respect to *border main roads*. For example, results show that 0% of the roads located in park 1, park 2, parking lot 1, and parking lot 2 were eliminated (see Table 1). Furthermore, few *main internal roads* were eliminated in the different functional areas of the map. Among these, 4.1% were observed in park 1, and 0% in parking lot 1 (see Table 2).

Table 1. Inclusion and elimination of *border main roads* with respect to the different functions in the map

%	eliminated	included
area 1	17.6	82.4
area 2	5.5	94.5
cemetery	4.4	95.6
park 1	0.0	100.0
park 2	0.0	100.0
parking lot 1	0.0	100.0
parking lot 2	0.0	100.0

Contrary to the previous results, high percentages of elimination of *secondary roads* were detected in most of the areas by which the map is classified. It was seen that 71% of the roads located in area 1, 70% of those located in park 1 were eliminated. However, a different situation was observed in the areas of the stadium and area 2, where only 7% and 3% of the *secondary roads* were eliminated, respectively (see Table 3). Moreover, regarding *dead ends*, results showed that 66.6% were eliminated in the area of the cemetery, and 80% in area 2.

Table 2. Inclusion and elimination of *main internal roads* with respect to the different functions in the map

%	eliminated	included
area 1	7.7	92.3
area 2	8.1	91.9
cemetery	8.8	91.2
park 1	4.1	95.9
park 2	7.2	92.8
parking lot 1	0.0	100.0
parking lot 2	6.6	93.4

Table 3. Inclusion and elimination of *secondary roads* with respect to the different functions in the map

%	eliminated	included
area 1	71.0	29.0
area 2	3.0	97.0
cemetery	68.0	32.0
park 1	70.0	30.0
park 2	62.0	38.0
stadium	7.0	93.0

The Process of Schematization.   Findings derived from the study on particular cognitive patterns of behavior identified during the schematization process are presented as follows. 73% of subjects began to construct the schematic map starting from the area of the stadium, whereas only 27% start from the map's borders. Furthermore, most subjects schematize the map by referring to hierarchical relationships between roads. 53% of subjects schematized *main border roads* before *main internal roads* (see Table 4), and 80% of the subjects schematized *main internal roads* before *secondary roads* (see Table 5). These results are valid at least for that side of the map at which subjects started schematization. On the other hand, only 13% of the subjects neither considered the *main border / main internal roads* hierarchy, nor referred to the *main internal / secondary roads* hierarchy in any sector of the map. By exploring the main strategies used in the process of schematization, we divided the map into two main sectors or sides. Both sides relate to the right or left of the stadium, which is the main destination of the wayfinding task located in the center of the map. The terms *horizontal* and *vertical* strategies used below refer to the pattern of displacement effected by subjects while schematizing the sides of the map. 40% of the subjects schematized one sector of the map, and then completed the other sector (vertical strategy). On the other hand, 60% of the subjects schematized both sides of the map in parallel, iterating from one side to the other (horizontal strategy).

Referring to the schematization of the whole map, 33% of those subjects who used the horizontal strategy failed to schematize from *main internal roads* to *secondary internal roads*, but all of those subjects who used the vertical strategy had a 100% success. Furthermore, 56% of those who use the horizontal strategy failed to schematize from *border roads* to *main internal roads*, but only 33% of those subjects who used the vertical strategy failed to do so. Moreover, it was seen that those subjects who used the vertical strategy were able to preserve branching points from the map (as considered in the taxonomy) by 12% more than those who used the horizontal strategy. It was also found that subjects who used the vertical strategy were able to preserve right-left relationships between branching points by 20% more than those who used the horizontal strategy. Finally, it was observed that those who used the vertical strategy eliminated roads of every type by 10% more than those who used the horizontal strategy.

Table 4. Hierarchies considered in the schematization process (from *border main roads (bmr)* to *main internal roads (mir)*)

%	from *bmr* to *mir*	from *mir* to *bmr* or oscillation
one side of the map	53.0	47.0
whole map	20.0	80.0

Table 5. Hierarchies considered in the schematization process (from *main internal roads (mir)* to *secondary roads (sr)*)

%	from *mir* to *sr*	from *sr* to *mir* or oscillation
one side of the map	80.0	20.0
whole map	40.0	60.0

5   Discussion

The Schematic Map. Results indicate that the taxonomy of branching points provided in the empirical task is helpful for carrying out the schematization of the original map. Subjects manage to construct the new map using the prototypical branching points supplied with the assigned taxonomy. Moreover, most branching points of the new map are schematized by considering their related prototypes in the given taxonomy. Although schematizing maps involves undesirable effects such as modifications of branching points, findings illustrate about the effectiveness of the taxonomy in general, and the qualitative angle constraint in particular, as appropriate tools for preserving important relationships between road connections in the environment and branching points depicted in the schematic map.

Another important finding shows that right / left distinctions between schematized branching points are largely preserved with respect to those of the original map; in other words, this means that in most cases we found that those branching points located to the left or to the right of another one, are fully preserved in the new map as well. Although some changes and distortions are likely to occur as a direct result of both the qualitative angle constraints and nature of the schematizing process, it is worth saying that right / left relationships are by no means part of such modifications.

However, it has to be argued that such relationships are seen as an important component for orientation in wayfinding.

Occasionally, distances between closely located branching points are simplified by merging two branching points into one. In some cases, this operation is also applied for merging three branching points into one. It must be noted, however, that the relationship between the environment and the map is subject to metrical scale reduction. The bigger this reduction, the larger the visual distortions between branching points. In specific cases in which schematic maps are represented in very small scales, relatively larger or medium distances will probably be wrongly perceived as shorter ones. Consequently, merged branching points might cause mismatches with the use of the schematic map in the real environment. In other words, although merged branching points in the schematic map are regarded as an effective mechanism of simplification, scale should also be referred to as an important parameter.

Regarding the simplification and schematization of roads, we have seen that one third of the (total) roads of the provided map is eliminated from the new map. Within these, we have found that almost half of the *external roads* are eliminated from the map. Furthermore, more than half of the cases of *secondary roads* and *dead ends* count as irrelevant for the purposes of wayfinding, and are therefore eliminated. However, *border main roads* and *internal main roads* are largely preserved in the new map. All these findings suggest possible relations between the relative importance of a road with respect to the specific wayfinding task, and its consequent inclusion or elimination in the schematic map. In order to gather more refined evidence about these arguments, we have also studied the relationship between road types and the functional areas upon which the map is organized.

Based on the previous findings, we have proposed a general classification in which both *border main roads* and *internal main roads* are part of a higher level hierarchy, and *secondary roads* and *dead end roads* are included in the lower levels of that hierarchy. The experiments indicate that roads being part of a high hierarchical level are generally preserved in the new map independently of the relative importance of the functional area in which they are located. This means that in the process of deciding whether to include a higher-level road in a schematic map or not, the meaning of functional areas may be irrelevant or of lesser importance.

However, a different conclusion is presented with respect to roads belonging to a low hierarchical level, where preserving or eliminating them mostly depends on the role played by the functional area in which any of these roads is located. This argumentation serves to explain why in contrast to what happened in most functional areas of the map, *secondary roads* are largely preserved in the stadium area (indispensable to reach the main destination), and in area 2 (vital to connect with the city train). On the other hand, it becomes reasonable to argue that since *dead ends* do not play any vital role for wayfinding to places outside those roads, they are largely eliminated in the map.

The Process of Schematization. The second part of the analysis focuses on the process of schematization. By analyzing the verbalizations of subjects, it is observed that most of them start constructing the schematic map from the stadium area. It seems reasonable to hypothesize that subjects felt more confident to start the schematizing process by first structuring the map from the stadium (the destination roads lead to) and then continue doing so with respect to the remaining areas of the map. It is worth saying that only a third of the subjects started to schematize the map from the *main border roads* to the inner areas of the map.

In addition, two main aspects related to the process of schematization and construction of the map are identified. While the first one deals with the consideration of hierarchies, the second one focuses on such hierarchies with respect to different schematization strategies.

First, it is observed that most subjects construct the schematic map by referring to hierarchical relationships between roads. The dominant pattern of behavior for the majority of subjects shows that at least for one side of the map, *main border roads* are schematized before *main internal roads*, and *main internal roads* are generally schematized before *secondary roads*. Observing the whole map, we found that almost half of the subjects schematize *main internal roads* before *secondary roads* in all sectors, but few of them schematize *main border roads* before *main internal roads*. On the other hand, it is worth saying that a minority of the subjects schematizes the map without referring to any hierarchical relationship at all. Considering the findings of this study, it is argued that thinking in hierarchical structures is generally used as a successful aid helping most subjects to construct the schematic map.

Second, two main strategies are identified in the process of map schematization. The first one shows that less than half of the subjects schematize first of all one side of the map, and thereafter complete the other sector (vertical strategy). In the second one, a little more than half of the subjects schematize both sides of the map in parallel, sequentially alternating from one side to the other (horizontal strategy).

From previous findings of this study, we see that few subjects who use the vertical strategy are unable to schematize the whole map from *main border roads* to *main internal roads*, and from *main internal roads* to *secondary roads*. It is claimed that most subjects who use the vertical strategy (from one side of the map to the other one) learn from their experiences acquired in one side of the map. Thus, when they schematize the other side of the map they are able to recall and re-apply such experiences in a systematic and hierarchical way. The same cannot be said regarding the horizontal strategy, which refers to the parallel schematization of both sides of the map. We saw that most subjects are able to schematize using the hierarchical relationship from *main internal roads* to *secondary roads*, but half of them fail to schematize from *main border roads* to *main internal roads*. Alternating between cyclical oscillations that go from one side of the map to the other, the horizontal strategy privileges the *main internal roads / secondary roads* hierarchy. Contrasting with findings related to the vertical strategy, using the horizontal strategy for this hierarchical relationship does not completely help in recalling experiences learned in one side of the map and systematically applying them to the other one.

By comparing the vertical and horizontal strategies, we conclude that the vertical strategy proves to be more helpful than the horizontal strategy. Schematizing one complete sector of the map, and thereafter proceeding with the other one contributes to: i) structure visual information according to hierarchies; ii) learn from experiences it in a systematic and organized way; iii) increase the preservation of branching points from the map as regarded in the taxonomy; iv) increase the elimination of unnecessary roads.

6   Conclusions

The focus of this research was set on the schematization of maps for wayfinding tasks. By considering main findings of this study, conclusions and possible implications for navigation in the real environment are presented. We have seen that subjects were able to establish strong relationships between the provided taxonomy and the map. Both branching points and right-left relationships between branching points were generally preserved in the schematized map. These results strengthen what was pointed out in Section 3.2, where we criticized the inaccurate and confusing official schematic map displayed at many locations in the stadium area. We claimed that mismatches between the physical environment and the wayfinding map were caused as a consequence of many important spatial relationships that were oversimplified. The fact that in our study subjects tend to avoid such kind of mismatches is a clear reference to the conceptual differences that exist between schematization and undesirable oversimplification. This aspect should not be ignored when constructing wayfinding maps.

It can also be argued that if subjects are able to construct schematic maps considering the assigned taxonomy, they are also potentially able to use these types of maps while performing wayfinding tasks in a real environment. Some light has been shed on the classical dilemma of what to include and what to eliminate in a schematic map. The high percentage of road elimination shows that not always all the existing roads in a real environment must be included in a schematic map. It was shown that this largely depends on the relative hierarchical level of the road, or on the importance of the functional area where the road is located represents regarding the wayfinding task. The latter suggests that decisions of what type of roads might be included in a map should not be taken on the mere basis of just geometrical considerations. Their relative meaning regarding wayfinding also counts. The finding that most subjects construct the schematic map by considering hierarchical relationships between roads could have important implications for navigation in the environment. This reference to hierarchies suggests that stressing relative relationships between roads could be seen as an important aid for wayfinding.

Acknowledgments

We wish to thank the students of the project seminar on knowledge representation techniques for spatial cognition for their field work around the Volksparkstadion. We thank the participants of the Hamburg *International Workshop on Maps and Diagrammatical Representations of the Environment* for helpful comments and fruitful discussions about this work. The valuable criticisms and suggestions by Gabi Janzen and the anonymous reviewers are gratefully acknowledged. Special thanks are due to Rik Eshuis and Barbara Kaup for valuable support in conducting the empirical investigations.

References

Arthur, P., & Passini, R. (1992). *Wayfinding: people, signs and architecture.* New York: MacGraw-Hill Ryerson.

Barkowsky, T., & Freksa, C. (1997). Cognitive requirements on making and interpreting maps. In S. Hirtle & A. Frank (Eds.), *Spatial information theory: A theoretical basis for GIS* (pp. 347-361). Berlin: Springer.

Berendt, B., Barkowsky, T., Freksa, C., & Kelter, S. (1998). Spatial representation with aspect maps. In C. Freksa, C. Habel, & K. F. Wender (Eds.), *Spatial cognition - An interdisciplinary approach to representing and processing spatial knowledge* (pp. 313-336). Berlin: Springer.

Correa de Jesus, S. (1994). Environmental communication: design planning for wayfinding. *Design Issues, 10*(3), 33-51.

Downs, R. M., & Stea, D. (1973). *Maps in minds: reflections on cognitive mapping.* New York: Harper & Row.

Frank, A. (1992). Qualitative spatial reasoning with cardinal directions. *Proceedings of the Seventh Austrian Conference on Artificial Intelligence, Vienna* (pp. 157-167). Berlin: Springer.

Freksa, C. (1991). Qualitative spatial reasoning. In D. Mark & A. U. Frank (Eds.), *Cognitive and linguistic aspects of geographic space* (pp. 361-372). Dordrecht: Kluwer.

Freksa, C. (1999). Spatial aspects of task-specific wayfinding maps. In J. S. Gero & B. Tversky (Eds.), *Visual and spatial reasoning in design* (pp. 15-32). University of Sydney: Key Centre of Design Computing and Cognition.

Freksa, C., Moratz, R., & Barkowsky, T. (2000). Schematic maps for robot navigation. In C. Freksa, W. Brauer, C. Habel, & K. F. Wender (Eds.), *Spatial Cognition II - Integrating abstract theories, empirical studies, formal models, and practical applications.* Berlin: Springer.

Furbach, U., Dirlich, G., & Freksa, C. (1985). Towards a theory of knowledge representation systems. In W. Bibel & B. Petkoff (Eds.), *Artificial Intelligence: Methodology, systems, applications* (pp. 77-84). Amsterdam: North-Holland.

Garling, T., Book, A., & Lindberg, E. (1986). Spatial orientation and wayfinding in the designed environment: A conceptual analysis and suggestions for postoccupancy evaluation. *Journal of Architectural and Planning Research, 3*, 55-64.

Golledge, R., Klatzky, R., & Loomis, M. (1996). Cognitive mapping and wayfinding by adults without vision. In J. Portugali (Ed.), *The construction of cognitive maps* (pp. 215-246). Netherlands: Kluwer.

Hernández, D. (1994). Qualitative representations of spatial knowledge. *Lecture Notes in artificial intelligence, 804* (pp. 25-54). Berlin: Springer.

Hillier, B., Hanson, J., & Peponis, J. (1984). What do we mean by building function? In J. Powell, I. Cooper, & S. Lera (Eds.), *Designing for building utilization* (pp. 61-71). New York: Spon.

Janzen, G., Herrmann, T., Katz, S., & Schweizer, K. (2000). Oblique angled intersections and barriers: Navigating through a virtual maze. In C. Freksa, W. Brauer, C. Habel, & K. F. Wender (Eds.), *Spatial Cognition II - Integrating abstract theories, empirical studies, formal models, and practical applications.* Berlin: Springer.

Kaplan, S. (1976). Adaptation, structure, and knowledge. In G. Moore & R. Golledge, *Environmental knowing* (pp. 32-46). Stroudsburg, Penn: Dowden, Hutchinson, and Ross.

Lawton, C., Charleston, S., & Zieles, A. (1996). Individual and gender-related differences in indoor wayfinding. *Environment and Behavior, 28*(2). 204-219.

Levine, M. (1982). You-are-here maps - Psychological considerations. *Environment and Behavior, 14*(2), 221-237.

Lynch, K. (1960). *The image of the city*, Cambridge: MIT Press.

MacEachren, A. (1995). *How maps work: representation, visualization, and design.* New York: The Guilford Press.

O'Neill, M. (1991a). Evaluation of a conceptual model of architectural legibility. *Environment and Behavior, 23,* 259-284.

O'Neill, M. (1991b). Signage and floor plan configuration. *Environment and Behavior, 23,* 553-574.

Passini, R. (1984). *Wayfinding in architecture.* New York: Van Nostrand Reinhold Company.

Passini, R. (1998). Wayfinding and dementia: some research findings and a new look at design. *Journal of Architectural and Planning Research, 15*(2), 133-151.

Palmer, S. E. (1978). Fundamental aspects of cognitive representation. In E. Rosch & B. B. Lloyd (Eds.), *Cognition and categorization* (pp. 259-303). Hillsdale, NJ: Lawrence Erlbaum.

Peponis, J., Zimring, C., & Choi, Y. (1990). Finding the building in wayfinding. *Environment and Behaviour, 22,* 555-590.

Raubal, M., & Egenhofer, M. (1998). Comparing the complexity of wayfinding tasks in built environments. *Environment and Planning B: Planning and Design, 25,* 895-913.

Suwa, M., Gero, J., & Purcell, T. (1998). The roles of sketches in early conceptual design processes. *Proc. 20th Annual Meeting of the Cognitive Science Society* (pp. 1043-1048), Hillsdale, NJ: Lawrence Erlbaum.

Timpf, S. (1998). *Hierarchical structures in map series.* Dissertation. Technical University Vienna.

Tolman, E. (1948). Cognitive maps in rats and men. *Psychological Review, 55,* 189-208.

Tversky, B., & Lee, P. U. (1999). Pictorial and verbal tools for conveying routes. In C. Freksa & D. M. Mark (Eds.), *Spatial information theory: cognitive and computational foundations of geographic information science* (pp. 51-64). Berlin: Springer.

Weisman, G. (1981). Evaluating architectural legibility: wayfinding in the built environment. *Environment and Behavior, 13,* 189-204.

Some Ways that Maps and Diagrams Communicate

Barbara Tversky

Department of Psychology, Stanford University
Stanford, CA 94305-2130
bt@psych.stanford.edu

Abstract. Since ancient times, people have devised cognitive artifacts to extend memory and ease information processing. Among them are graphics, which use elements and the spatial relations among them to represent worlds that are actually or metaphorically spatial. Maps schematize the real world in that they are two-dimensional, they omit information, they regularize, they use inconsistent scale and perspective, and they exaggerate, fantasize, and carry messages. With little proding, children and adults use space and spatial relations to represent abstract relations, temporal, quantitative, and preference, in stereotyped ways, suggesting that these mappings are cognitively natural. Graphics reflect conceptions of reality, not reality.

1   Introduction

One candidate for an intellectual achievement separating humankind from other kinds is the creation of cognitive artifacts, of external devices that extend the human mind. They range from using fingers for counting or fashioning bends in trees to mark trails to powerful computers or global positioning systems. Cognitive tools augment the mind in two major ways: they reduce memory load by externalizing memory, and they reduce processing load by allowing calculations to be done on external rather than internal objects and by externalizing intermediate products. Of course, external representations have other benefits as well. They take advantage of people's facility with spatial representations and reasoning, they are more permanent than thoughts or speech, they are visible to a community (cf. Tversky, in press, b). Written language is prominent among useful cognitive tools, but graphics of various types preceded written language and serve many of the same functions. What renders graphics privileged is the possibility of using space and elements in space to express relations and meanings directly, relations and meanings that are spatial literally as well as metaphorically.

Early graphics, found in cultures all over the world, include not only lines on paper or bark, but also inscriptions on trees, paintings in caves, incisions in bones, carvings in wood, scarifications on bodies, and more. They portrayed things that took up space, animals, objects, and events, actual or imagined, or spaces, the prime example being maps. It was only in the late 18th century that graphics began to be used, in the West, to portray non-spatial, abstract information, most notably, economic data, such as

Ch. Freksa et al. (Eds.): Spatial Cognition II, LNAI 1849, pp. 72-79, 2000.
© Springer-Verlag Berlin Heidelberg 2000

balance of trade over time (Beniger and Robin, 1958; Tufte, 1990). I will first characterize ways that graphics convey the essentially visual through a discussion of maps, the prime example of these graphics. Then I will characterize ways that diagrams visualize the non-visual.

2   Characterizing Maps

One of the most typical and ubiquitous of graphics is a map. Nowadays, when people think of maps, they think of the world map in school classrooms or the road map in the glove compartment of the car or the city map in the tourist guide. Yet the maps that have been invented across cultures throughout human existence are both far more and far less than these. They seem to share two features, though by no means strictly. Maps typically portray an *overview,* and they reduce and roughly preserve *scale.* As we shall see, in practice, maps are not limited to overviews, and scale is not always consistent. Here, we analyze those maps, characterize what they do and do not do, and bring those insights to the study of diagrams in general.

2.1.   Maps are Two-Dimensional

Although the worlds that maps typically represent are three-dimensional, maps are typically two-dimensional. This is for a number of converging reasons. Obviously, it is easier to portray a two-dimensional space on a piece of paper than a three-dimensional space. But, in addition to the constraints of the medium, there are cognitive reasons for portraying maps as two-dimensional overviews. First, it seems that people readily conceive of three-dimensional environments as two-dimensional overviews, in itself a remarkable cognitive achievement. This is attested by the invention of two-dimensional overview maps by diverse and dispersed cultures as well as their spontaneous invention by children. Next, three-dimensional diagrams are difficult to construct and difficult to comprehend (e. g., Cooper, 1984; Gobert, 1999). Architects and other designers, for example, prefer to first construct two-dimensional plans or overviews and two-dimensional elevations before integrating them into three-dimensional sketches or models (Arnheim, 1977). As Arnheim points out, the considerations important for plans differ from those important for elevations. Like maps, plans display the spatial relations among large structures, providing information useful for navigating among them. Plans, then, are a presentation useful for evaluating function. Elevations provide information about what the structures look like, important for recognition of them. Elevations are a presentation useful for evaluating aesthetics. In addition to these reasons, for many purposes, three-dimensional information about environments is simply not needed, and may even interfere; the spatial relations among the large features is sufficient information.

2.2.   Maps Omit Information

The next thing to notice about maps is that they omit information. One of the reasons for this has to do with the very nature of mapping. Borges' invented fable of an Empire where the Art of Cartography was perfected so that a map of the Empire the size of the Empire could be created was just that, a fable, an absurdity (Borges, 1998). The very usefulness of a map comes from its reduction of space. Reductions in size require reductions in information to be useful. An aerial photograph does not make a good map. Maps omit information because much of the information in space is not only not relevant, but also gets in the way of finding the essential information. Maps are typically designed for a communicative purpose; that purpose determines what information should be kept and what information can be eliminated. Consider, for examples, two kinds of maps created by seafaring cultures (Southworth & Southworth, 1982). Coastal Eskimos carried carved wood outlines of the coastline with them in their canoes to guide them in their travels. The Marshall Islanders in the Pacific, who navigate among islands too distant to be seen for much of their voyages, constructed maps out of bamboo sticks and shells. The shells indicated islands and the sticks ocean currents, identifiable from the flotsam and jetsam that accumulates along them. For more familiar examples, consider the information useful for a map to guide drivers in the city or, alternatively, a map to guide hikers in the mountains. Details of types of roads and intersections are important to the former, whereas topographical details are important to the latter.

2.3.   Maps Regularize

Yet another characteristic of maps in practice is that they simplify and regularize information. A classic example is the London subway map, which has served as a model for subway maps all over the world. The London subway system, like many subway systems, is quite complex, with many different lines and intersections. The information important to travelers includes the general direction of the lines, the stops, and the intersections with other lines. The specific directions, path curvatures, and distances are not usually critical. So the lines on the London subway map are presented as straight lines, oriented vertically, horizontally, or diagonally, ignoring the finer distinctions of local curves and orientations. This simplification, however, facilitates computing the desired information, the general directions and the connections, and the distortions produced by the regularization do not cause sufficient errors to create problems.

2.4.   Maps Use Inconsistent Scale and Perspective

Road maps illustrate another common feature of maps. They use inconsistent scale. Indeed, roads, rivers, railroads, and other important environmental information portrayed in maps would simply not be visible if scale were consistently adopted. In addition, many maps violate consistent perspective. Consider, for example, a popular kind of tourist map. These present an overview of the city streets superimposed with frontal views of frequently visited landmarks, such as churches and public buildings. Presenting both perspectives in a single map is a boon to tourists. It allows them to navigate the streets in order to find the landmarks, and then to recognize the landmarks when they see them. Maps with mixed perspectives are by no means a modern invention. For example, maps portraying overviews of paths and roads and frontal views of structures and living things are clearly visible in petroglyphs dating back more than 3000 years in northern Italy (Thrower, 1996).

2.5.   Maps Exaggerate, Fantasize, and Carry Messages, Aesthetic, Political, Spiritual, and Humorous

In 1916, King Njoya presented the British with a map of his kingdom, Banum, in northwestern Cameroon. To impress the Europeans with his modernity, he put 60 surveyors to work for two months to construct the map. While fairly accurate, the map exaggerates the size of the capital, and locates it, incorrectly, smack in the center of the kingdom (Bassett, 1998). Maps influenced by the political, mythical, historical, or fantastic are common all over the world. In medieval Europe, T-O maps were popular. They were called that because they looked like T's embedded in O's, the circular background for the world. East, the direction of the (presumed) beginning of the world, Adam and Eve, the rising sun, was at the top (hence the word "oriented" from *oriens* or east). The top bar of the T was formed by the Nile on the south or the left and the Dan on the north or the right. The Mediterranean formed the vertical bar. Such maps portrayed religious beliefs and reflected elegant geometry and symmetry more than actual geography. They also added decorative depictions, of Adam and Eve in the Holy Land, of the four winds, and more. Maps mixing geography, beliefs, and history are not unique to Europeans. T-O maps appeared in Asia (e. g., Tibbetts, 1992), Similar maps appeared in preColumbian Mesoamerica, for example, a map showing the imagined or real migrations of the ancestors superimposed on a geographic map (Mundy, 1998). Maps of the heavenly spheres appeared in both Europe and Asia (Karamustafan, 1992).

Humorous maps enliven newspapers, books, and journals. Perhaps best known are the "New Yorker's View of the World" maps of Steinberg that graced the covers of the New Yorker as well as many dormitory rooms. Such maps take a local perspective so that close distances loom larger than far distances. They also include landmarks likely to be of interest to the New Yorker and omit those of less interest. A more recent example from the New Yorker was a map of New York City as the palm of a hand, with Broadway as the lifeline and the boroughs as fingers. These are but a few of many, many examples of maps that are designed to convey far more than geography, and that sacrifice geographic accuracy for other messages.

Put briefly, maps, those produced by professionals as well as amateurs, are schematic (for a related view, see Freksa, Moratz, and Barkowsky, this volume). Schematic maps are created for a specific goal or goals, usually communicative, and they distill and highlight the information relevant to those purposes. They eliminate extraneous information to remove clutter, making the essential information easier to extract. They simplify and even exaggerate this information. People's minds also schematize spatial and other information. In fact, many of the ways that minds schematize correspond to the way that maps schematize. Internal representations of environments omit and regularize information, they mix perspectives, reduce dimensionality, and exaggerate. This can lead to internal "representations" that are impossible to realize even as a three-dimensional world (e. g., Tversky, 1981; in press, a). Matching external schematizations to internal ones may also facilitate processing information from maps. Of course, the match between internal schematizations of environments and external schematizations of maps is no accident; both are products of human minds. Moreover, these same processes, omission, use of inconsistent information, regularization, exaggeration, politicization, beautification, and more, appear in other depictions and external representations.

3   Characterizing Graphics

When people talk or think about abstract concepts, they often do so in terms of spatial concepts (e. g., Clark, 1973; Lakoff and Johnson, 1980). There is good reason for this. Spatial competence appears early in life and is essential for survival (e. g. Tversky, in press, a). Bootstrapping abstract thought onto spatial thought should allow transfer of spatial agility to abstract agility. Graphic visualizations of abstract concepts and processes such as those in economics, biology, chemistry, physics, mathematics, and more should pay similar benefits. An examination of graphics produced by children and adults throughout history and across cultures reveals some general characteristics of the way they use space and the elements in it to convey meaning (Tversky, 1995; in press, b).

3.1.   Spatial Relations

Graphics, such as maps, consist of elements and the spatial relations among them (for a broader view of this analysis, see Tversky, in press, b). Whereas maps use space to represent space, graphics can use space to represent other concepts, such as time, quantity, and preference. Underlying the use of space is a simple metaphor, distance in depictive space reflects distance in real space, or, in the case of abstract graphics, distance on some other dimension or feature. The spatial mapping from the represented world to the visualization can convey information at different levels, categorical, ordinal, interval, or ratio. Even written alphabetic languages use graphic devices in addition to symbolic ones to convey meaning. At the categorical level, only groupings are meaningful; elements in one group share a feature or features not shared by elements in other groupings. Organizing names of students by classes, clubs, or dormitories are examples of categorical groupings. Separating the lists spatially is a rudimentary use of space to indicate groupings. There was a time when words were not separated in writing; the current practice of putting spaces between words is also a rudimentary use of space to indicate groupings. Parentheses, boxes, and frames are visual devices that accentuate spatial separation in conveying categorical information. In the case of ordinal mappings, the order of the groups is meaningful. A rudimentary way to indicate order is spatial (or temporal) order of a list: ordering children by age or ordering a shopping list by the route through the supermarket. The spatial order of the list reflects some other order. Indentation as in paragraphs or outlines is another simple way of indicating order. Networks and trees convey order by augmenting the spatial devices with visual ones. For interval mappings, the distance between the groupings as well as the order is meaningful, representing the distance on some other dimension. X-Y graphs are familiar examples of using space to express interval relations. Graphs are also common for ratio mappings where the zero point is meaningful so ratios are meaningful. For the prototypical map, the sort produced by government agencies, mapping is ratio. For the typical map, the sort produced for special purposes like driving or tourism, the mapping from actual space to depictive space may be, as we have seen, complex and even inconsistent.

Direction of Spatial Mappings. Although direction in space is objectively neutral, cognitively, it is not neutral, even for young children. Children from cultures where language is written left to right as well as children from cultures where language is written right to left were asked to place stickers on a square sheet of paper to represent items that could be ordered by time, quantity or preference (Tversky, Kugelmass, and Winter, 1991). On one trial, they were asked to represent the time they get up in the morning, the time they go to school, and the time they go to bed. On another trial, they were asked to represent a TV show they loved, one they didn't like at all, and one they sort of liked. Some of the youngest children did not put all the items on a line; that is, their representations were categorical, not interval. Most of the children's mappings preserved ordinal information, but only children around 11 years of age preserved interval information. As for direction, children from all language cultures mapped increases in quantity or preference from left to right, from right to left. and from down to up. The one direction they avoided for mapping increases was from up to down. The only mappings to follow the order of writing were the temporal ones.

The bias to map up to more or better or stronger is not restricted to children. A survey of common visualizations in college text books, such as those portraying evolution and geological eras, revealed that time is usually conveyed vertically, with the present time, the pinnacle of evolution, at the top (Tversky, 1995). Words like "pinnacle" indicate that these biases are present in language as well. We say someone's at the "head" or "top" of the class or below the mean. For gestures as well, up is generally good or strong or powerful or successful, and down the opposite. Thus vertical mappings seem loaded, with the good/more/powerful pole at the top and the bad/less/weak pole at the bottom. In contrast, as the politicial consensus attests, the horizontal left/right axis is more neutral.

3.2. Elements

Graphics portraying information that is not spatial, but rather abstract, use elements in space as well as space to convey that information. Think, for example, of an evolutionary tree. It uses nodes and lines and perhaps icons to convey the development of species. Or consider a diagram of the workings of a machine or system, such as a car or a power plant. Like the evolutionary tree, it uses lines, nodes, and icons to convey the system parts and their causal relations. Icons can depict literally or figuratively. More literal icons are widespread in simple contemporary graphics, such as road and airport signs, where a curve in a sign indicates a curve in the road and a place setting indicates a restaurant stop. They also occur throughout history in petroglyphs, cave paintings, and ideographic scripts. Ideographic scripts as well as contemporary icons make use of "figures of depiction" as well. In ideographic scripts, ideographs for animals, for example, often included only part of the animal, the head or horns, and ideographs for rulers often portrayed an association to the role, such as a staff or crown. Icons common in graphical user interfaces do likewise; scissors can be used to cut unwanted text and a trash can to dispose of unwanted files.

Expressing Meanings Using Space. Space, then, as well as visuospatial devices, can be used to convey abstract concepts in cognitively natural ways. Starting from abstract concepts and examining the devices used to convey them emphasizes the point. Similarity can be represented by similar appearance, such as font or color, and difference by different appearances. Groupings, based of course on similarity, can be represented by proximity in space, and emphasized by devices that suggest enclosures, such as circles, boxes, frames, and parentheses. Connections between groups are readily conveyed by devices that indicate paths, especially lines, connected or broken. Orderings may be conveyed in a variety of ways including order in space and order on visual dimensions such as brightness or size. A popular device to indicate direction is an arrow. Arrows may be cognitively compelling for two reasons. Arrows as weapons fly in the direction of the arrowhead. Water and other substances flow downward in rivulets that converge, forming arrows in the direction of the flow. The concept of extent can be readily represented by spatial extent, length or area, and the concept of proportion can be readily represented by spatial proportion, as in pie charts.

4   In Sum

Depictions reflect conceptions of reality, not reality. This holds for depictions of things in the world as well as things in the mind. Maps, drawings, graphs, and diagrams are forms of communication. As such, they are inherently social devices used in social interactions. As in most social interactions, veridicality may not be the primary concern. Rather the primary concern may be affecting the cognitions, emotions, and ultimately, the actions of those for whom the communication is intended and designed.

References

Arnheim, R. (1977). The dynamics of architectural form. University of California Press: Berkeley.

Bassett, T. J. (1998). Indigenous mapmaking in intertropical Africa. In Woodward, D. and Lewis, G. M. (Editors). History of cartography. Vol. 2 Book 3: Cartography in the traditional Africa, America, Arctic, Australian, and Pacific societies. Pp. 24-48. Chicago: Chicago Press.

Beniger, J. R. & Robyn, D. L. (1978). Quantitative graphics in statistics. The American Statistician, 32, 1-11.

Borges, J. L. (1998). On the exactitude of science. Collected Fictions. P. 325. Translated by Andrew Hurley. New York: Penguin.

Clark, H. H. (1973). Space, time, semantics, and the child. In T. E. Moore (Ed.), Cognitive development and the acquisition of language. Pp. 27-63. New York: Academic Press.

Cooper, L. A. (1989). Mental models of the structure of visual objects. In B. Shepp & S. Ballesteros (Eds.), Object perception: Structure and process. (pp. 91-119). Hillsdale, N. J.: Erlbaum.

Freksa, C., Moratz, R. and Barkowsky, T. (2000). Schematic maps for robot navigation. In C. Freksa, W. Brauer, C. Habel, & K.F. Wender (Eds.), *Spatial Cognition II – Integrating abstract theories, empirical studies, formal models, and practical applications*. Berlin: Springer.

Gobert, J. D. (1999). Expertise in the comprehension of architectural plans. In J. Gero and B. Tversky (Editors), Visual andspatial reasoning in design. Pp. 185-205. Sydney, Australia: Key Centre of Design Computing and Cognition.

Karamustafan, A. T. (1992). Cosmographical diagrams. Celestial mapping. In Harley, J. B. and Woodward, D. (Editors) The history of cartography. Vol. 2. Book One Cartography in the traditional Islamic and South Asian Societies. Pp. 71-89. Chicago: University of Chicago Press.

Lakoff, G. & Johnson, M. (1980). Metaphors we live by. Chicago: University of Chicago Press.

Mundy, B. E. (1998). Mesoamerican cartography. In Woodward, D. and Lewis, G. M. (Editors). History of cartography. Vol. 2 Book 3: Cartography in the traditional Africa, America, Arctic, Australian, and Pacific societies. Pp. 183-256. Chicago: University of Chicago Press.

Southworth, M. and Southworth, S. (1982). Maps: A visual survey and design guide. Boston: Little, Brown and Company.

Thrower, N. J. W. (1996). Maps and civilization: Cartography in culture and society. Chicago: University of Chicago Press.

Tufte, E. R. (1990) Envisioning Information. Cheshire: Graphics Press.

Tversky, B. (1995). Cognitive origins of graphic conventions. In F. T. Marchese (Editor). Understanding images. Pp. 29-53. New York: Springer-Verlag.

Tversky, B. (1981). Distortions in memory for maps. Cognitive Psychology, 13, 407-433.

Tversky, B. (In press, a). Levels and structure of cognitive mapping. In R. Kitchin and S. M. Freundschuh (Editors). Cognitive mapping: Past, present and future. London: Routledge.

Tversky, B. (In press, b). Spatial schemas in depictions. In M. Gattis (Editor), Spatial schemas and abstract thought. Cambridge: MIT Press.

Tversky, B., Kugelmass, S. and Winter, A. (1991) Cross-cultural and developmental trends in graphic productions. Cognitive Psychology, 23, 515-557.

Spatial Communication with Maps: Defining the Correctness of Maps Using a Multi-Agent Simulation

Andrew U. Frank

Dept. of Geoinformation
Technical University Vienna
frank@geoinfo.tuwien.ac.at

Abstract. Maps are very efficient to communicate spatial situations. A theoretical framework for a formal discussion of map production and map use is constructed using a multi-agent framework. Multi-agent systems are computerized models that simulate persons as autonomous agents in a simulated environment, with their simulated interaction. A model of the process of map production and map use is constructed based on a two-tiered *reality and beliefs model*, in which facts describing the simulated environment and the simulated agents' beliefs of this environment are separated. This permits to model errors in the persons' perception of reality.

A computerized model was coded, including all operations: the observation of reality by a person, the production of the map, the interpretation of the map by another person and his use of the knowledge acquired from the map for navigation, are simulated as operations of agents in a simulated environment.

1 Introduction

Daily experience tells us that maps are a very efficient and natural way to communicate spatial situations. Small children produce maps spontaneously and maps are among the earliest human artifacts. However, we seem not to have a good understanding how maps communicate spatial situations. Formal models for the processes of map production and use are missing. This leaves judgment of map quality to a large degree subjective, as map construction and map reading are both implying intelligent human interpretation. A person using a map knows the general morphology of the terrain and uses this knowledge to draw appropriate conclusions from the graphical signs on the map. Unfortunately, this general assumption of intelligent interpretation breaks down in unfamiliar terrain when a map is most needed. A more objective measure for correctness of a map, which does not rely on additional knowledge, is required. So far, we can only define consistency of a database as the absence of internal contradiction in a data quality. A formal definition for correctness, i.e., the correspondence between data and reality, cannot be constructed, as it would need to bridge between reality and the formal representation.

Formal methods to define correctness of geographic data are urgently needed in the emerging business with geographic information. It is necessary to assess the quality of geographic data collections and compare them. For example, we must be capable of

Ch. Freksa et al. (Eds.): Spatial Cognition II, LNAI 1849, pp. 80-99, 2000.

comparing the quality of competing data providers for In-Car Navigation Systems and point out errors based on objective criteria and not just based on anecdotes. The unfortunate adventure of a car driver following the advice of his In-Car navigation system to cross a river over a bridge was widely published. Too late, when the car was already floating in the river, he noticed that there was no bridge but only a ferry!

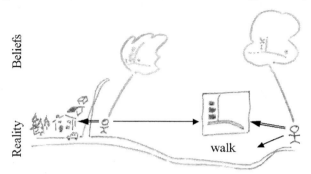

Fig.1. An agent producing a map and another agent using a map for navigation

In this paper a multi-agent formalism is used to produce a model of map production, map communication and map use. Multi-agent systems are computerized formal models, which contain a modeled environment and autonomous agents, which interact with this environment (Ferber 1998; Weiss 1999). The model formalizes the processes involved (Figure 1); it is not intended to be used for actual navigation in a city. I use here a simple task to make the discussion concrete; namely, the production and use of a small street network map for navigation. The model constructed simulates:

- The environment, which is constructed after the example of a small part of downtown Santa Barbara (Figure 4);
- A map-maker who explores the environment and collects information, which he uses to construct a map of the area; and
- A map-user who acquires this map to gain knowledge, which he uses to navigate in this environment.

The environment represents the world in which persons live and the agents represent the persons who make and use maps (Figure 2). The simulation includes multiple agents – at least one map-making agent and one or several map-using agents. In the simulated environment, it is possible to define what it means that a map is correct and how to compare the effectiveness of map communication with verbal communication. The definitions point out the strong connection between correctness of a map and the intended use of a map or spatial data set.

Multi-agent systems have been used previously for map generalization (Baeijs, Demazeau et al. 1995), where different agents apply rules to a given map graphics to produce a generalized map. The approach here is very different; we do not intend to model a part of the map production process, but to model the complete process, which starts with data collection in reality, produces the map and then the process of map use: reading the map to navigate in an unknown territory.

Braitenberg has introduced computational models in psychology (Braitenberg 1984) and demonstrated how insight can be gained from a fully simulated (synthetic) model.

The agents used here are nearly as simple as Braitenberg's "vehicles"; they are sufficient to contribute to our understanding of correctness and effectiveness of maps. *Correctness* of the map is judged within the model as the success of the agents in navigating in the environment. *Effectiveness* of the map can be judged by comparing the size of different representations to communicate the same information between agents. One can demonstrate that verbal descriptions are equally effective if one has to communicate a single route, but are inefficient to communicate a complex spatial situation, e.g., the street segments in a downtown area.

To construct a computational model for map production and map use is novel. The model uses a two-tiered *reality and beliefs representation*, in which reality (facts) and the agents' cognition (beliefs) are represented separately (Figure 1). Errors in the agent's perception of reality or errors in the production or reading of artifacts like maps, representing and communicating an agent's (possibly erroneous) beliefs can be modeled. It is possible to include imaginary or contrafactual maps as well. The formalization is in an executable functional language, using Haskell (Hudak, Peyton Jones et al. 1992; Peterson, Hammond et al. 1997; Peyton Jones, Hughes et al. 1999). The code is available from the web (http://www.geoinfo.tuwien.ac.at) and extensions to investigate similar questions are possible.

2   Computational Models Separating Reality and Beliefs

Computational models are formalized methods to describe our understanding of complex processes. They have been extremely successful in many areas, especially modeling aspects of our physical environment where models of reality and models of processes are linked (Burrough and Heuvelink 1988). They have been less successful in the information domain, in my opinion because the linkage between the static data describing reality and the processes of data observation, data collection and data use has not been achieved (Frank 1998). The described two-tiered *reality and beliefs computational model*, where the simulation contains separate representations of the environment and representations of the agent and its knowledge about the environment as two separate data sets, overcomes this problem, because the observation and data use processes are explicitly included. Following an AI tradition, the agent's knowledge is called 'belief' to stress the potential for differences between reality and the agent's possibly erroneous beliefs about reality (Davis 1990).

Real World Situation	**Multi-Agent Model**
Real World Situation	Model
World	Environment
Person	Agent
Map-maker	Map-making agent
Map user	Map using agent
Fact	Belief

Fig.2. Mapping from reality to model

The model takes into account many of the often-voiced critiques against formal models of cognition. In terms of Warfield and Stich (Stich and Warfield 1994, p. 5ff) models of cognition must have three properties:

- Naturalness: The semantics of the mental representations are linked to the operations of the agent observing the environment and acting in it. These observation operations are part of the model and their properties described.
- Misrepresentation is possible, as the model contains separate representations for the data, which stand for reality, and the data, which represent an agent's beliefs. The models of observation processes may produce errors; the actions may not use the information the agent has correctly represented.
- Fine-grained meanings are achieved, as concepts and what they are linked to in reality are separate. It is possible that the agent maintains beliefs about two different concepts, only later to find out that the two are the same.

The model constructed here has these properties:

2.1 Naturalness and Semantics

The semantics of the mental operations on the beliefs are directly connected to the person's bodily actions (Johnson 1987): mentally following a street segment's mental representation is given meaning through the correspondence with the physical locomotion of the agent along a street segment. This correspondence is kept in the model; the simulated mental operations of the agents are linked to the simulated bodily actions of the agents. The model is therefore not disembodied AI (Dreyfuss 1998) because the linkage between bodily actions of the agents and their mental representation is direct and the same as in persons (Lakoff and Johnson 1999).

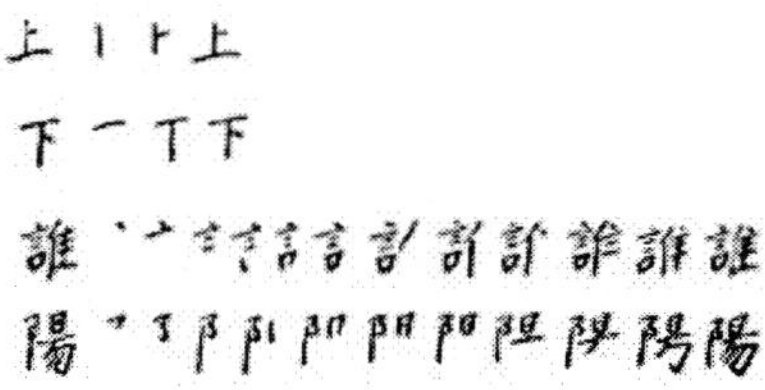

Fig. 3. Instructions how to draw Chinese characters (two simple and two complex ones from Tchen 1967)

"Information itself is nothing special; it is found wherever causes leave effects" (Pinker 1997, p. 65-66). The map product can be seen as the sequence of drawing steps (the causes) the map-maker follows to produce it and the map-reader does retrace these steps in his map reading process. It is instructive to observe that Chinese characters are not learned as figures but as a sequence of strokes (Figure 3). This is not only important for production, but also for recognition of signs created at different levels of fluidity. Westerners often copy Chinese characters as a picture and produce images, which are difficult to recognize.

In the multi-agent model, the structure of the operations for locomotion along a street segment, for drawing a street segment or for following a drawn street segment and for mentally following the belief about a street segment can be coded as the same polymorphic operation, applicable to different data structures; e.g., maps, real streets, etc. (not stressed in this presentation).

2.2 Misrepresentation

Persons – both the map-maker and the map-user – can make errors in the perception and form erroneous beliefs about the environment. The maps produced can also have errors or the map reading operation can include errors into the beliefs map-users form about the environment. Such errors or imprecisions can be modeled in the beliefs of the agents. Eventually, agents are prohibited to achieve 'impossible' states of the environment and are stopped in the model from executing impossible actions; e.g., to travel along a street not present in the environment.

2.3 Fine-Grained Meaning

Concepts can have various levels of detail – they can be 'read' from a map and therefore have no experience, e.g., a visual memory associated, or can have a partial knowledge, e.g., a street segment can have a known start but a not yet known end. It is possible to realize later that two different concepts are linked to the same real object, e.g., the intersection where 'Borders' is and the intersection of 'State Street' and 'Canon Perdido Street', which is the same in Santa Barbara (Figure 4). This is possible in multi-agent models, but not included in the simple model presented here.

3 Focused Discussion Based on an Example Case

The investigation is focused with a specific set of tasks in a concrete environment, namely finding a path between named intersections in a city street network. Research in cartography usually concentrates on transformations applied to maps – mostly discussions of map generalization (Weibel 1995) – situated in a diffuse set of implied assumptions about the intended map use and the environment represented (Lechthaler 1999). Concentration on a very specific example avoids this problem. I select here the communication with maps about a city environment for the purpose of navigation. The environment and the tasks are fixed. Then the construction of a computational model becomes possible.

The *environment* is maximally simplified to make this paper self-contained. It includes street segments, which are connected at street intersections (Figure 4). Many interesting aspects of real cities are left out: for example, one-way streets are excluded as well as turn-restrictions at the intersections; cost of travel is proportional to distance; agents at a node can recognize for all street segments which node they are connected to, etc. The model of the environment is static, as no changes in the environment are assumed; only the position and beliefs of the agent change in the model. Nevertheless, the model retains the important aspect of exploring an environment and navigating in it using the knowledge collected by others. Even from this generalized model, interesting conclusions can be drawn.

Agents are located in this environment at a street intersection oriented to move to a neighboring intersection. They can turn at an intersection to head a desired street segment and can move forward to the end of the street segment they are heading. Agents recognize intersections and street segments connecting them by labels without error. This follows roughly a simplification of the well-known TOURS model (Kuipers and Levitt 1978; Kuipers and Levitt 1990). Operations of the agents simulate the corresponding operations of persons in the real world. For example, "observe" applied to agents always means the simulated execution in the model.

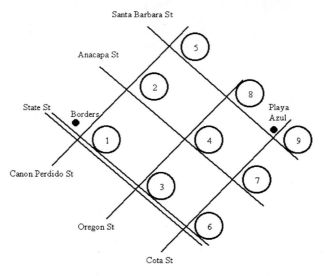

Fig. 4. A small subset of streets of downtown Santa Barbara (with node Ids as used in the code)

The *map-making agent* is exploring the modeled reality and constructs knowledge of each segment traveled and accumulates this knowledge in its memory. From this knowledge, a map is produced as a collection of lines and labels, placed on paper. This map, which looks much like Figure 4 as well, is then given to the agent that represents the map user.

The task *the map-using agent* is carrying out is to navigate between two named street intersections. The agent is constructing knowledge from the map drawn by the map-making agent and then plans the shortest path to the destination using the knowledge gained from the map.

4 Multi-Agent Theory

Multi-agent systems are a unifying theory for a number of developments in computer science. They have interesting applications in robotics, e-commerce, etc., but they also provide a fruitful model to discuss questions of communication and interaction, for example, in Artificial Life research (Epstein and Axtell 1996). Multi-agent systems consist of a simulated environment with which one or more simulated actors interact (Figure 5).

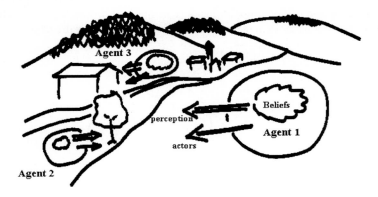

Fig.5. Environment and actors

Actors perceive through sensors the environment and have effects on the environment through their actors. The agents are part of the environment and are situated in it. Most authors include direct communication among agents (Ferber 1998; Weiss 1999), but the approach used here models communication between agents as the exchange of artifacts (i.e., a map). The operations of the actors can be described as algebras.

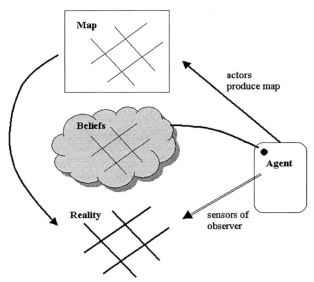

Fig.6. The different kinds of representations

The representation of this computational model is two-tiered: it separates completely the facts which stand for the environment, i.e., modeled *reality*, and the set of *beliefs* the agents hold about the environment (Figure 6). Maps are produced by the agents and exist in the environment. They can be used ('read') by the agents; they encode the knowledge the map-making agent has constructed during its exploration of the environment and communicates it to other agents.

The environment, which represents the world in the model, is encoded by a data structure, which represents the street graph and the locations of the intersections with coordinates. This representation could be extended to include labels for street names and the address range for each side (following the DIME model, which is a widely used representation (Corbett 1975; Corbett 1979)). This constitutes in the model 'reality' – it is therefore by definition complete and correct (and assumed here as static).

Simulated agents observe this environment and form a set of *beliefs* about it. The agents' beliefs may be incomplete, imprecise or even wrong; they are the results of the specific observation process, which is part of the computational model. The agents do usually not have knowledge of the coordinates of locations. For simplicity, only incompleteness of knowledge will be considered here, but investigating the effects of imprecise or vague spatial knowledge (Burrough and Frank 1995; Burrough and Frank 1996) and comparing strategies to compensate for missing information is possible. The agent can produce artifacts, which represent their knowledge. They simulate *Maps* in this environment.

5   Correctness of Maps

Formalizing environment, agents and maps allows us formally define correctness of a map following Tarski as correspondence between the map and the environment, which the map should represent. A representation of reality is correct, when operations in reality have results, which correspond to the results of corresponding operations in the representation. Applied to navigation: a map is correct, if the person using it plans the same path as a person, who knows the environment well, would walk.

This is best expressed as a homomorphism diagram (Figure 7) as usually drawn in category theory (Pierce 1993). It is based on a mapping f between two domains (the domain of agent's representation and the environment) and two corresponding operations sp and sp' (one to plan a shortest path in the agents' mind and the other walking the shortest path in the environment), such that first mapping the input from one to the other domain and then applying the operation, or first applying the operation and then mapping the result is the same:

$$f(sp\,(l)\,) = sp'\,(f\,(l)).$$

In general, formalization of Tarski semantics is not possible – the chasm between the world and the representation cannot be bridged. In the multi-agent model, however, both the environment and the agents' beliefs are part of the model and represented (but we gave up the restriction that the environment in the model is representing exactly some real-world situation). The correspondence between an agent's beliefs and the environment is definable as a homomorphism (Figure 7), where objects and operations are set into correspondence: the (simulated) mental act of an agent determining a path and the actual (simulated) walking of the path must be homomorphic. This means that the agent's beliefs about distances must be precise enough to determine the correct shortest path.

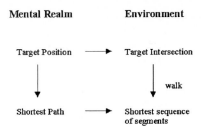

Fig. 7. Homomorphism between Real and Mental Representation

The processes of map making and map use are combinations of homomorphisms (Figure 8). The construction of a map is based on the (correct) mental representation of the map-maker gained through exploration of the environment. The mental representation of the map user is then constructed while reading the map. If each square in Figure 8 is a homomorphism, then a homomorphism from begin to end applies (Walters 1991). In all cases, the homomorphism maps not only between the simulated objects of the environment (street segments, intersections, respectively lines and points on the map), but also between the corresponding simulated operations of the agents (walking a street segment, imagine walking the street segment, drawing a line, etc.).

For complex decisions it may be difficult to show that these mappings are homomorphic. Inspecting the algorithm used to determine the shortest path (Dijkstra 1959; Kirschenhofer 1995), one finds that only two operations access the representation of the data. Initially, all the nodes in the graph must be found (other forms of the algorithm need the nodes connected to a given node) and the distance between two nodes. It is, therefore, only necessary to show that these operations are mapped correctly between the domains (see section 8).

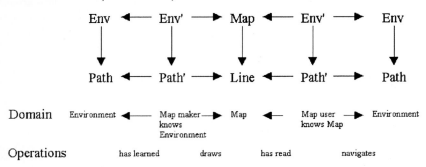

Fig.8. Combinations of homomorphisms

6   Formalization

For a multi-agent model, we have to represent the environment and the agents together with their interactions. We select here an algebraic approach and define classes with operations. To implement the model, data representations are also given. The formalization uses the Haskell language syntax (Peyton Jones, Hughes et al. 1999) to describe algebras (as classes) with the operations defined for the parameterized types and representations (as data or type).

The next subsections show the abstractions selected for street environment, the agents, map-makers, maps and map-users. For each type of object, the necessary operations are described. The representations selected here are given for illustration purposes and the goal is simplicity of the presentation. They document what information must be available, but any other representation for which the algebras given can be implemented would serve as well. No claim is made that the representations given here resemble the representations used in human mental operations.

6.1 Environment

The simulation is in an environment, which contains the agents with their beliefs, the street network, and, for simplicity, a single map. It can be represented as a data structure, consisting of

```
data Env = Env [Intersection] [Agent] Map
```

The next subsection define now the data for these three parts:

6.2 Static Street Environment

The street network consists of street segments (*edges*), which run from an intersection to the next. The intersections are called *nodes* and the street network is represented as a graph. The algebra for the street-network must contain operations to determine the position of a node as a coordinate pair (*Vec2* data type), test if two nodes are connected and find all nodes, which can be reached from a given node (operations *connectedNodes*); the shortest path algorithm requires to find all nodes and to get the distance between two nodes. Two operations to add a node and to add a connection to the network are also included.

```
class Streets node env where
    position :: node -> env -> Vec2
    connected :: node -> node -> env -> Bool
    travelDistance :: node -> node -> env -> Float
    connectedNodes :: node -> env -> [node
    allNodes :: env -> [node]
    addNode :: (node, Vec2) -> env -> env
    addConnect :: (node, node) -> env -> env
```

Nodes are just numbered (Figure 4) and Intersections consist of the Node (the node number as an ID), the position (as a coordinate pair) and a list of the connected node numbers.

```
data Intersection = IS Node Vec2 [Node]
data Position = Position Node Vec2
data Node = Node Int | NoNode
data Vec2 = V2 Float Float
```

6.3 Agents

The agents have a position at a node and a destination node they head to. They can either move in the direction they head or can turn to head towards another destination. They are modeled after Papert's Turtle geometry (Papert and Sculley 1980; Abelson and Disessa 1986)). After a move, the agent heads to the node it came from (Figure 9). This behavior can be defined with only four axioms:

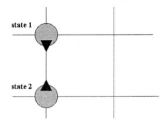

Fig. 9. The position of an agent before (state1) and after a move (state2)

1. Turning (*changeDestination*) does not affect the position:
   pos (a, (changeDestination (a,n,e)) = pos (a,e)
2. Moving brings agent to the node that was its destination:
   pos (a, move (a,e)) = destination (a,e)
3. The destination after a move is the location the agent was at before the move:
   destination (a, move (a,e)) = pos (a,e).
4. Turning (changeDestination) makes the agent's destination the desired intersection:
   destination (a, changeDestination (a, n, e)) =
      if n elementOf (connectedNodes (pos (a, e) e) then n
      else error ("not a node")

The agent constructs knowledge about the environment while it moves. The operation learnConnection constructs the belief about the last segments traveled (start and end intersection and its length) and accumulates these beliefs about the environment. The operation exploreEnv lets an agent systematically travel all connections in the environment and accumulate complete knowledge about it. Agents can determine the shortest path (here simulated with the algorithm given by Dijkstra) to a destination based on their knowledge and move to a desired target following the planned path using moveAlongPath (using single steps of moveOneTo).

```
class Agents agent env where
    pos :: agent -> env -> Node
    destination :: agent -> env -> Node

    move :: agent -> env -> env
    changeDestination :: agent -> Node -> env -> env

    moveOneTowards :: agent -> Node -> env -> env
    learnConnection :: agent -> env -> env
    exploreEnv :: agent -> env -> env

    moveAlongPath :: [Node] -> agent -> env -> env
    pathFromTo :: agent -> Node -> Node -> env -> [Node]
    moveTo :: agent -> Node -> env -> env
```

Positions are tuples with a node and a coordinate pair; an edge connects two nodes with a certain cost, which is encoded as a real number. A possible data structure for agents contains the beliefs as a list of edges and position recordings, which are used only by map-makers:

```
data Agent = Agent AId Node Node [ConnectionCost] [Position]

data AId = AId Int deriving (Show, Eq)
type ConnectionCost = Edge Node

data Cost = Cost Float | CostMax
data Edge n = Edge n n Cost
```

An ordinary agent after having traveled over some segments has a knowledge, which is represented as (using the codes from Figure 4):

```
Agent A1 at Node 4 destination Node 2 beliefs
    Node 4 to Node 2 dist 3.20156
    Node 2 to Node 1 dist 1.41421
```

6.3.1   Map Making Agent. The map-making agent is a specialized agent and explores first the environment and then draws a map. In addition to the observation of connections, any agent is capable; it can observe the coordinate values of his current position. The map-maker can draw a map based on his current knowledge or can draw a sketch of a path between two nodes (used in section 9, Figure 10).

```
class MapMakers agent environment where
    isMapMaker :: agent -> environment -> Bool
    getCoords :: agent -> environment -> Vec2
    learnPos :: agent -> environment -> environment
    drawMap :: agent -> environment -> environment
    drawPathMap :: Node -> Node -> agent -> environment -> environment
```

A map-making agent after having visited node 1,2 and 5 has also coordinates for these nodes (using again the codes from Figure 4):

```
Agent A1 at Node 5 destination Node 8 beliefs
    Node 8 to Node 5 dist 3.60555
    Node 3 to Node 5 dist 5.09902
    Node 5 to Node 2 dist 2.5
    Node 4 to Node 2 dist 3.20156
    Node 2 to Node 1 dist 1.41421
    Node 3 to Node 1 dist 3.20156
  visited
    Node 5:(5.0/8.0)
    Node 2:(3.0/6.5)
    Node 1:(2.0/5.5)
```

6.3.2   Map-Using Agents. The map-using agents have the task of moving from the node they are located at to another node in the environment. Their locomotion operations are the same as for all agents. They intend to travel the shortest path (minimal distance). A map-user first reads the map (using *readMap*) and adds the knowledge acquired to his set of beliefs about the environment before he plans the shortest path to his destination node.

```
class MapUsers agent environment where
    readMap :: agent -> environment -> environment
```

6.4 Maps

Maps are artifacts, which exist in the environment (for simplicity, only one map is present in the model at any given time). The map-making agent produces the map usually after he has collected all beliefs about the environment. The map represents these beliefs in a (simulated) graphical format.

Maps are simulated in the model as a list of line segments (with start and end map coordinates) and labels at the intersection coordinates; one can think of this as suitable instructions for drawing a map with a computerized plotter. The map, in the form of the drawing instructions, is then read by the map-using agent and translated into a list of beliefs. This representation of the map avoids the need to simulate drawing a bitmap and then using pattern recognition to analyze it; it leaves out the graphical restrictions of map-making and map reading.

Maps can be drawn and read, as well as sketches of a path (Figure 10):

```
class Maps aMap where
    drawTheMap :: [ConnectionCost:] -> [Position] -> aMap
    drawAPath :: [Node] -> [Position] -> aMap
    readTheMap :: aMap -> [ConnectionCost]
```

They are represented as

```
data Map = Map [Line] [Label]
data Line = Draw Vec2 Vec2
data Label = Label Node Vec2
```

7 Coding

The formalization has been coded in the functional notation of Haskell (Peyton Jones, Hughes et al. 1999). In a purely functional language, values cannot change and each movement of an agent is recorded as a new snapshot of the environment and not as a destructive change of the representation. This allows the use of standard mathematical logic, especially reasoning with substitution, and does not force to use temporal logic as would be necessary to reason with imperative programming languages.

Using a functional notation with some restrictions allows constructing executable prototypes (models for the abstract algebras constructed). These help to check that a formal system captures correctly our intentions. The following test starts with two agents "Jan" and "Dan" (more would be possible) in an environment with the streets from the center of Santa Barbara (with the coding shown in Figure 4). Jan is a "map-maker" and explores the environment. We can ask him for the path from Node 1 to Node 9 and get the shortest path. The same question to Dan gives no answer, as he has no knowledge yet. If Jan draws a map (*env2*) and Dan reads it (*env3*), then Dan can give the correct answer as well. This answer is the same as if Dan had explored the environment himself (*env1a*). The simulated system exhibits this behavior and confirms that our intuition about maps and the formalization correspond as explained in section 5. The following text shows a sequence of `code` and the *responses* from the system:

```
-- readable names for the agents:
jan = AId 1
dan = AId 2
```

```
-- create two agents at node 1 destination in direction of node 2
jan0 = Agent jan (Node 1) (Node 2) [] []
dan0 = Agent dan (Node 1) (Node 2) [] []

env0 = Env santaBarbara [jan0, dan0] emptyMap

--the two agents with the streets of Santa Barbara (figure 9)
env1' = learnPos jan env0
env1 = exploreEnv jan env1'

-- the positions of jan and dan
janpos1 = pos jan env1
danpos1 = pos dan env1
```

```
    test input> janpos1
            Node 3
    test input > danpos1
            Node 1
```

```
-- the path from 1 to 9
janpath1 = pathFromTo jan (Node 1) (Node 9) env1
danpath1 = pathFromTo dan (Node 1) (Node 9) env1
```

```
    test input> janpath1
            [Node 1,Node 2,Node 4,Node 7,Node 9]
    test input> danpath1
            []
```

```
-- jan draws map and dan reads it
env2 = drawMap jan env1
env3 = readMap dan env2

danpath3 = pathFromTo dan (Node 1) (Node 9) env3
```

```
    test input> danpath3
            [Node 1,Node 2,Node 4,Node 7,Node 9]
```

```
-- this path is the same as
--      if dan had explored the environment itself:

env1a = exploreEnv dan env0
danpath1a = pathFromTo dan (Node 1) (Node 9) env1a
env2a = drawPathMap (Node 1) (Node 9) jan env1
env3a = readMap dan env2a

danpath3a = pathFromTo dan (Node 1) (Node 9) env3a
```

```
    test input> danpath3a
            [Node 1,Node 2,Node 4,Node 7,Node 9]
```

8 Definition of Correctness of a Map

In this environment, a formal and stringent definition for a map to be a correct representation of reality is possible. A map is correct if the result of an operation based on the information acquired from the map is the same as if the agent would have explored the world to gain the same information. The proof is in two steps: completeness and correctness. Completeness assures that all relevant elements – here nodes and segments – are transformed between the respective representations. Correctness requires that the transformations preserve the properties important for the decision (here the determination of the shortest path).

8.1   Completeness: Collecting All Observations into Beliefs

The operations to explore the environment and gradually learn about it or the exploration of a map are a repeated application of an operation '*learnConnection*', which is applied to all segments in the environment, respectively, the map. The construction of the beliefs of an agent about the environment can then be seen as a transformation between two data structures: the data structure which represents the environment is transformed into the internal structure of the beliefs. Similarly is the construction of the map a transformation between the data structure of the agent's beliefs into the list of drawing instructions; reading the map is the transformation of the data element of the map into beliefs.

We have to show that these transformations are applied to all elements and nothing is 'overlooked'. The *exploreEnv* operation is quite complex. It explores a node at a time, learning all segments, which start at this node, and keeps a list of all nodes ever seen. The environment is completely explored if all nodes where completely explored.

Drawing the map is a transformation procedure; coded with the second order function *map*, which applies a transformation to each element in a list. The transformation changes the belief into a drawing instruction. Reading the map is a similar function, taking line after line from the map and building a list of beliefs.

8.2   Correctness: Transformations Preserve the Important Properties

The different transformation for individual objects must preserve the properties necessary for the correct determination of the shortest path.
- A street segment is added to the beliefs after it is traveled; having traveled the segment ensures that the segment is viable and the cost is the cost just observed. Surveyors correctly observe the coordinate values for intersections.
- Map-makers translate each segment into a line drawn. The positions are based on the observed coordinate values for intersections.
- Map-users read the drawn line as viable segments and use the length of the line as an indication of the cost.

These operations guarantee that beliefs about viable street segments by the map-maker are communicated to the map-users. The (relative) cost is communicated correctly if the cost function is based on distance only. These transformations could be more realistic and include systematic and random errors and we could then observe the effects on the determination of the shortest path.

8.3   Discussion

In this example, where the observation and the use of the map are based on the same operation, nothing can go wrong. The model, however, indicates the potential for errors in communication. Here two examples:

8.3.1   Problems with the Classification of Elements. The world contains different classes of pathways, which can be driven, biked or walked, and not all segments can be passed with all vehicles. The classification of the road must be included in the map to allow use of the map for car drivers, bikers and persons walking. These problems seem trivial, but some of the current In-Car Navigation systems recommend paths, which include segments of a bike path!

In the simulation, if the exploring agent uses the same mode of locomotion as the map user, then correct communication is assured. If the exploring agent rides a (simulated) bike and the map using agent drives a (simulated) car, one may discover that the shortest path determined is using segments of a bike path the car driving agent cannot travel on or may find that a shorter route using an interstate highway is not found, because the map-making agent could not travel there and did not include it.

In general, the map-makers are not using the same operation that the map-user executes. The correctness of the map then depends on the composition of the transformation functions from observations of the map-maker to beliefs in the map user. The same criteria must be used during observation when the coding of an object is fixed. For example, while classifying roads using air photographs only road width, but not police regulations, are available to decide on the coding. This may classify some wide road segments which are closed for traffic as viable.

Users with different tasks may require different maps (or at least careful coding). A map for a hiker must be different from the map for driving – and indeed road maps for car driving are published separately from the maps for bikers or hiking maps. If a geographic database should be constructed for multiple purposes, then the properties which differentiate uses of objects must be recorded separately: the physical width and carrying capacity of a road must be recorded separately from the traffic regulations for the same road. It becomes then possible to establish the particular combinations of classifications, which simulate the intended type of use.

8.3.2   Problems with the Transformation. If the function to draw the map is using one of the many map projections, which do not preserve distances, then the representation of distances on the map is not representative of the distance between the nodes (but systematically distorted). The map-reader's naïve approach to link the distance between two nodes on the map with the cost for travel is then wrong and can lead to an error in determining the shortest path. More questions arise if the travel cost is a complex function of distance and other elements, e.g., the Swiss hiker's rule:

$$time\ (h) = distance(km)/5 + total\ ascent\ (m)/300 + total\ descent\ (m)/500$$

9   Effectiveness of Maps to Communicate Spatial Information

In this context, one may address the question why maps are so effective to communicate information about a complex environment in comparison to verbal descriptions. Take the small part of downtown Santa Barbara in Figure 4 and imagine communicating the information verbally: it would read as a long list, describing each segment, with the intersection it starts and ends:

The first segment of State St runs from Canon Perdido St to Ortega St, the next segment runs from Ortega St to Cota St. The first segment of Anacapa St runs from

Canon Perdido St to Ortega St, etc., etc. This list contains a total of 12 segment descriptions, is tedious and does not communicate well. Alternatives would use the naming of 9 nodes and 24 incidence relations.

For areas where streets are regularly laid out, abbreviations could be invented. For example, in large parts of Santa Barbara, it is sufficient to know which streets run (conventionally) North-South and which East-West and to know the order in which they are encountered. This does not work for areas where the street network is irregular and a detailed description, for example, for areas, where an Interstate highway or a railway line intersect and distort the regular grid.

A verbal description for a street network is tedious and verbose, because it must create communicable identifiers for each object; for example, a name must be given to each intersection, such that another street segment starting or ending at the same location can refer to it. A graphical communication uses the spatial location to create references for the locations and does not need other names. The incidence is expressed as spatial position on paper and picked up by the eye. The information retained is the same, but the communication is more direct, using the visual channel. It is curious to note that American Sign Language, which is a well-documented natural language, uses a similar device of 'location' used as references. The speaker may designate a location in the (signing) space before him to stand for a person or place he will later refer to. A later reference to this person or place is then made by simply pointing to the designated location, using the location as a reference to the objects (Emmorey 1996).

The situation is different when only a specific path should be communicated. The list of instructions is shorter and simpler than the sketch (Figure 10). The instructions for a path from Borders (Intersection Canon Perdido St and State St) to Playa Azul (Intersection of Santa Barbara St with Cota St):

> Follow Canon Perdido Street to the East for one block,
> Turn right and follow Anacapa Street for two blocks
> Follow Cota St to the East for one block

In the language of the agents, a list of nodes as the shortest path is communicate as:

    [Node 1,Node 2,Node 4,Node 7,Node 9]

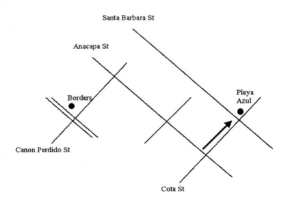

Fig. 10. Sketch for path from Borders to Playa Azul

Each of the representations is short and can be communicated using a linear channel, e.g., verbally). A sketch map would be somewhat more complex as the following example of a simulated map demonstrates

```
env4 = drawPathMap (Node 1) (Node 9) jan env1

line:(2.0/5.5), (3.0/6.5),line:(3.0/6.5), (5.0/4.0),
line:(5.0/4.0), (6.0/3.0),line:(6.0/3.0), (8.0/4.0),
label Node 4 (5.0/4.0),label Node 7 (6.0/3.0),label Node 9 (8.0/4.0),
label Node 2 (3.0/6.5),label Node 1 (2.0/5.5),
```

10   Conclusion

A framework for the formalization of the production and use of maps and carto-graphic diagram is described. The model is two-tiered; it contains a representation of what stands for *reality* and what stands for the *beliefs* of multiple, simulated agents about reality. The model is natural, allows misrepresentation and is fine-grained. It goes beyond current models, as it permits to model the observation processes of the agents and the agents' actions, which use the information collected. It can include errors in these processes or in the information stored.

The production and use of maps or diagrams for navigation can be described in this computational model, which includes processes for exploring the environment while traveling, casting the information collected by the agent into a graphical form, which can be communicated to and be used by another agent. The semantics of the map is directly related to the processes that observe reality or use the data. Correctness of a map can be established in this formalization. It is directly related to the connection between the operations used for observing and representing reality and the operations the map users intend to perform.

Using an executable functional language to construct multi-agent models allows experimenting. The code is very compact and takes only 5 pages, plus 3 pages for the graph related utilities, including the shortest path algorithm. The code is available from http://www.geoinfo.tuwien.ac.at.

In this multi-agent framework, other related questions can be explored. For exam-ple, the effects of incomplete street maps on navigation can be simulated and tested, how much longer the path traveled becomes and what are the best strategies for users to cope with the incomplete information. One can also explore different strategies for map users to deal with observed differences between the map and the environment.

Acknowledgements

This contribution is based on research I carried out at the University of California, Santa Barbara. I thank both the National Center of Geographic Information and Analysis, and the Department of Geography for the hospitality and the support they have provided. I have greatly benefited from the discussion with the students in the course 'Formalization for GIS' I taught here. Discussions with Dan Montello, Helen Couclelis, Mike Goodchild, Waldo Tobler, and Jordan Hastings were invaluable to sharpen my arguments.

I also thank Werner Kuhn, Martin Raubal, Hartwig Hochmair, Damir Medak and Annette von Wolff for the patience to listen to and comment on these ideas in various stages of progression. The contribution to help clarify my ideas made by Haskell is invaluable, and I thank the Haskell community, especially Mark Jones, for the Hugs implementation. The comments from two unknown reviewers were extremely useful for the revision of the paper. Roswitha Markwart copy-edited the text and Hartwig Hochmair improved the figures.

References

Abelson, H. and A. A. Disessa (1986). *Turtle Geometry : The Computer As a Medium for Exploring Mathematics*. Cambridge, Mass., MIT Press.

Baeijs, C., Y. Demazeau, et al. (1995). *SIGMA: Approche multi-agents pour la generalisation cartographique*. CASSINI'95, Marseille, CNRS.

Braitenberg, V. (1984). *Vehicles, experiments in synthetic psychology*. Cambridge, MA, MIT Press.

Burrough, P. A. and A. U. Frank (1995). "Concepts and paradigms in spatial information: Are current geographic information systems truly generic?" *International Journal of Geographical Information Systems* **9**(2): 101-116.

Burrough, P. A. and A. U. Frank, Eds. (1996). *Geographic Objects with Indeterminate Boundaries*. GISDATA Series. London, Taylor & Francis.

Burrough, P. A. W. v. D. and G. Heuvelink (1988). *Linking Spatial Process Models and GIS: A Marriage of Convenience or a Blossoming Partnership?* GIS/LIS'88, Third Annual International Conference, San Antonio, Texas, ACSM, ASPRS, AAG, URISA.

Corbett, J. (1975). *Topological Principles in Cartography*. 2nd International Symposium on Computer-Assisted Cartography, Reston, VA.

Corbett, J. P. (1979). Topological Principles of Cartography, Bureau of the Census, US Department of Commerce.

Davis, E. (1990). *Representation of Commonsense Knowledge*. San Mateo, CA, Morgan Kaufmann Publishers, Inc.

Dijkstra, E. W. (1959). "A note on two problems in connection with graphs." *Numerische Mathematik*(1): 269-271.

Dreyfuss (1998). *What Computers Still Cannot Do*. Cambridge, Mass., The MIT Press.

Emmorey, K. (1996). The cofluence of space and language in signed language. *Language and Space*. P. Bloom, M. A. Peterson, L. Nadel and M. F. Garett. Cambridge, Mass., MIT Press: 171 - 210.

Epstein, J. M. and R. Axtell (1996). *Growing Artificial Societies*. Washington, D.C., Brookings Institution Press.

Ferber, J., Ed. (1998). *Multi-Agent Systems - An Introduction to Distributed Artificial Intelligence*, Addison-Wesley.

Frank, A. U. (1998). *GIS for Politics*. GIS Planet'98, Lisbon, Portugal (September 9-11, 1998), IMERSIV.

Hudak, P., S. L. Peyton Jones, et al. (1992). "Report on the functional Programming Language Haskell, Version 1.2." *ACM SIGPLAN Notices* **27**(5): 1-164.

Johnson, M. (1987). *The Body in the Mind: The Bodily Basis of Meaning, Imagination, and Reason*. Chicago, University of Chicago Press.

Kirschenhofer, P. (1995). The Mathematical Foundation of Graphs and Topology for GIS. *Geographic Information Systems - Material for a Post Graduate Course*. A. U. Frank. Vienna, Department of Geoinformation, TU Vienna. **1**: 155-176.

Kuipers, B. and T. S. Levitt (1990). Navigation and Mapping in Large-Scale Space. *Advances in Spatial Reasoning*. S.-s. Chen. Norwood, NJ, Ablex Publishing Corp. **2**: 207 - 251.

Kuipers, B. J. and T. S. Levitt (1978). "Navigation and mapping in large-scale space." *AI Magazine* **9**(2): 25-43.

Lakoff, G. and M. Johnson (1999). *Philosophy in the Flesh*. New York, Basic books.

Lechthaler, M. (1999). "Merkmale der Datenqualitaet im Kartographischen Modellbildungsprozess." *Kartographische Nachrichten* **49**(6): 241-245.

Papert, S. and J. Sculley (1980). *Mindstorms: Children, Computers and Powerful Ideas*. New York, Basic Books.

Peterson, J., K. Hammond, et al. (1997). "The Haskell 1.4 Report." *http://haskell.org/report/index.html*.

Peyton Jones, S., J. Hughes, et al. (1999). Haskell 98: A Non-strict, Purely Functional Language.

Pierce, B. C. (1993). *Basic Category Theory for Computer Scientists*. Cambridge, Mass., MIT Press.

Pinker, S. (1997). *How the Mind Works*. New York, W. W. Norton.

Stich, S. P. and T. A. Warfield, Eds. (1994). *Mental Representation*. Cambridge, Mass., Basil Blackwell.

Tchen, Y.-S. (1967). *Je parle Chinois*. Paris, Librairie d'Amérique et d'Orient.

Walters, R. F. C. (1991). *Categories and computer science*. Cambridge, UK, Carslaw Publications.

Weibel, R. (1995). "Map generalization in the context of digital systems." *CaGIS* **22**(4): 259-263.

Weiss, G. (1999). *Multi-Agent Systems: A Modern Approach to Distributed Artificial Intelligence*. Cambridge, Mass., The MIT Press.

Schematic Maps for Robot Navigation[1]

Christian Freksa, Reinhard Moratz, and Thomas Barkowsky

University of Hamburg, Department for Informatics,
Vogt-Kölln-Str. 30, 22527 Hamburg, Germany
{freksa,moratz,barkowsky}@informatik.uni-hamburg.de

Abstract. An approach to high-level interaction with autonomous robots by means of schematic maps is outlined. Schematic maps are knowledge representation structures to encode qualitative spatial information about a physical environment. A scenario is presented in which robots rely on high-level knowledge from perception and instruction to perform navigation tasks in a physical environment. The general problem of formally representing a physical environment for acting in it is discussed. A hybrid approach to knowledge and perception driven navigation is proposed. Different requirements for local and global spatial information are noted. Different types of spatial representations for spatial knowledge are contrasted. The advantages of high-level / low-resolution knowledge are pointed out. Creation and use of schematic maps are discussed. A navigation example is presented.

1   Introduction: A Robot Navigation Scenario

We describe a scenario consisting of an autonomous mobile robot and a structured dynamic spatial environment it lives in. The robot is equipped with rudimentary sensory abilities to recognize the presence as well as certain distinguishing features of obstacles that may obstruct the robot's way during navigation. The robot's task is to move to a given location in the environment.

This task – that appears so easy to humans – is a rather difficult task for autonomous robots. First of all, the robot must determine where to go to reach the target location; thus it needs knowledge about space. Next, the robot must determine what actions to take in order to move where it is supposed to go; thus it needs knowledge about the relation between motor actions and movements and about the relation between movements and spatial locations.

In theory, we could provide the robot with detailed information about the spatial structure of its environment including precise distance and orientation information as well as information about its own location in the environment. The robot then could compute a route through unobstructed space from its current location to the target location. Consequently, some route following procedure could traverse this route.

---

[1] Support by the Deutsche Forschungsgemeinschaft, the International Computer Science Institute, and the Berkeley Initiative in Soft Computing is gratefully acknowledged.

Ch. Freksa et al. (Eds.): Spatial Cognition II, LNAI 1849, pp. 100-114, 2000.

In practice, however, this approach does not work. What are the problems? First, it is very hard to provide the robot with detailed knowledge about its spatial environment in such a way that this knowledge actually agrees with the encountered situation in the environment at a given time in all relevant aspects. Even if it agrees, it is impossible to get the robot to carry out actions that correctly reflect the computed result. Second, the real world is inherently dynamic: knowledge about the state of the world at a given time does not guarantee the persistence of that state at a later time.

Why is autonomous robotics so difficult? The general problem a robot must cope with when acting in the real world is much harder than the problem a computer[2] must deal with when solving problems. The reason is that autonomous robots live in two worlds simultaneously while computers only must deal with a single world. Autonomous robots live in the physical world of objects and space and in the abstract world of representation and computation. Worst of all: these two worlds are incommensurable, i.e., there is no theory that can treat both worlds in the same way (Palmer, 1978; Dirlich et al., 1983).

Computers act entirely in a formalized computational (mental) world: their problems are given in formalized form, they compute on the basis of formalized procedures, and the results come out as formal statements. The physical existence and appearance of computers are not essential for the solution of the formal problem. Autonomous robots, on the other hand, are not only superficially submerged in the physical world; they are essential physical parts of their own physical environment. When a robot moves, the physical world changes. In addition to their physical existence, autonomous robots have an important mental facet: autonomous robots are controlled by computers that compute the decisions about the robots' actions in their physical environment.

We can take at least two views regarding the relationship between the physical robot and its controlling computer: (1) We can consider the computer as just a piece of physical circuitry that connects sensor inputs to motor outputs in a more or less complex way. In this view, we do not need to consider representations and mental processes; all issues can be addressed in the physical domain. (2) We acknowledge that formal theories about physical space are required for intelligently acting in a physical environment. Then we have two options: (a) we believe that these theories can be made sufficiently precise to describe all that is needed to perform the actions on the level of the representations; this option corresponds to the classical AI approach. Or (b) we recognize that it is unfeasible to employ a global theory that accounts for all aspects the robot may be confronted with in physical space. Then we can formalize a theory that deals with some aspects of the physical world and leaves other aspects to be dealt with separately – for example in the manner suggested by the first view.

The first view was brought forward most prominently by Brooks (1985). It works well on the level of describing reactive behavior and for modeling adaptation behavior of insects and robots in their environments (Braitenberg, 1984). However, it has not been possible to describe purposeful proactive behavior in this paradigm, so

---

[2] We use the term 'computer' to designate the abstract reasoning engine and the term 'robot' to designate a physical device with sensors and effectors that interact with the environment and with a computer that interprets the sensor data and controls the actions.

far. To describe and model intelligent planning behavior, a representation of knowledge about the world is necessary.

In board games or other domains that are defined entirely within a formal framework, a representation with suitable inference procedures is all that is needed to provide appropriate solutions. For tasks and problems that are given in the physical world, however, formal representations must be set in correspondence with the physical world and can only approximate actual situations. This is true not only for robots but also for people and other living beings. Biological systems cope with this general representation problem so well, that the extent of this correspondence problem has been underestimated for a long time. Through the use of robots we have become aware of the severeness of this problem and by using robots we can thoroughly study mappings between the physical world and its mental representation.

An example of information that typically will not be available from a world model is information about an object that happens to have entered the scene due to unpredictable reasons. Another example is the information to which degree a certain location of the robot environment will be slippery and cause a given robot wheel to slip (at a particular angle, at a given force, speed, temperature, etc.). Such situations can be dealt with reactively through perception and adaptation in the environment. In summary, the autonomous robot requires a suitable combination of represented and directly perceived knowledge.

2   A Robot that Communicates by Means of Maps

Our robot is designed to be autonomous to a certain extent: A navigation task is given to the robot and it must find the specified destination autonomously (cf. Röfer, 1999; Musto et al., 1999). Given the situation as described in the previous section, the robot must interact in two directions: (1) it must communicate with the instructor who specifies the task and checks its solution, and (2) it must interact with the environment to master the task. For a human instructor there are three natural modes to communicate spatial information: by deictic means (looking and/or pointing at spatial locations); by a description of spatial locations or objects in natural language; by using a spatial medium to convey spatial information in an analogical manner. Frequently these modes are combined to make use of their respective advantages.

As the robot must interact with its spatial environment to master its navigation task, communication by means of a spatial medium appears particularly advantageous and interesting. Common spatial media to communicate about space are sketches or maps. Maps may serve as interaction interfaces between people and their environment, between robots and their environment, but also between people and robots. In the present paper we explore the communication with robots by means of schematic maps.

The power of maps as representation media for spatial information stems from the strong correspondence between spatial relations in the map and spatial relations in the environment. This allows for reading spatial relations directly off the map that have not explicitly been entered into the representation, without engaging inference processes (Freksa & Barkowsky, 1999). When maps are used to convey spatial information, spatial relations in the map can be directly applied to the environment and vice versa, in many cases. All maps distort spatial relations to some extent, the most obvious distortion being the distortion due to scale transformation (Barkowsky & Freksa, 1997). Most spatial distortions in maps are gradual distortions. No translation of spatial information through symbol interpretation is required as in the case of natural language descriptions.

The strong spatial correspondence between maps and spatial environments has specific advantages when dealing with spatial perception; in our case the robot is equipped with sensors that determine the spatial location of objects to perform its navigation task. The distortions obtained in the sensor readings may share properties with the distortions we get in map representations; thus, the same interpretation mechanisms may be used for the interpretation of the maps and of the sensor readings.

In the setting described, maps can be constructed from the spatial relations in the environment by a human overlooking the environment or by a robot moving through the environment. The human can convey instructions to the robot using maps. In solving its task, the robot can match spatial relations in the map against spatial relations in the environment. And the robot can communicate back to the human instructor by using a map. This provides us with a rich environment to study formal properties of different maps and practical map use. Figure 1 indicates the communication relations between the human and the robot on one hand and the spatial correspondence between the environment and the map on the other hand.

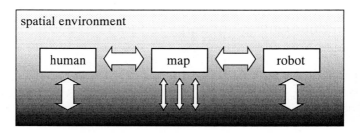

Fig. 1. Spatial communication relations between human, map, robot, and environment (thick arrows). Thin arrows indicate spatial correspondence relations between map and spatial environment

3   Qualitative Spatial Knowledge for Navigation Tasks

Depending on the class of tasks to be performed, different abstractions of spatial knowledge may be useful. To determine what type of knowledge may be most useful to solve navigation tasks, let us consider two extreme cases: (1) the robot knows everything about its spatial environment and (2) the robot knows nothing about its spatial environment. In the first case, the robot does not require perception as it can navigate entirely on the basis of the available knowledge. (We can dismiss this case on the basis of unattainability of complete correct knowledge, in particular in dynamic environments). In the second case, the robot must get all the information to solve its navigation task directly through perception of the environment. (We dismissed this case as unsuitable for developing intelligent navigation strategies.)

Between the two extremes, we should find an appropriate combination of information to be provided externally through the map and information extracted directly from the environment. The information in the environment is superior over information in a map in several respects: (a) it is always correct and (b) it does not require an additional medium. Information provided externally by a map is superior in other respects: (c) it may be available when perception fails (for example at remote locations) and (d) it may be provided at a more suitable level of abstraction for a given task. These considerations suggest that information about the local situation preferably should be obtained directly from the environment through perception while the information about the global spatial situation should be provided externally to allow for developing suitable plans and / or strategies to solve the navigation task.

This division between the primary sources of information also suggests the levels of abstraction the respective information sources should deal with: the information in the environment is very concrete; the perception processes must make it just abstract enough for the decision processes to be able to act on it. The externally provided global information, on the other hand, should preferably be abstract to allow for efficient route planning processes; however, it must be concrete enough to be easily matched to the actual spatial environment.

A suitable level of abstraction for these requirements is the level of qualitative spatial knowledge (Zimmermann & Freksa, 1996). Qualitative spatial knowledge abstracts from the quantitative details of precise distances and angles, but it preserves the information relevant to most spatial decision processes. Navigation then can be carried out in two phases: a coarse planning phase relying mainly on externally provided qualitative global knowledge and a detailed execution phase in which the plan is confronted with the actual details of reality in the local surroundings of the robot (cf. Sogo et al., 1999). This requires that the two sources of knowledge for the robot can be brought into close correspondence.

4   Schematic Maps

Maps to convey spatial relations come in different varieties. Depending on the scale, on the objects to be represented, and on the symbols to be used, they can be more or less veridical with respect to the spatial relations depicted. Scaled-down maps (i.e. in particular all geographic maps) distort spatial relations to a certain degree due to representational constraints (Barkowsky & Freksa, 1997). For many purposes, it is desirable to distort maps beyond the distortions required for representational reasons to omit unnecessary details, to simplify shapes and structures, or to make the maps more readable. This latter type of map we will refer to as 'schematic map'. Typical examples of schematic maps are public transportation maps like the London underground map or tourist city maps. Both types may severely distort spatial relations like distances or orientations between objects.

Schematic maps are well suited to represent qualitative spatial concepts. The orientation of a line on the map may correspond to a general orientation (or category of orientations) in the nature; a distance on the map may correspond to the number of train stops, rather than to the metric distance in nature, etc. (Berendt et al., 1998).

If we consider abstract mental concepts of the spatial world as constituting one extreme in a hypothetical continuum of representations and the concrete physical reality itself as the other extreme, it is interesting to determine where different types of representations of the world would be located in this continuum. Mental concepts can be manifested most easily by verbal descriptions (in fact, some researchers believe that we cannot think what we cannot express in words - Whorfian hypothesis, Whorf, 1956). When we move in the hypothetical continuum closer to the physical manifestation of the world, we can put concepts of spatial objects and relations into a sketch map to convey selected spatial relations. Sketch maps tend to have close correspondences to verbal descriptions and they are used to augment verbal descriptions by spatial configurations that correspond to spatial configurations in the physical world.

Moving from the other extreme, the physical reality, we obtain a mild abstraction by taking a visual image (e.g. a photograph) that preserves important spatial relations. Moving a few steps further towards concept formation, we may get a topographic map in which objects have been identified and spatial relations from the real environment are maintained. Further abstraction may lead to a schematic map as suggested above. Figure 2 depicts this abstraction scheme.

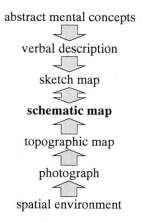

Fig. 2. Abstraction levels between conceptual-linguistic and physical-spatial structures

In this framework, schematic maps differ from sketch maps in that they are derived from topographic maps that are meant to represent a certain part of the environment completely at a given granularity level. Sketch maps, on the other hand, usually correspond to the linear flow of speaking and drawing and frequently to the temporal sequence of route traversal (Habel & Tappe, 1999). Thus, schematic maps provide information about a region while sketch maps more typically provide information about a single route or about a small set of routes. However, there is no sharp boundary between schematic maps and sketch maps as schematic maps may be incomplete and sketch maps may be unusually elaborate.

5   Using Schematic Maps for Robot Instruction

Schematic maps provide suitable means for communicating navigation instructions to robots: they can represent the relevant spatial relationships like neighborhood relations, connectedness of places, location of obstacles, etc. Humans can construct schematic maps rather easily, as the necessary qualitative relations to be encoded are directly accessible to human perception and cognition. But autonomous robots also can construct schematic maps by exploring their environment and by keeping track of notable entities (cf. Fox, 1998; Fox et al., 1999; Thrun, 1998; Thrun et al., 1999); thus, schematic maps can be used for two-way communication between humans and robots.

In Fig. 3 we give a simple example of an initial schematic map of an indoor office environment that may be provided by a human instructor to an autonomous robot. It consists of three rooms, three doors connecting the rooms, and the robot that is located in one of the rooms. This example may serve as reference for the following discussion.

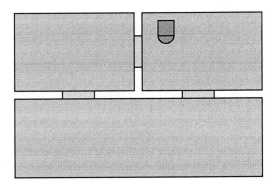

Fig. 3. Schematic map of a simple indoor environment consisting of three rooms, three doorways, and one autonomous robot

Schematic maps can be encoded in terms of qualitative spatial relations. They preserve important ordering information (Schlieder, 1996) for identifying spatial configurations. Qualitative spatial reasoning (Freksa & Röhrig, 1993; Cohn, 1997) can be used to infer relationships needed for solving the navigation task.

To use schematic maps for actual robot navigation, a correspondence between entities and relations in the schematic map and entities and relations in the spatial environment must be established. As we have argued above, this is very difficult to do on the level of high-resolution information. However, we believe that this task can be much more easily performed on coarser, low-resolution information (Zadeh, 1999).

One of the reasons for this is that we can expect a larger number of rare or unique configurations on the coarser and higher level of representation. This should make the approach rather robust against perturbations due to incomplete, imprecise and even partially conflicting knowledge. When spatial relations found in the map and in the spatial environment do not match perfectly, conceptual neighborhood knowledge (Freksa, 1992a, b) can be used to determine appropriate matches.

Furthermore, in realistic settings suitable reference information usually will be available to simplify the problem of matching the map to the environment. Like in route instructions to fellow human beings we can inform a robot about its own location on the map and possibly about its orientation in the environment. Other locations may be indicated by unique landmarks or rare objects that the robot should be able to recognize. These measures help control the number of possible matches between map and environment.

This leads us to the problem of object recognition. Here we adopt a coarse, qualitative approach, as well. Rather than attempting to recognize objects from details in visual images, our strategy is to identify configurations through rather coarse classification and by employing knowledge as to how these configurations may be distorted by the perception and by matching processes. For example, we use coarse color and distance information to identify landmarks in our indoor office scenario. We may relate our approach to Rosch's findings of linguistic categories in human communication (Rosch, 1975). Rosch found that the basic conceptual categories people use in communication tend to be neither the very specific nor the very general categories but intermediate categories that may be most suitable for object identification and concept adaptation.

Multimodal information[3], for example a combination of color, distance, and ordering information, can support the identification process on the level of high-level conceptual entities and structures considerably, as the use of different feature dimensions helps select appropriate matching candidates.

5.1   Creating Schematic Maps

Schematic maps can be created in at least three different ways: (1) by a human observer / instructor; he or she can acquire knowledge about the spatial layout of the environment through inspection and can put down relevant relationships in a schematic map. The actual layout of that map can be supported by a computerized design tool that creates a simple regularly structured map and helps making sure the depicted relations can be interpreted in the intended way; (2) by the robot itself; in its idle time, the robot can explore its environment, note landmarks, and create a schematic map that reflects notable entities and their spatial relationships as discovered from the robot's perspective; (3) from a spatial data base: for artificial environments data about the kinds of objects and their locations may be specified in a data base; this information can be fed into a computerized design tool to create a schematic map, as well.

---

[3] We use the term 'multimodal' in a rather general sense. It refers to conceptual as well as to perceptual categories.

5.2   Navigation Planning and Plan Execution Using Schematic Maps

The initial schematic map (cp. Fig. 3) provides the robot with survey knowledge about its environment. The robot extracts important features from the map for identification in the environment. The robot can enter discoveries into the map that it made during its own perceptual explorations in the environment. It produces a coarse plan for its route using global knowledge from the map and local knowledge from its own perception. Details of a planning procedure that we use are described in the next section. The resulting plan is a qualitative plan comparable to what people come up with when giving route instructions to a fellow human being: it indicates which roads to take but does not specify in precise quantitative terms where to move on the road.

During plan execution, the robot will change its local environment through locomotion. This enables it to instantiate the coarse plan by taking into account temporary obstacles or other items that may not be present in the map. Also, the local exploration may unveil serious discrepancies between the map and the environment that prevent the instantiation of the plan. In this case, the map can be updated by the newly accumulated knowledge and a revised plan can be generated.

5.3   Communication and Negotiation Using Schematic Maps

The robot may not be able to generate a working plan for its task due to incompleteness or incorrectness of the map or due to constraints that lead the robot to believe that it will not be able to move to the destination. Rather than just stopping its actions, the robot should get in touch with its instructor, in such a situation. Using the schematic map, the robot should be able to indicate to the instructor what kind of problem it has in plan generation or plan execution. The human instructor then can inspect the schematic map to evaluate the problem and revise his or her instructions. Figure 4 summarizes the different interaction pathways discussed.

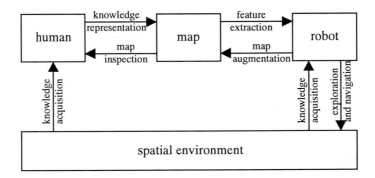

Fig. 4.   The interaction pathways between the human instructor, the schematic map, the autonomous robot, and its spatial environment

6   A Simulation Scenario

Currently we develop our scenario such that a simulated robot solves navigation tasks in a simple world with the aid of a schematic map. The schematic map depicts selected spatial aspects of the environment as well as the position and the orientation of the robot and the target location for the navigation task. An example is presented in Fig. 5. The map depicts a few spatial aspects of the three-room office environment in a qualitative manner. Specifically, walls, room corners, and doorways are represented. Other aspects are neglected. For example the thickness of the walls is not depicted. Also, distances and angles need not be to scale, in the depiction.

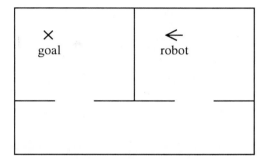

Fig. 5.   Schematic map of robot environment including goal location and location and orientation of robot

The sensors of the simulated robot simulate two laser range finders, each covering a range of 180 degrees. Together these sensors yield a panorama view of 360 degrees. Using the schematic map, the robot determines a 'qualitative path', i.e. a path specified only in terms of the route to be taken, not in terms of metrically specified locations. To compute this path the free regions depicted on the map are partitioned into convex regions (Fig. 6a) (Habel et al., 1999).

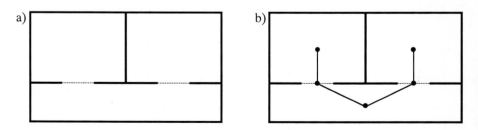

Fig. 6.   Qualitative path generation using a schematic map: a) partitioning the free region into three convex cells; b) connecting the cell centers with the centers of their adjacent cell transition lines to obtain the path graph for the route to be traversed

A robot can overlook a convex cell entirely from any location in that cell with a single panorama view. To make use of this feature, the algorithm partitions the free regions in the schematic map into convex cells. Each concave corner is transformed into two convex corners by converting it into a corner of two different regions. A qualitative path graph is constructed by connecting the cell centers with the centers of their adjacent cell transition lines (Fig. 6b). In this graph, the path from start to goal is found by a simple graph search (Wallgrün, 1999).

The simulated robot environment is spatially more veridical than the schematic map. Here, the thickness of the walls is represented and distances and angles reflect the distances and angles of the actual robot environment (Fig. 7).

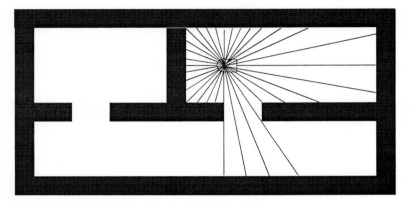

Fig. 7. The simulated robot environment. The spatial dimensions reflect the actual robot environment to scale. The simulated laser range finders measure the distances from the robot to the closest obstacles (walls)

The robot determines a path in the map. Consequently, it will traverse the path in the simulated world. To do this, the robot must establish a correspondence between the schematic map and the world (Fig. 8). This is done by mapping multimodal configurations detected in sensor space to configurations in the schematic map. The mapping task is supported by the use of qualitative spatial relations. This approach promises to be more reliable than isolated feature matching, as high-level feature configurations are less likely to be confused in restricted contexts.

The robot first matches the current sensor percepts with the cell that is marked in the map as its own location (see Fig. 8). The marked cell then is translated into a qualitative spatial representation in terms of vector relative position relations ("double cross calculus" – Fig. 9) (Freksa, 1992b). These relations allow for effective qualitative spatial inferences suitable for wayfinding (Zimmermann & Freksa, 1996). The potential complexity of spatial relationships can be restricted by organizing locations hierarchically (cf. Allen, 1983) and/or by limiting encoding of spatial relations to neighboring entities (Zimmermann, 1995).

Now, the qualitative description of the relevant map area can be matched with the sensor percept produced by the simulation. Since only the qualitative relations are represented the corresponding corners in the simulated world typically have the same relations like the ones on the map. Therefore the correct mapping between the entities on the map and in the world can be identified. Next, the transition line that is on the goal path can be determined in the simulated world. The midpoint of the transition line is the next intermediate goal of the simulated robot. At this point, the neighboring cell is entered, and the new local region is subject of a new map matching process. With this iterative procedure the target location in the simulated environment is reached.

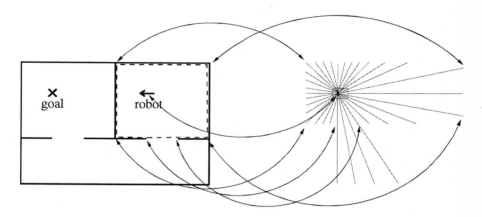

Fig. 8. Correspondence between spatial features in the schematic map (Fig. 5) and spatial features derived from the simulated sensor readings (Fig. 7). The arrows depict the correspondence relations between the corners in the schematic map and those derived from the sensory input, as well as the correspondence relations between the robot's location in the schematic map and that in the simulated world

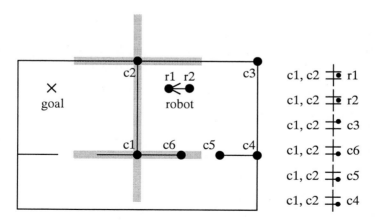

Fig. 9. Spatial relations in local robot region expressed in terms of vector relative position relations suitable for qualitative spatial inferences. Only the relations with reference to the room corners c1 and c2 are presented as an example

7   Conclusion and Outlook

We have presented an approach to high-level interaction between humans and robots on one hand and between robots and their environment on the other hand by means of schematic maps. The approach is based on the presumption that meaningful interaction requires an appropriate level of abstraction for intelligently solving tasks in a given domain. In the domain of wayfinding in a structured environment, a representation of space on the abstraction and granularity levels of decision-relevant entities is considered appropriate. Schematic maps are found to be suitable (1) for representing spatial knowledge on this level, (2) for qualitative spatial reasoning, (3) for human-robot interaction, and (4) for robot-environment interaction.

In pursuing this approach, we proceed in three stages: (1) conceptual design taking into account (a) spatial properties of the perceptual apparatus and the environment, (b) representational tools, and (c) inference methods; (2) implementation and experimentation in a simulation environment emphasizing the spatial reasoning aspects; and (3) implementation in a physical robot environment. The three stages are not carried out in a purely sequential manner; instead we have rather strong interactions between the stages during their development. As the three stages can be developed independently of one another to a large extent, we gain interesting insights about the transitions between the static analytic theory and the dynamic simulation environment on one hand and between the idealized perception / action model in the simulation environment and the real perception / action situation in the physical environment, on the other hand.

The work we are reporting on is work in progress. We have developed a formal system for qualitative spatial reasoning, a platform for simulation studies for spatial reasoning, and we have carried out experiments in physical robot navigation on the basis of qualitative spatial knowledge. The main focus of our present work is on the simulation environment that we build on the basis of our existing qualitative spatial reasoning theories. In parallel, we carry out perception studies in the physical robot environment to determine the type of landmarks we can use best for the navigation task.

In our future work on this project we will particularly focus on issues of dealing with incomplete sensor and map information, exploiting neighborhood and other spatial structures, matching descriptions of different granularity, and integrating information sources of different modality. We plan to experiment in our simulation environment successively with additional features and modalities to better understand and reduce the gap between the simulation and real environments.

Acknowledgments

We thank Sven Bertel, Stefan Dehm, Steffen Egner, Steffen Gutmann, Alexander Klippel, Kai-Florian Richter, Jesco von Voss, Jan Oliver Wallgrün, and Diedrich Wolter for discussions and contributions. We gratefully acknowledge very helpful comments by Thomas Röfer, Christoph Schlieder, and an anonymous reviewer. We apologize for not having incorporated all suggestions.

References

Allen, J. F. (1983). Maintaining knowledge about temporal intervals. *Communications of the ACM, 26*(11), 832-843.

Barkowsky, T., & Freksa, C. (1997). Cognitive requirements on making and interpreting maps. In S. Hirtle & A. Frank (Eds.), *Spatial information theory: A theoretical basis for GIS* (pp. 347-361). Berlin: Springer.

Berendt, B., Barkowsky, T., Freksa, C., & Kelter, S. (1998). Spatial representation with aspect maps. In C. Freksa, C. Habel, & K. F. Wender (Eds.), *Spatial cognition - An interdisciplinary approach to representing and processing spatial knowledge* (pp. 313-336). Berlin: Springer.

Braitenberg, V. (1984). *Vehicles - Experiments in synthetic psychology*. Cambridge, MA: MIT Press.

Brooks, R. A. (1991). Intelligence without representation. *Artificial Intelligence, 47*, 139-159.

Cohn, A. G. (1997). Qualitative spatial representation and reasoning techniques. In G. Brewka, C. Habel, & B. Nebel (Eds.), *KI-97: Advances in Artificial Intelligence* (pp. 1-30). Berlin: Springer.

Dirlich, G., Freksa, C., & Furbach, U. (1983). A central problem in representing human knowledge in artificial systems: the transformation of intrinsic into extrinsic representations. *Proc. 5th Cognitive Science Conference*. Rochester.

Dudeck, G. L. (1996). Environment representation using multiple abstraction levels. *Proc. IEEE, 84*(11), 1684-1705.

Fox, D. (1998). *Markov localization: A probabilistic framework for mobile robot localization and navigation*. Dissertation, Bonn.

Fox, D., Burgard, W., Kruppa, H., & Thrun, S. (1999). Collaborative multi-robot localization. In W. Burgard, T. Christaller, & A. B. Cremers (Eds.), *KI-99: Advances in Artificial Intelligence* (pp. 255-266). Berlin: Springer.

Freksa, C. (1992a). Temporal reasoning based on semi-intervals. *Artificial Intelligence, 54* (1-2), 199-227.

Freksa, C. (1992b). Using orientation information for qualitative spatial reasoning. In A. U. Frank, I. Campari, & U. Formentini (Eds.), *Theories and methods of spatio-temporal reasoning in geographic space* (pp. 162-178). Berlin: Springer.

Freksa, C., & Barkowsky, T. (1999). On the duality and on the integration of propositional and spatial representations. In G. Rickheit & C. Habel (Eds.), *Mental models in discourse processing and reasoning* (pp. 195-212). Amsterdam: Elsevier.

Freksa, C., & Röhrig, R. (1993). Dimensions of qualitative spatial reasoning. In N. Piera Carreté & M. G. Singh (Eds.), *Qualitative reasoning and decision technologies, Proc. QUARDET'93, CIMNE Barcelona 1993* (pp. 483-492).

Habel, C., Hildebrandt, B., & Moratz, R. (1999). Interactive robot navigation based on qualitative spatial representations. In I. Wachsmuth & B. Jung (Eds.), *KogWis99 - Proc. d. 4. Fachtagung der Gesellschaft für Kognitionswissenschaft, Bielefeld* (pp. 219-224). St. Augustin: Infix.

Habel, C., & Tappe, H. (1999). Processes of segmentation and linearization in describing events. In R. Klabunde & C. von Stutterheim (Eds.), *Representations and processes in language production* (pp. 117-153). Wiesbaden: Deutscher Universitäts-Verlag.

Musto, A., Stein, K., Eisenkolb, A., & Röfer, T. (1999). Qualitative and quantitative representations of locomotion and their application in robot navigation. In T. Dean (Ed.), *Proceedings IJCAI-99* (pp. 1067-1072). San Francisco, CA: Morgan Kaufmann.

Palmer, S. E. (1978). Fundamental aspects of cognitive representation. In E. Rosch & B. B. Lloyd (Eds.), *Cognition and categorization* (pp. 259-303). Hillsdale, NJ: Lawrence Erlbaum.

Röfer, T. (1999). Route naviation using motion analysis. In C. Freksa & D. M. Mark (Eds.), *Spatial information theory - Cognitive and computational foundations of geographic information science* (pp. 21-36). Berlin: Springer.

Rosch, E. (1975). Cognitive representations of semantic categories. *Journal of Experimental Psychology: General, 104*, 192-233.

Schlieder, C. (1996). Qualitative shape representation. In P. Burrough & A. Frank (Eds.), *Geographic objects with indeterminate boundaries* (pp. 123-140). London: Taylor & Francis.

Sogo, T., Ishiguro, H., & Ishida, T. (1999). Acquisition of qualitative spatial representation by visual observation. In T. Dean (Ed.), *Proceedings IJCAI-99* (pp. 1054-1060). San Francisco, CA: Morgan Kaufmann.

Thrun, S. (1998). Learning metric-topological maps for indoor mobile robot navigation. *Artificial Intelligence, 99*, 21-71.

Thrun, S., Bennewitz, M., Burgard, W., Cremers, A. B., Dellaert, F., Fox, D., Hähnel, D., Rosenberg, C., Roy, N., Schulte, J., & Schulz, D. (1999). MINERVA: A tour-guide robot that learns. In W. Burgard, T. Christaller, & Cremers, A. B. (Eds.), *KI-99: Advances in Artificial Intelligence* (pp. 14-26). Berlin: Springer.

Wallgrün, J. (1999). *Partitionierungsbasierte Pfadplanung für autonome mobile Roboter.* Studienarbeit, Fachbereich Informatik, Universität Hamburg.

Whorf, B. L. (1956). *Language, thought, and reality; selected writings.* Cambridge, MA: Technology Press of Massachusetts Institute of Technology.

Zadeh, L. A. (1999). From computing with numbers to computing with words - From manipulation of measurements to manipulation of perceptions. *IEEE Trans. Circuits and Systems - I: Fundamental Theory and Applications, 45*(1).

Zimmermann, K. (1995). Measuring without measures - The Delta-Calculus. In A. U. Frank & W. Kuhn (Eds.), *Spatial information theory - A theoretical basis for GIS* (pp. 59-67). Berlin: Springer.

Zimmermann, K., & Freksa, C. (1996). Qualitative spatial reasoning using orientation, distance, and path knowledge. *Applied Intelligence, 6*, 49-58.

From Motion Observation to Qualitative Motion Representation

Alexandra Musto[1], Klaus Stein[1], Andreas Eisenkolb[2], Thomas Röfer[3], Wilfried Brauer[1], and Kerstin Schill[2]

[1] Technische Universität München, 80290 München, Germany
{musto,steink,brauer}@in.tum.de
[2] Ludwig-Maximilians-Universität München, 80336 München, Germany
{amadeus,kerstin}@imp.med.uni-muenchen.de
[3] Universität Bremen, Postfach 330440, 28334 Bremen, Germany
roefer@tzi.de

Abstract. Since humans usually prefer to communicate in qualitative and not in quantitative categories, qualitative spatial representations are of great importance for user interfaces of systems that involve spatial tasks. Abstraction is the key for the generation of qualitative representations from observed data. This paper deals with the conversion of motion data into qualitative representations, and it presents a new generalization algorithm that abstracts from irrelevant details of a course of motion. In a further step of abstraction, the shape of a course of motion is used for qualitative representation. Our approach is motivated by findings of our own experimental research on the processing and representation of spatio-temporal information in the human visual system.

1 Introduction

It is a matter of fact that humans are capable of storing transient data, e.g., for tracking an object through occlusion [14], or for the reproduction of a path formed by a moving black dot on a computer screen (see section 4.1). For this prima facie trivial capability a bunch of problems has to be solved by the human visual system. Here we are concerned with the integration of representations with different time stamps. This general logistic problem arises especially when complex spatio-temporal information has to be classified, e.g., a sophisticated gesture in Japanese Kabuki Theater, but is immanent to formal-technical applications too. As an example, a locomotion-capable agent who has explored route knowledge about some environment and who has to communicate a possible route to a human listener in a human format has to transform the raw data (obtained by odometry) into some *qualitative* format (see [1] for an approach on a similar problem).

When representing an observed course of motion qualitatively, the most important step is abstraction. This paper presents a new possibility of how to abstract from irrelevant details of a course of motion in order to represent the coarse structure qualitatively. A further step of abstraction is the segmentation

Ch. Freksa et al. (Eds.): Spatial Cognition II, LNAI 1849, pp. 115–126, 2000.
© Springer-Verlag Berlin Heidelberg 2000

and classification of the gained qualitative representation. In this step any detail of the course of motion concerning spatio-temporal variation is omitted and it is represented merely as a sequence of motion shapes. Both generalization and segmentation are essential aspects of the qualitative modeling of motion representation. This qualitative stage is part of an integrated approach [13] which ranges from the modeling of qualitative representation to the modeling of spatio-temporal information on earlier stages of the human visual system. In this context, we also propose a memory model on the level of visual concept formation which tries to explain human capabilities of reproducing the spoor of a moving object.

2   Qualitative Motion Representation

Representing a course of motion qualitatively means abstracting from irrelevant details to get only the big directional changes which can then be translated in some qualitative categories for direction. Therefore, an observed course of motion should be simplified by some generalization algorithm (e. g., the one described in section 3) that suppresses small zigzags and deviations (fine structure) and returns a course of motion that contains only the big directional changes and the overall shape information (coarse structure).

The directional changes can then be represented qualitatively, as well as the distance between each directional change. We can do this by mapping the numeric values for distance and angle into a corresponding interval, thus collapsing numerical values which we do not want to distinguish into a single qualitative value.

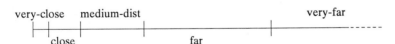

Fig. 1. A discretization of the distance domain

Figure 1 and 2 show how this can be done. In the direction domain, we model only positional change, so we can use the well known orientation relations from qualitative spatial reasoning. A discretization with intervals of equal size can be systematically derived on any level of granularity as shown in [6]. Possibilities of how to discretize the distance domain are shown in [2]. The qualitative values for distance can e. g., be named *very-close, close, medium-dist, far, very-far*, the values for direction *north, south, east, west*.

The resulting qualitative representation of a course of motion is a sequence of qualitative motion vectors (QMVs). The vector components are qualitative representations of direction and distance as shown above. A qualitative representation of motion speed can be added by simply mapping the ratio of time

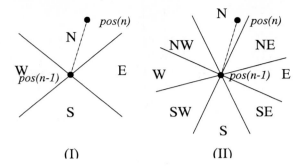

Fig. 2. I: 4 direction relations – II: 8 direction relations

elapsed from one directional change to the next and the distance covered in this time into qualitative intervals.

The mapping of the observed course of motion into a QMV sequence should best take place after a suitable generalization algorithm has been applied. However, it is also possible to do it the other way round, that is to generate a QMV representation containing coarse and fine structure of the course of motion and to apply the generalization algorithm to this QMV sequence. All developed algorithms (see [7] and section 3) can be used with numerical vector sequences as well as with QMV sequences. In fact, the algorithm presented here only smoothes the course of motion when applied to a numeric vector sequence, but generalizes only when applied to a QMV sequence, as we shall see.

3   Generalization

By generalization, a motion track is simplified and the fine structure is suppressed, stressing on the coarse form, containing the information on the major directional changes and the overall shape of the course of motion. Such an operation can be useful when processing motion information with a limited memory capacity, e.g. in biological systems, but also in technical systems like the Bremen Autonomous Wheelchair, where generalization is used for navigation on basis of recorded locomotion data [10].

Analytical generalization algorithms as used by, e.g., cartographers [3,5], are mostly not suitable for our purpose, since the computational effort of computing a smooth curve is not necessary with our discrete representation, but so considerable that it would perhaps not be possible to generalize "on the fly" in a real time environment. Other algorithms like the one described in [8] need the complete motion path, and therefore they are also not suitable for processing on the fly. So we developed two incremental generalization algorithms, the ε- and Σ-Generalization which are described in [7]. An alternative approach is based on a spatio-temporal memory model introduced in [12], which is a necessary prerequisite for the representation and processing of intricate motion information in the biological visual system.

In contrast to the sequential approaches described in [7] it enables the parallel access to all motion information gathered within the last few time cycles, i. e. to a sequence (of a certain length l) of successive motion vectors, in order to extract and generalize information about a course of motion. The following algorithms are based on the main idea of this model. Parallel access to l successive vectors is equal to a window of size l sliding on the motion track:

$$\text{Step } n \quad\ : \text{Access to } v_n, v_{n+1}, \ldots, v_{n+l-1}$$
$$\text{Step } n + 1 : \text{Access to } v_{n+1}, v_{n+2}, \ldots, v_{n+l}$$
$$\cdots \qquad\qquad \cdots$$

Technically this is a FIFO-buffer of size l.

Linear Sliding Window Generalization. A simple idea of using the local information of few successive vectors for generalization is to calculate the averages of k successive vectors (here $k = l$: compute the average of all parallel accessible vectors):

$$w_t = \frac{1}{k} \sum_{i=s}^{s+k-1} v_i, \qquad w_{t+1} = \frac{1}{k} \sum_{i=s+1}^{s+k} v_i, \qquad \cdots,$$

where w_t is the corresponding result vector for the sliding window from v_s to v_{s+k-1}. The index t marks the middle of the sliding window. $t = s + \frac{k-1}{2}$ holds with odd k. Thus we get a new vector sequence where each w_t is the average of k successive vectors v_i.

Nested Sliding Window Generalization. [9] used the idea of the spatiotemporal storage and parallel access for a different approach: the motion vectors $v_n, v_{n+1}, \ldots, v_{n+l-1}$ are used to compute the motion vector $(u_{n,n+l-1})$ for the whole motion of the observed object within the time slice of the spatio-temporal memory:

$$u_{n,n+l-1} = \sum_{i=n}^{n+l-1} v_i, \qquad u_{n+1,n+l} = \sum_{i=n+1}^{n+l} v_i, \qquad \cdots$$

Thus we get a sequence of overlapping vectors: $u_{n,n+l-1}, u_{n+1,n+l}, \ldots$ In this sequence, v_{n+l-1} is covered by the combined vectors $u_{n,n+l-1}$ to $u_{n+l-1,n+2l-2}$, v_{n+l} by $u_{n+1,n+l}$ to $u_{n+l,n+2l-1}, \ldots$

Now each w_t is computed as the average of all the $u_{a,b}$ covering v_t:

$$w_t = \frac{1}{k'} \sum_{a \leq t \leq b} u_{a,b} = \frac{1}{k'} \sum_{i=t}^{t+l-1} u_{i-l+1,i},$$

where k' is the number of input vectors: Each vector $u_{a,b}$ is the sum of l vectors v_i, and l vectors $u_{a,b}$ are added, therefore $k' = l \cdot l = l^2$.

Generic Sliding Window Generalization. Having a closer look on the algorithm above shows two nested sliding windows used:

$$w_t = \frac{1}{k'} \sum_{i=t}^{t+l-1} u_{i-l+1,i} = \frac{1}{l^2} \sum_{i=t}^{t+l-1} \sum_{j=i-l+1}^{i} v_j$$

$$= \frac{1}{l^2}(v_{t-l+1} + 2 \cdot v_{t-l+2} + \ldots + l \cdot v_t + (l-1) \cdot v_{t+1} + \ldots + v_{t+l-1}).$$

The inner sliding window (sum of the v_i) is of size l, the outer one used for computation of the average of size $k = 2l - 1$. Rewriting indices leads to

$$w_t = \frac{4}{(k+1)^2}(v_1 + 2 \cdot v_2 + \ldots + 2 \cdot v_{k-1} + v_k),$$

a weighted sum over k successive vectors v_i.

In general any set of weights can be used, so the generic form is

$$w_t = \frac{1}{\sum \alpha_i}(\alpha_1 v_1 + \alpha_2 v_2 + \ldots + \alpha_k v_k),$$

with a sliding window of size k, and with the linear and nested sliding window generalizations as special cases with $\alpha_i = 1$ (linear) and $\alpha_1 = 1$, $\alpha_2 = 2$, $\ldots$, $\alpha_{k-1} = 2$, $\alpha_k = 1$ (nested).

Processing of Start and End of a Motion Track. The start and end of the track have to be processed by an adapted (shortened) sliding window for starting the window with the first k vectors would shorten the whole track: Imagine a track of length n processed with a sliding window of length n, then the whole track would be shortened to $\frac{1}{n}$th of its length.

Properties. The sliding window generalization algorithm does not (as the ε- and Σ-generalization, see [7]) transform a vector sequence with many small movement vectors into one with a few large vectors to simplify the motion track, because the output track contains the same number of vectors as the input.

original   linear   nested

Fig. 3. Sliding Window Generalization with size k=3

The track is smoothed due to calculating the average of neighboring vectors, so sudden small but rapid changes in direction are lessened as shown in Figure 3 with linear sliding window of size 3.

If we use this algorithm on vector sequences with qualitative directions, the track is not only smoothed, but generalized: the mapping from numeric vectors in qualitative intervals gives a threshold below which all successive vectors are mapped to the same direction. If the track is sufficiently smoothed, little deviations are so totally suppressed. So and by fine-tuning the parameters α_i, the sensitivity of the generalization can be chosen. As can be seen in Figure 3, the gradient of the smoothed track is determined by choice of the α_i: With $\alpha_1 = \alpha_2 = \alpha_3 = 1$ (linear), the gradient is the same at all neighboring points of the original kink. With $\alpha_1 = \alpha_3 = 1, \alpha_2 = 2$ (nested), the gradient is the steepest at the point, where the kink was in the original track. With an arbitrary choice of the α_i, other operations can be done on the track. E.g., with $\alpha_1 = 1, \alpha_2 = \alpha_3 = 0$, all vectors in the window would be pushed one position forward.

4   Segmentation

4.1   Human Visual Perception

When asked, humans can reproduce the motion of single black dot moving along a (invisible) path on an otherwise white background of a computer screen (*target path*). To our knowledge there is no experimental data as to how precise such a stimulus can be reproduced. As a first step, we devised an experiment to examine as to how well humans can reproduce simple trajectories. In an experiment subjects (Ss) had to watch a target path and subsequently reproduce its spoor by moving the finger along the surface of a touch screen. First results of the experiments are shown in Figure 4.

Given the state of research this subsection has necessarily a speculative character. Some of the trajectories seem to be reproduced better than others (compare target path P13 and P18). It also seems that some patterns are adequately "generalized" and classified (P18: square loop, round loop) whereas others provide more difficulties, like most reproductions of P9 which can be considered "over-generalized". It may be the case that certain sub-patterns of a motion stimulus can be matched "on the fly" with existing internal templates, like "loop" or "kink". It is then possible that a dynamic structuring based on forms, i.e. *segmentation* can be performed whereas in other cases no template is found and the resulting representation is poor.

However, a difference in reproduction quality may mean that for some patterns it is simply easier to generate the respective motor commands whereas others are difficult to "express". It may also mean that the motor commands are of the same level of difficulty but the internal representation of some patterns is worse than others. If the template is in an eidetic form, i.e., there is no "higher" description based for instance on a form vocabulary, there is no reason why the curved pattern P13 in Figure 4 is reproduced worse than P18. It is in most cases a nontrivial task to determine the quality of the reproduction. Also, we know of no way to exclude the influence of the motor component in our reproduction task.

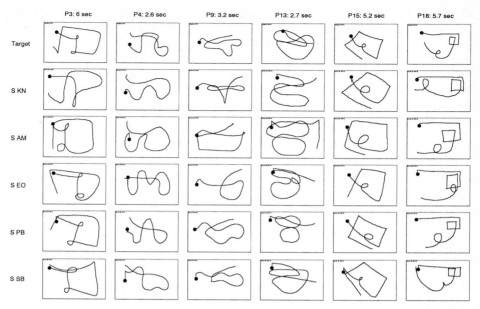

Fig. 4. Ss saw a target path (first row). Immediately after it had finished they had to reproduce its spoor by moving the finger along the screen surface (touch screen). The performance of 5 Ss for 6 target paths (columns) is depicted

4.2   Segmentation Algorithm

Inspired by the idea that courses of motion might be segmented into pieces that are matched with existing templates and so classified, we developed an algorithm that segments a QMV sequence such that it can be represented very compactly by a sequence of motion shapes. The basic step for this is to find out where one shape ends and the next begins. The algorithm described below will transform a QMV sequence to the basic shapes straight line, 90° curves to right and left and backward turns in different sizes, where the size of a curve is its radius. In case of eight direction relations, we also have 45° curves to right and left as well as 135° curves to right and left. More complex shapes[1] are built by combination of these basic ones. We restrict our illustrating example to the use of four direction relations. The case of eight relations is analogous.

The input for the algorithm is a generalized QMV sequence, so we do not have to worry about minimal movements; all movements below a minimal size are filtered by generalization. Thus, we can take each turn in the QMV sequence to be a curve in the shape representation. Since we do not know where a curve starts and ends, we would not know, for example, whether a QMV sequence

---

[1] from a predefined shape vocabulary, containing shapes like *loop-left, loop-right, u-turn-left, u-turn-right, s-curve-left, ...*

has to be segmented into a *left-turn, straight-line* and a *left-turn*, or into an *u-turn-left*. To get smooth shapes, we therefore made the decision to give every curve the maximum size possible, distributing the distances among all the curves starting with the smallest. Hence, we work on the representation of our QMV sequence as a track in the QMV linear space as described in [7]. This track is made up of horizontal and vertical straight lines as shown in Figure 5 (I).

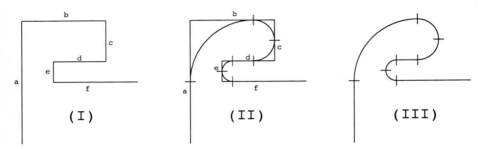

Fig. 5. Segmentation of a QMV sequence

The segmentation of each line will contain two curves, except for the first and the last line. We sort the lines by size, every line with two curves counts only half its length, for it will be split into two. In our example we get e, c, d, b, f, a. Then we start at the smallest line e. It has two curves, so it is split in the middle and the attached lines d and f are shortened by this amount (half size of e, as shown in Figure 5 (II)). Then the changed lines are sorted in with their new length (d now has just one curve from c to d and so is sorted in with full length whereas e and f have no curves left and are not sorted in at all), and we start from the beginning (for our example this means c, b, d, a). If we proceed on a line with only one curve left we give the curve the maximum size possible, as shown at line b. We are done when no lines are left.

Figure 5 (III) shows the graph of our shape representation of the QMV sequence. It reads

```
<straight-line middle-size> <right-turn big> <right-
turn middle-size> <right-turn middle-size> <straight-line small>
<left-turn small> <left-turn small> <straight-line middle-size>.
```

Now we can build more complex shapes. Two attached 90° turns of same size and direction (like at the lines c and e) can easily be combined to an u-turn, four attached 90° turns of same size can be combined to a loop, etc.

In a further step, we can suppress straight lines with very small size compared with the adjacent curves and take curves with similar size as equal. So, we can build complex shapes from simple shapes that are not matching the template completely. The synthesis of some shapes in the example sequence results in

```
<straight-line middle-size> <right-turn big> <u-turn-
right middle-size> <straight-line small> <u-turn-left
small> <straight-line middle-size>.
```

This segmentation algorithm also works well when processing a motion sequence on the fly with a lookahead of 2 to 3 vectors. The only problematic case for processing on the fly is when the vectors in the sequence get constantly shorter, which is a rare case. It can nevertheless be processed with small error, as shown in [15].

4.3   Memory Model

Can the shape representation be used to explain how humans reproduce the spoor of a course of motion? Given the preliminary character of the research results in section 4.1, this section is surely speculative, too. Nevertheless, we suggest a model that, given a segmentation of a course of motion in basic shapes, constructs more complex shapes on the fly and with a limited memory capacity.

We assume that at this level of processing, we have a similar memory model for parallel access like the model proposed in [12] for an earlier stage, with a few modifications:

- We assume not a simple FIFO-Buffer like used in the generalization algorithm, but more a kind of a stack: Incoming shapes on top of stack are "popped" and combined to a more complex shape as soon as possible, and the more complex shape is then again "pushed" on top of the stack.
- Our preliminary results strongly suggest that, if the course of motion is too "long" to be memorized correctly[2], not the beginning is forgotten, but features that are not very salient or pronounced. Therefore, in our model, if the motion sequence becomes too long for being held in the buffer, not the oldest shapes are removed, but the shapes that are least salient. This is the reason why parallel access must be granted.

To find out what "salient" means, experiments have to be conducted. For the moment, we assume that a shape is the more salient, the more complex it is. Complex means here, that the shape is constructed from simpler shapes. The more basic shapes participate in a complex shape, the higher is its complexity, but also the more information is coded in a single buffer element. Therefore it seems to be a reasonable strategy to supersede the simple shapes that carry the least information. This basic measurement of saliency can be weighted by several factors, e. g., age of the shape in the buffer and size of the shape. It seems also that shapes with pronounced corners are more salient than others. Another factor seems to be that the first and the last shape are remembered better due to primacy/recency effects, which should also be considered.

---

[2] Where too long means not the temporal duration, but the number of shapes that have to be held in memory

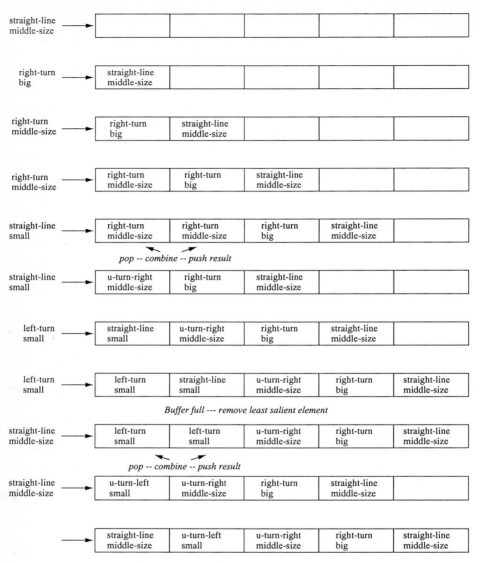

Fig. 6. Memory buffer for shape representation

Figure 6 shows the construction of the complex shape sequence from the course of motion in Figure 5 in a memory buffer of size 5.

This memory structure with parallel access to all elements also allows that a higher processing level extracts global features from the elements stored in the buffer, for example the shape of the overall arrangement of the shape sequence in space, similar as described in [13] for a lower processing stage.

A problem with our approach is that with the QMV representation we are unable to distinguish between a smooth curve and a sharp corner. That means that very important information is lost. Of course we could work with a numeric representation of the course of motion at the lower level, but then segmentation is much more difficult than with QMVs. Another possibility is to store additional information about the smoothness of a directional change with the QMVs.

5   Results and Future Work

In [9], the ε- and the nested sliding window generalization were applied concurrently to the same navigation approach in a simulation of the Bremen Autonomous Wheelchair [11]. The generalization was there used for navigation of the wheelchair (a semi-autonomous robot) along pre-taught routes. The track of the taught route was generalized and stored, and the track of the actual route was generalized on the fly and compared to the stored track. Thus, it could be determined whether the wheelchair has found its way correctly or made a mistake in following the route. In this work, both algorithms were applied to a semi-qualitative representation of the tracks, i. e., directions were represented qualitatively and distances numerically. With the qualitative directions, the results of both algorithms were satisfying and very similar, only the sliding window generalization always produces smooth tracks without sharp corners which is not the case with the ε-generalization.

Up to now, we have no experimental data whether our generalization algorithms correspond to anything that takes place in motion perception. This is a topic for future work. Neither do we have experimental results on the memory buffer model so far. In this context, we have to conduct first experiments on the saliency of motion shapes. From these findings, we can then define a saliency function to determine which of the shapes in the buffer can be forgotten first. Then we can apply our algorithm to the same trajectories human subjects have to reproduce in experiments to see whether similar results are produced, which would validate our model.

References

1. Elisabeth André, Guido Bosch, Gerd Herzog, and Thomas Rist. Characterizing trajectories of moving objects using natural language path descriptions. In *Proceedings of the 7th ECAI*, volume 2, pages 1–8, Brighton, UK, 1986.
2. Eliseo Clementini, Paolino Di Felice, and Daniel Hernández. Qualitative representation of positional information. *Artificial Intelligence*, 95(2):317–356, 1997.
3. Douglas and Peucker. Algorithms for reduction of the number of points required to represent a digitized line or its caricature. *The Canadian Cartographer*, 10/2, 1973.
4. Christian Freksa and David M. Mark, editors. *Spatial Information Theory. Cognitive and Computational Foundations of Geographic Information Science. International Conference COSIT'99*, volume 1661 of *Lecture Notes in Computer Science*, Berlin, Heidelberg, New York, August 1999. Springer.

5. Emmanuel Fritsch and Jean Philippe Lagrange. Spectral representations of linear features for generalisation. In Andrew U. Frank and Werner Kuhn, editors, *Spatial Information Theory. A Theoretical Basis for GIS. European Conference, COSIT'95*, volume 988 of *Lecture Notes in Computer Science*, pages 157–171, Berlin, Heidelberg, New York, September 1995. Springer.

6. Daniel Hernández. *Qualitative Representation of Spatial Knowledge*, volume 804 of *Lecture Notes in Artificial Intelligence*. Springer, Berlin, Heidelberg, New York, 1994.

7. Alexandra Musto, Klaus Stein, Kerstin Schill, Andreas Eisenkolb, and Wilfried Brauer. Qualitative Motion Representation in Egocentric and Allocentric Frames of Reference. In Freksa and Mark [4], pages 461–476.

8. Jonas Persson and Erland Jungert. Generation of multi-resolution maps from run-length-encoded data. *International Journal of Geographical Information Systems*, 6(6):497–510, 1992.

9. Hans Raith. Methoden zur Analyse von Bewegungsinformationen in technischen Systemen und Implementierung eines biologischen Modells zur Repräsentation spatio-temporaler Informationen. Diploma Thesis, 1999. Institut für Informatik, Technische Universität München.

10. Thomas Röfer. Route navigation using motion analysis. In Freksa and Mark [4], pages 21–36.

11. Thomas Röfer and Axel Lankenau. Architecture and applications of the Bremen Autonomous Wheelchair. In P. P. Wang, editor, *Proceedings of the Fourth Joint Conference on Information Systems*, volume 1, pages 365–368. Association of Intelligent Machinery, 1998.

12. Kerstin Schill and Christoph Zetzsche. A model of visual spatio-temporal memory: The icon revisited. *Psychological Research*, 57:88–102, 1995.

13. Kerstin Schill, Christoph Zetzsche, Wilfried Brauer, Andreas Eisenkolb, and Alexandra Musto. Visual representation of spatio-temporal structure. In B. E. Rogowitz and T. N. Pappas, editors, *Human Vision and Electronic Imaging. Proceedings of SPIE*, pages 128–137, 1998.

14. Brian J. Scholl and Zenon W. Pylyshyn. Tracking multiple items through occlusion: Clues to visual objecthood. *Cognitive Psychology*, (38):259–290, 1999.

15. Klaus Stein. Generalisierung und Segmentierung von qualitativen Bewegungsdaten. Diploma Thesis, 1998. Institut für Informatik an der TU München.

Lexical Specifications of Paths[*]

Carola Eschenbach, Ladina Tschander, Christopher Habel, Lars Kulik

University of Hamburg, Department for Informatics
Vogt-Kölln-Str. 30, D-22527 Hamburg
{eschenbach, tschander, habel, kulik}@informatik.uni-hamburg.de

Abstract. Natural language descriptions of motion frequently combine verbs of motion and directional prepositions. This article presents an analysis of German expressions from these two classes and their interaction. The focus is on the spatial structures (called *path*) that link them. Paths and the relations accessing them are formally specified in a geometric framework and with reference to the situation structure. The analysis distinguishes verbs of motion from other kinds of verbs that combine with directional adverbs. In addition, it provides a basis for explaining certain restrictions in combining verbs of motion with directional prepositional phrases and for comparing different approaches to the same kind of expressions.

1   Introduction

Natural language descriptions of motion frequently involve two kinds of expressions that connect to spatial structure: a verb of motion (such as *go, turn, enter, put, throw*) and a directional adverb or a directional prepositional phrase (such as *into the zoo, through the park, back, straight on*). Combinations of verbs of motion and directional prepositional phrases are characteristic for a group of languages that Talmy (1975) calls *manner languages*. Their linguistic and spatial constraints are of specific interest in the context of route descriptions and route instructions (cf. Habel, Moratz & Hildebrandt 1999). The objective of this article is to specify the general structures that underlie and sanction such combinations. On the linguistic level we give schemata for lexical entries. On the spatial level we present the geometry of bounded linear oriented structures, called *paths* for short.

It is generally assumed that the main verb determines several aspects of the syntactical and logical structure of the sentence. Its semantic content characterizes the situation described by the whole sentence. The verb assigns specific roles to the referents

---

[*] The research reported in this article has been carried out in the context of the project 'Axiomatik räumlicher Konzepte' (Ha 1237/7) that is part of the priority program on 'Spatial Cognition' of the Deutsche Forschungsgemeinschaft (DFG).
Many thanks to Christina Kiehl, Claudia Maienborn, Reinhard Moratz, Thomas Röfer, Barry Smith, Afra Sturm, and an anonymous reviewer for their comments on an earlier version of the article. Unfortunately but due to space restrictions, we were not able to react on all their comments.

Ch. Freksa et al. (Eds.): Spatial Cognition II, LNAI 1849, pp. 127-144, 2000.

of the noun phrases with which it is combined. Syntactic features such as case and word order mediate the assignment of roles. In (1) the verb *throw* is combined with two noun phrases (*Bill, the ball*) and a directional preposition (*into the pond*).

(1)   Bill threw the ball into the pond.

The verb assigns the role of the *actor* to Bill, and the role of the *bearer of motion* to the ball. The prepositional phrase specifies the *path* the ball moves along. The path of motion is also specified by the verb of motion. This can be concluded from sentences with verbs of motion that do not contain a directional adverb or prepositional phrase. The assignment of the two roles of bearer of motion and path is characteristic for verbs of motion. Only sub-groups of verbs of motion single out an actor explicitly. In (2) the actor is not distinguished from the bearer of motion. The verb *walk* assigns both roles to Peter. (This is our reason to use the term bearer of motion instead of theme or patient; cf. Levin 1993, Ehrich 1996.)

(2)   Peter walked out of the pond.

The semantics of the verbs of motion is responsible for connecting the path with the bearer of motion. Directional prepositions characterize the spatial embedding of the path. The prepositions *into, onto, to* specify the path as having a final position or goal. The prepositions *from* and *out of*, in contrast, specify the path by relating it to its starting position or source.[1] The verbs *enter* and *leave* also specify goal and source, respectively. Talmy (1975) points out that languages systematically differ in the assignment of semantic and spatial aspects to the syntactic categories. However, the following discussion focuses on manner languages and verbs of motion that combine with directional prepositional phrases.

Verbs of motion and directional prepositions are traditionally considered in the analysis of how language describes motion (cf. Talmy 1975, Jackendoff 1990, Maienborn 1990, Kaufmann 1995). On the informal level most approaches seem to agree on which meaning components have to be assumed for these expressions. However, the agreement is easily hidden under different names for the operators used in the symbolic versions of the proposals offered. For example, to compare the approaches of Jackendoff (1990) and Kaufmann (1995) one has to take account of the fact that the component Jackendoff calls GO corresponds to Kaufmann's MOVE, whereas Jackendoff's MOVE corresponds to Kaufmann's MOD.

One goal of the following discussion is to provide a formal description of path-based concepts for natural language analysis. Therefore, we focus on expressions that refer to motion, direction or paths and on combinations of them. The formal description of the concepts and their interrelations is a basis for comparing various approaches offered in the literature. It enables further work on more detailed questions

---

[1] The notions of path, source and goal employed here are technical ones, following common uses in the linguistic literature. In this sense of path, the term *the street* does not denote a path, but the directional prepositional phrase *along the street* does.

of the spatial-semantic structure of the linguistic elements involved in the description of motion and paths.

The outline of the article is as follows: It starts with a closer look at the German verbs of motion and identifies four sub-classes. The directional prepositions are divided into sub-classes that interact with the verb classes. We offer proposals for the structures of the lexical entries according to the verb classes. The different structures agree in how the bearer of motion and the path of motion are related. The article concentrates on the combination of German verbs of motion with different prepositional phrases and focuses on the path concept. It summarizes several aspects of analyses given by other linguists, which are then compared to our own proposal. The concepts that relate to paths and that are used in this proposal are specified using a geometric characterization of paths and their interaction with the situation structure.

2   Verbs of Motion and Directional Prepositions

2.1   General Characteristics

While there are various views of the exact extension of the field of verbs of motion, these differences can be reduced by distinguishing between relevant types of verbs of motion. The types we will consider in this article are based on the assignment of the role of actor and the focus on motion. Examples from German with approximate English translations are listed in Table 1. *Verbs of locomotion* express that the bearer of motion moves along a linearly extended path. *Verbs of change of position* focus on the change into the final position, and neglect the rest of the path. *Verbs of transport* and the *verbs of caused change of position* express the distinction between the bearer of motion and the actor.

Table 1. Sub-classes of verbs of motion

verbs of locomotion	schlendern, wackeln, rennen, gehen, rollen, kommen, …
	stroll, wobble, run, walk, roll, come, …
verbs of change of position	treten, eintreten, betreten, ankommen, verlassen, …
	step, (step into), enter, arrive, leave, …
verbs of transport	werfen, tragen, bringen, …
	throw, carry, bring, …
verbs of caused change of position	stellen, setzen, legen, hängen, …
	put, place, lay, hang, …

A linguistic characteristic of verbs of motion is that they combine with spatial prepositional phrases referring to a path. Thus, if a verb combines with directional prepositional phrases, it is a candidate of the field of verbs of motion. However, several perception verbs and static verbs can combine with directional prepositional phrases as well (consider *zur Tür blicken / look towards the door* or *nach Norden zeigen / to point north*). But they systematically differ from verbs of motion because they do not assign the role of the bearer of motion. In section 2.4 we offer a way to model this difference by different lexical structures of the verb classes.

The German verbs of change of position are mostly morphologically complex (*ein-treten / step into, an-kommen / arrive, be-treten / enter*) and syntactically heterogeneous. Three sub-classes of verbs of change of position demand different syntactic means for the specification of the goal of the change. Depending on the verb, the goal has to be specified via a directional prepositional phrase (*in das Haus eintreten / step into the house*), a local prepositional phrase (*auf dem Bahnhof ankommen / arrive at the station*), or a direct object (*das Haus betreten / enter the house*). In the following we will consider only the first of these sub-classes. Thus, we will exclusively consider verbs of motion that combine with directional prepositional phrases.

Directional prepositions differ regarding the spatial specification of the paths. *Goal prepositions* specify the end of the path, *source prepositions* give the start of the path, and *course prepositions* characterize the intermediate course of the path. They either indicate an *intermediate* place (*durch / through, über / over*) or the *shape* of the path (*um / around, längs / along*).

Table 2. Sub-classes of German directional prepositions[2]

goal prepositions	in, an, auf, vor, hinter, über, unter [+Acc]; zu
	into, to, onto, in front of, behind, over, under, to
source prepositions	von, aus; from, out of
course prepositions (intermediate place)	durch, über [+Acc]; through, across
shape prepositions	um, längs, entlang; around, along, along

Some combinations of verbs of motion and directional prepositional phrases are given in Table 3. It turns out that the sub-classes of verbs of motion differ in their restrictions to combine with directional prepositional phrases. Such restrictions have to be considered in the representations of the lexical entries for the verbs of motion.

Verbs of locomotion and verbs of transport can be combined with all kinds of directional prepositional phrases.[3] Verbs of caused change of position are best with goal phrases and worst with shape phrases. Verbs of change of position, again, show an irregular (and idiosyncratic) pattern of combinations.

---

[2] The German prepositions *in, an, auf, vor, hinter, über, unter* form local prepositional phrases when combined with a dative noun phrase, and directional prepositional phrases when combined with an accusative noun phrase. *In dem Haus* specifies the region inside the house, whereas *in das Haus* specifies the path leading into the house. *Entlang* can be used both as a preposition and as a postposition (*entlang der Straße, die Straße entlang / along the street*). All examples in this article employ *entlang* as a postposition.

[3] Exceptions in the class of verbs of locomotion are *schweifen / rove* or *streifen / roam*. They are atelic and cannot be combined with prepositions specifying goal or source. These verbs deserve specific attention and are excluded from the following discussion.

Table 3. Combinations of verbs of motion and directional prepositional phrases. (*) signals that the combination hardly makes sense or seems ungrammatical, whereas (?) signals that the combination is unusual but can be interpreted in certain contexts. Obviously, the judgements of the German examples cannot directly be transferred to the English translations provided here

verb preposition	locomotion	transport
goal	Tim geht an den Bach. Tim walks to the brook.	Tim wirft den Ball an die Wand. Tim throws the ball against the wall.
source	Tim geht aus dem Haus. Tim walks out of the house.	Tim wirft den Ball aus dem Teich. Tim throws the ball out of the pond.
course	Tim geht durch die Tür. Tim walks through the door.	Tim wirft den Ball durch die Tür. Tim throws the ball through the door.
shape	Tim geht den Bach entlang. Tim walks along the brook.	Tim wirft den Ball die Wand entlang. Tim throws the ball along the wall.

verb preposition	change of position	caused change of position
goal	Tim tritt in das Haus ein. Tim steps into the house. *Tim tritt an die Tür ein. Tim enters against the door.	Tim stellt den Eimer an die Wand. Tim puts the bucket against the wall.
source	?Tim tritt aus dem Garten ein. Tim enters out of the garden.	?Tim stellt den Eimer aus dem Haus. Tim puts the bucket out of the house.
course	Tim tritt durch die Tür ein. Tim enters through the door.	?Tim stellt den Eimer durch die Tür. Tim puts the bucket through the door.
shape	*Tim tritt das Haus entlang ein. Tim enters along the house.	*Tim stellt den Eimer die Wand entlang. Tim puts the bucket along the wall.

2.2   The Format of Lexical Representations

A lexical entry consists of different components connecting several levels of linguistic structure such as syntax, semantics, and phonology (cf., Bierwisch 1997, Eschenbach 1995, pp. 28–35). For our purpose two components are relevant. The first is the *semantic structure*. It specifies the meaning of a lexical item in the form of a combination of basic predicates and parameters. In the following we use expressions in the standard format of first order predicate logic for representing the semantic structure. In (3) the semantic structure of a transitive verb is represented as $VERB'(s, x, y)$.[4]

(3) $\lambda y\, \lambda x\, \lambda s\, [VERB'(s, x, y)]$

---

[4] The prime indicates that the predicate or relation is not analyzed further.

The second component of interest here is the *argument structure*. It specifies the syntactic realization of the semantic parameters and forms the main connection between the syntactic and the semantic level. We represent the argument structure as a prefix formed via λ-abstractors, where syntactic features can be added to each abstractor. The argument structure of (3) is $\lambda y \, \lambda x \, \lambda s$. Three types of arguments are generally distinguished in lexical entries of verbs. *Internal arguments* (y) are regularly syntactically realized as (direct or indirect) objects or by prepositional phrases. The *external argument* (x) is realized as the subject. The *referential argument* (s) refers to the situation described and serves as an anchor point for additional specifications by adverbs.[5]

The investigation presented in this article joins the tradition of representing lexical meaning in a decompositional manner (cf. Bierwisch 1997, Wunderlich & Herweg 1991, Wunderlich 1991, Kaufmann 1995) along with specifying the semantic interrelation of basic predicates by meaning postulates or axioms (cf. Hilbert 1918). The semantic structure of a lexeme can be analyzed as a combination of different predicates. The combination of lexical entries yields the complex meanings according to general rules. Axioms specify the semantic interrelations among the basic predicates. The goal is to use and formally interrelate a small number of predicates, such that the semantic content of verbs of motion and directional prepositions can be grasped.

2.3   Lexical Entries of Directional Prepositions

Rauh (1997) argues that the argument structure of spatial prepositions is similar to that of transitive verbs. It also includes an internal, an external and a referential argument. The main difference is that the referential argument of a spatial preposition specifies a spatial entity. We assume that the referential argument of directional prepositions is a path (w), whereas the referential argument of a local preposition is a place (p). (4) shows the general format of directional prepositions. (5) gives the specific case of the lexical entry for the German directional preposition *in*.

(4) $\lambda y_{[CASE]} \, \lambda x \, \lambda w \, [\text{PATH-RELATION}(w, \text{PLACE-FUNCTION}(y)) \wedge D(x, w)]$

(5) $\lambda y_{[+Acc]} \, \lambda x \, \lambda w \, [\text{TO}(w, \text{IN}^*(y)) \wedge D(x, w)]$

The direction given by the preposition is specified as the PATH-RELATION based on a region. In the case of TO, the path (w) ends in the region. The internal argument (y) and the PLACE-FUNCTION determine the region. IN* is the characteristic function that is part of the meaning of the preposition *in* (cf. Wunderlich & Herweg 1991, p. 773). IN*(y) refers to the region that is the interior of y. This part of our proposal directly corresponds to Jackendoff's (1990) analysis. The relation D relates the external argument (x) and the path. Thus, (5) says that the goal of the path (w) is a dimension of the external argument (x) and located in the interior of the internal argument (y). The internal argument of a preposition is a noun phrase with the case specified by the preposition,

---

[5] The notion of situation is used according to Mourelatos (1978). The situation structure includes events, processes and states.

accusative in the case of the directional preposition *in*. (6) represents the structure of the phrase *in ein Haus / into a house*.[6]

(6) $\lambda x \, \lambda w \, [TO(w, IN^*(y)) \wedge D(x, w) \wedge HAUS'(y)]$

It relates the external argument (x) with the path (w) and w is specified as having its goal in a house.[7] As a result of combining the two components, y becomes a free parameter in the semantic structure. Free parameters of this type can be compared to discourse referents in DRT (Kamp & Reyle 1993).

2.4   Lexical Entries of Verbs of Motion

It is broadly accepted that the path is an element of the lexical entry of directional prepositions (cf. Bierwisch 1988, Jackendoff 1990, Maienborn 1990, Kaufmann 1995, Wunderlich 1991). However, views vary mainly regarding the question whether verbs of motion have to provide the path explicitly. Bierwisch (1988), Kaufmann (1995) and Wunderlich & Herweg (1991) do not integrate the path parameter in the lexical entry of the verb but leave the integration to further mechanisms of the interpretation of the complex expression. In contrast, Jackendoff (1990), Maienborn (1990), and Ehrich (1996) use the path parameter as the connecting element between verbs and prepositions. This guarantees that the movement that the verb refers to is constituted on the path given by the prepositional phrase. We will follow the latter approach (for a syntactic argument cf. Tschander (1999)).

Assuming that directional prepositional phrases and adverbs figure as internal arguments of verbs of motion, (7) shows the semantic representation of verbs of locomotion. The phrase *in ein Haus / into a house* (6) can specify the internal argument position P of the verb yielding the representation in (8). Thus, the assumptions guarantee that the path given by the verb is identified with the path given by the preposition.

(7) $\lambda P \, \lambda x \, \lambda s \, [VERB'(s, x, w) \wedge P(x, w)]$

(8) $\lambda x \, \lambda s \, [VERB'(s, x, w) \wedge TO(w, IN^*(y)) \wedge D(x, w) \wedge HAUS'(y)]$

As mentioned before, the structure of verbs of motion is similar to the structure of perception verbs that combine with directional prepositional phrases (consider *zur Tür starren / to stare towards the door*). But verbs of motion and perception verbs

---

[6] The lexical entries for nouns have the form $\lambda x \, [NOUN'(x)]$, where x is both the external and the referential argument. See Eschenbach (1995) for details on principles of combining lexical entries to form representations of complex phrases.

[7] The role of the external argument x and the relation D will not be important in the following discussion. They are relevant for modeling uses of prepositional phrases such as *die Fahrt nach München / the jouney to Munich, der Pfeil zum Bahnhof / the arrow to the station* (cf. Kaufmann 1995, Wunderlich 1991, Wunderlich & Herweg 1991).

systematically differ in the embedding of the path. In the first case, the bearer of motion traverses the path, whereas in the second case an entity orients its gaze according to the path. This is reflected in the representation of verbs of motion by assuming the meaning component OCCURS-ON(s, GO(x, w)) to be part of the meaning of every verb of motion. It says that s is a situation in which x traverses w. (The formalization of this is given in section 4.3.) This meaning component assigns the role of the bearer of motion in the situation s to x. Following Jackendoff (1990) we assume that perception verbs include the component ORIENT(x, w) specifying the path. However, an analysis of verbs of perception and ORIENT cannot be presented here.

Several German verbs of locomotion specify the manner of motion in addition to the path (cf. Talmy 1975). The relation VERB-MOD'(s, x) is a place holder for the specific condition on the manner of motion encoded in the verb. Accordingly, we assume (9) to be the general form of the lexical entries of verbs of locomotion.

(9) $\lambda P \lambda x \lambda s$ [OCCURS-ON(s, GO(x, w)) $\wedge$ VERB-MOD'(s, x) $\wedge$ $P(x, w)$]

Verbs of transport systematically distinguish an actor and the bearer of motion. We simply use the expression ACTOR(x, s) to symbolize that x is the actor in s, because the question how the actor is distinguished in the semantic form of a verb is not the focus of our discussion. This expression will not be specified further. The verbs of transport differ in assigning the manner to the actor (*werfen* / *throw*), to the bearer of motion (*rollen* / *roll*), or to their interaction (*schieben* / *push*) (cf. Ehrich (1996) for a more elaborate discussion). Therefore, (10) is the general structure of verbs of transport allowing the assignment of manner to both or either of the participants.

(10) $\lambda P \lambda y \lambda x \lambda s$ [ACTOR(x, s) $\wedge$ OCCURS-ON(s, GO(y, w)) $\wedge$ VERB-MOD'(s, x, y) $\wedge$ $P(y, w)$]

Verbs of transport and verbs of locomotion refer to the complete course of the path. Verbs of caused change of position additionally specify that the bearer of motion arrives at a certain position in a certain manner. Correspondingly, section 2.1 showed that verbs of caused change of position (*stellen, legen* / *put, lay*) can best be combined with goal expressions. We propose (11) as the basic structure of the lexical entries of verbs of caused change of position.

(11) $\lambda P \lambda y \lambda x \lambda s$ [ACTOR(x, s) $\wedge$ OCCURS-ON(s, GO(y, w)) $\wedge$
                    BECOME(s, VERB-MOD'(y) & BE AT(y, p)) $\wedge$ $P(y, w)$]

On the one hand the entries include the component OCCURS-ON(s, GO(y, w)) characteristic for verbs of motion. On the other hand, BECOME(s, VERB-MOD'(y) & BE AT(y, p)) expresses that y arrives at place p in a certain manner (VERB-MOD') of being located (for example, *stand* vs. *lie*). The location p is not characterized further in the semantic structure of the verbs. It has to be specified on the conceptual level using background knowledge because it is not directly accessible for specification by linguistic items.

When (11) is combined with the representation of a goal prepositional phrase (such as *in ein Haus stellen / put into a house*) the result is (12).

(12) $\lambda y\, \lambda x\, \lambda s$ [ACTOR(x, s) ∧ OCCURS-ON(s, GO(y, w)) ∧ BECOME(s, VERB-MOD'(y) & BE AT(y, p))
∧ TO(w, IN*(z)) ∧ D(y, w) ∧ HAUS'(z)]

The semantic structure of (12) can be verbalized as follows: x is the actor in the situation s, in s, y moves along the path w that ends in the interior of some object z, which is a house, and s thereby establishes a situation in which y is at a position p in manner VERB-MOD'. Obviously, the final point of the path is the best candidate for specifying the goal position p. This can explain the clear preference for goal phrases, because they can directly specify the path w and indirectly specify the position p. The formal analysis of OCCURS-ON(s, GO(y, w)) and BECOME(s, VERB-MOD'(y) & BE AT(y, p)) in section 4.3. ensures that the final point is the initial position of the bearer of motion in the directly following situations specified by BECOME.

We do not assume that the lexical entries of the verbs of change of position have a common structure, because they form a heterogeneous class in several respects. However, the lexical entry of the verb *eintreten / (step into)* might look like (13).

(13) $\lambda P\, \lambda x\, \lambda s$ [OCCURS-ON(s, GO(x, w)) ∧ BECOME(s, BE AT(x, p)) ∧ OCCURS-ON(s, STEP'(x)) ∧
$p \subseteq$ IN*(d) ∧ P(x, w)]

In this case, OCCURS-ON(s, STEP'(x)) encodes the manner of motion of x making a step. BECOME(s, BE AT(x, p)) says that the situation in which the motion takes place is followed by a situation of x being at position p. This position is included in the interior of some object (d) that is not specified further in the semantic structure of the verb. However, if (13) combines with *in ein Haus / into a house*, the result is (14). In this structure, the house (z) is a good candidate for specifying the free parameter d.

(14) $\lambda x\, \lambda s$ [OCCURS-ON(s, GO(x, w)) ∧ BECOME(s, BE AT(x, p)) ∧ OCCURS-ON(s, STEP'(x)) ∧
$p \subseteq$ IN*(d) ∧ TO(w, IN*(z)) ∧ D(x, w) ∧ HAUS'(z)]

Summarizing the proposals given, the component OCCURS-ON(s, GO(., w)) figures as the characteristic component of verbs of motion. It connects the bearer of motion to the path which it traverses. The actor of the situation is characterized by ACTOR(x, s). If the further analysis reveals the necessity, intransitive verbs can include the component ACTOR related to the bearer of motion as well.

The discrimination between sub-groups of the verbs is done based on two features. The first critical point is whether the actor is distinguished from the bearer of motion. The second point is whether the verb specifies the establishment of a position for the following situation, symbolized as BECOME(s, BE AT(., p)). The connection of the position p and the path w in these cases is not done by the verb but on the conceptual level of interpreting the complex structures.

In addition, we saw that the verbs include characteristic modes of motion. However, we did not have a closer look at the details of these meaning components. This is based on the thesis that the manner is completely independent from the path. But this is a matter of further investigations.

3   Comparison to Other Approaches

The semantic components used in our analysis are related to proposals for lexical representations given in the literature. The similarities and differences in the representations of verbs of motion and directional prepositional phrases are presented in this section.

3.1   Semantic Structures of Verbs of Motion

The lexical representations proposed for verbs of motion (Jackendoff 1990, Maienborn 1990, Kaufmann 1995, Ehrich 1996) agree regarding the components of the lexical representation of verbs of motion, although different forms of symbolization are used. The proposals are listed in Table 4 and Table 5. Slight alternations of these formulae from the original formulations mainly concern the naming of variables.

Table 4. Proposals for the semantic structures of verbs of locomotion and transport

	verbs of locomotion
this article	OCCURS-ON$(s, GO(x, w)) \wedge$ VERB-MOD'$(s, x) \wedge P(x, w)$
Jackendoff (1990)	[Event GO([Thing]x, [Path]w)]s [Event MOVE([Thing]x)]s
Maienborn (1990)	s INST [DO$(x$, MOVE$(x, w))] \wedge$ MOD(s)
Kaufmann (1995)	MOVE$(x) \wedge$ MOD(x)
Ehrich (1996)	GO$((x, w)$ BY MOD$(x))$

	verbs of transport
this article	ACTOR$(x, s) \wedge$ OCCURS-ON$(s, GO(y, w)) \wedge$ VERB-MOD'$(s, x, y) \wedge P(y, w)$
Jackendoff (1990)	[Event CAUSE([Thing]x, [Event GO([Thing]y, [Path]w)]s)]
Ehrich (1996)	[AFFECT$((x, y)$ BY MOD$(x))$] CAUSE [GO(y, w)] [AFFECT(x, y)] CAUSE [GO$((y, w)$ BY MOD$(y))$]

The proposals for verbs of motion have in common that the representations contain a relation of dynamic localization and a specification of the manner of motion. The relation of dynamic localization refers to the bearer of motion, which covers the path during a situation. This is symbolized as GO (Jackendoff 1990, Ehrich 1996) or as MOVE (Kaufmann 1995, Maienborn 1990). Jackendoff (1990) assumes different forms of representation for two classes of verbs of motion. The first expresses motion along a path (as *go, enter, approach*) and the second manner of motion (as *wiggle, dance, wave*). In contrast to all other approaches, he does not assume that both components are combined in a single lexical entry. Kaufmann (1995) deviates from the others by omitting the path in the representation of the verbs.

The manner component is connected to the localization relation either with a conjunction (Maienborn 1990, Kaufmann 1995) or with a BY-component (Ehrich 1996). The representations of Ehrich (1996) result from an investigation of verbs of transport. She takes into account that transitive verbs of motion differ in the assignment to the manner component to the actor or the bearer of motion. Maienborn (1990) in contrast assigns manner to the complete situation.

Maienborn (1990) analyzes verbs of caused change of position similar to verbs of locomotion. In her representation the final location of the bearer of motion is conveyed in the dynamic localization relation and the manner component. Hence, the distinction between an extended path and a change of location is not represented explicitly. Jackendoff (1990) and Kaufmann (1995) propose representations of the verbs of (caused) change of position that employ predicates corresponding to the BECOME-operator and the relation of an object being at a place, called BE by Jackendoff and $KONT(T(y), p)$ by Kaufmann. These representations do not entail a path parameter.

Table 5. Proposals for the semantic structures of verbs of (caused) change of position

	verbs of change of position
this article	OCCURS-ON$(s$, GO$(x, w)) \wedge$ BECOME$(s$, BE AT$(x, p)) \wedge P(x, w) \wedge \dots$
Jackendoff (1990)	[Event INCH([State BE([Thing]x, [Place]p)])]s

	verbs of caused change of position
this article	ACTOR$(x, s) \wedge$ OCCURS-ON$(s$, GO$(y, w)) \wedge$ BECOME$(s$, VERB-MOD'(y) & BE AT$(y, p)) \wedge P(y, w)$
Maienborn (1990)	s INST[CAUSE$(x$, MOVE$(y, w))] \wedge$ MOD(s)
Kaufmann (1995)	CAUSE$(x$, BECOME(KONT$(T(y), p)$ & MOD$(y)))$

Jackendoff (1990) uses INCH for representing verbs of change of position. He says 'the Event reading describes a change taking place whose final state is the State reading— the familiar inchoative relation. [...] INCH maps its argument into an Event that terminates in that State' (Jackendoff 1990, p. 75). According to this definition the INCH-function acts in the same way as the BECOME-relation. Thus, Jackendoff and Kaufmann analyze these verbs according to resultative verbs. This obscures the kinship with the other verbs of motion.

While our approach specifies both, the movement along the path and the establishment of a final position, the other authors selected one or other of them in their proposals. As mentioned before, the double specification in our approach can model that these verbs combine with directional adverbs and thereby prefer (specific) goal expressions.

3.2 Semantic Structures of Directional Prepositions

Jackendoff (1990) and Kaufmann (1995) propose representations for directional prepositional phrases. Both use path descriptions that relate paths to positions and specify whether the position is source, goal, or an intermediate position of the path.

Table 6. Path descriptions

	Jackendoff (1990)	Kaufmann (1995)	this article
goal	[Path TO[Place]$_p$]w	CHANGE(w, $\lambda r(r \subseteq p)$)	TO(w, p)
source	[Path FROM[Place]$_p$]w	kE(w) $\subseteq p$ CHANGEu(w, $\lambda r(r \subseteq p)$)	FROM(w, p)
direction$_{goal}$	[Path TOWARDS[Place]$_p$]w	FIN(OS(w) $\subseteq p$)	TOWARDS(w, p)
direction$_{source}$	[Path AWAY_FROM[Place]$_p$]w		AWAY-FROM(w, p)
intermediate	[Path VIA[Place]$_p$]w		VIA(w, p)

Table 6 summarizes the proposals for directional prepositions given by Jackendoff (1990) and Kaufmann (1995) and the names of the relations used in the following.[8] The correspondences yielding from the explanations of the authors are also represented in this list. The path function in the description of Jackendoff elaborates a path by mapping a reference object or place into a related path. Jackendoff represents a path directed to a goal with TO. This entails the change of location of an object traversing the path. With the CHANGE-relation, Kaufmann makes the change of location explicit. However, the relations shall not necessarily be interpreted as change in time. For a preposition referring to a starting point of a path, Kaufmann uses two different forms. The function TOWARDS specifies paths that are directed towards the goal but need not reach it. Similarly, paths AWAY-FROM a source need not start at the source. The VIA-function refers to a part of a path which is localized (as in *Tim travels via Bremen to Hamburg*).

4   A Geometric Foundation for the Representation of Paths

The path forms the semantic link between verbs of motion and directional prepositional phrases. Both, prepositions and verbs can include restrictions on the path. The verb is responsible for embedding the path in the description of the situation. The prepositions lead to the spatial embedding of the path.

This section gives a formal account on paths, their geometric properties that are the basis for defining path functions, and their interrelation with the situation structure. The first sub-section specifies the paths and the basic relations on and between them, the second sub-section specifies path functions and the third one the interaction between paths and situations of traversing them.

---

[8] Unfortunately we have to omit discussing Kaufmann's additional proposals for modeling shape prepositions.

4.1   The Geometry of Paths

Although there is agreement in the literature on which spatial aspects are expressed, this agreement can mainly be attributed to the common use of technical terms such as source and goal. One aim of this article is to provide a geometric foundation of the spatial terms that are used in the specification of paths. The geometric description is based on the geometric framework given by Eschenbach, Habel and Kulik (1999) and Kulik and Eschenbach (1999).

Paths are linear, directed and bounded entities. The basic relation for paths is a linear ordering relation. The symbolized form $\prec(w, Q, Q')$ can be read as *point Q precedes point Q' on path w*. The term 'precedence' signifies the order induced by the path's direction. It is not defined with reference to time. The relation of precedence is the basis for defining other relations and functions. In the course of giving the axioms for paths we introduce some of them as abbreviations. We continue to use the following conventions for sorted variable names: w for paths (ways), p for position or place of an object, which can be a point or a region (cf. Eschenbach et al. 1998), Q for points.

A *point is on a path* ($Q \iota w$), iff it is related by precedence to another point on the path (D1). The ordering relation on paths is transitive (A1) and irreflexive (A2). It relates every pair of different points on the path (A3). Therefore, paths do not include circles, branches or unrelated parts.

(D1) $Q \iota w$ $\qquad$ $\Leftrightarrow_{def} \exists Q' [\prec(w, Q, Q') \lor \prec(w, Q', Q)]$

(A1) $\forall w\, Q_1\, Q_2\, Q_3$ $\qquad$ $[\prec(w, Q_1, Q_2) \land \prec(w, Q_2, Q_3) \Rightarrow \prec(w, Q_1, Q_3)]$

(A2) $\forall w\, Q$ $\qquad$ $[\neg \prec(w, Q, Q)]$

(A3) $\forall w\, Q\, Q'$ $\qquad$ $[Q \iota w \land Q' \iota w \land Q \neq Q' \Rightarrow (\prec(w, Q, Q') \lor \prec(w, Q', Q))]$

A *path is a sub-path of another path* ($w \sqsubseteq w'$), iff precedence on w implies precedence on w' (D2). *Point Q_2 is between point Q_1 and point Q_3 on path w* ($\beta(w, Q_1, Q_2, Q_3)$), iff Q_1 precedes Q_2 and Q_2 precedes Q_3, or Q_3 precedes Q_2 and Q_2 precedes Q_1 (D3). According to its definition, the sub-path relation is transitive and reflexive. In addition, it is anti-symmetric (A4). Therefore, paths are uniquely characterized by the ordering relation they induce on points. Furthermore, we assume that sub-paths are convex with respect to the enclosing path (A5), i.e., if a point Q_2 is between two points Q_1 and Q_3 on path w', then it is on every sub-path w of w' that includes both Q_1 and Q_3.

(D2) $w \sqsubseteq w'$ $\qquad$ $\Leftrightarrow_{def} \forall Q\, Q' [\prec(w, Q, Q') \Rightarrow \prec(w', Q, Q')]$

(D3) $\beta(w, Q_1, Q_2, Q_3)$ $\qquad$ $\Leftrightarrow_{def} (\prec(w, Q_1, Q_2) \land \prec(w, Q_2, Q_3)) \lor (\prec(w, Q_3, Q_2) \land \prec(w, Q_2, Q_1))$

(A4) $\forall w\, w'$ $\qquad$ $[w' \sqsubseteq w \land w \sqsubseteq w' \Rightarrow w = w']$

(A5) $\forall w\, w'\, Q_1\, Q_2\, Q_3$ $\qquad$ $[\beta(w', Q_1, Q_2, Q_3) \land w \sqsubseteq w' \land Q_1 \iota w \land Q_3 \iota w \Rightarrow Q_2 \iota w]$

The starting point of a path (stpt(w)) is on the path and precedes every other point on the path (D4). *The final point of a path* (fpt(w)) is on the path and it is preceded by every other point on the path (D5). The axioms introduced so far justify the definition of starting point and final point as function symbols, because both conditions can be fulfilled by one point per path at most. The boundedness of paths is given by the existence of both a starting point (A6) and a final point (A7). If two paths have the same points, then the starting point and the final point of one of them are also starting point or final point of the other path (A8). In addition, if paths have the same points and the same starting point then they are identical (A9).

(D4) $Q = \text{stpt}(w)$ $\Leftrightarrow_{\text{def}}$ $Q \iota w \wedge \forall Q' [Q \neq Q' \wedge Q' \iota w \Rightarrow \prec(w, Q, Q')]$

(D5) $Q = \text{fpt}(w)$ $\Leftrightarrow_{\text{def}}$ $Q \iota w \wedge \forall Q' [Q \neq Q' \wedge Q' \iota w \Rightarrow \prec(w, Q', Q)]$

(A6) $\forall w \exists Q$ $[Q = \text{stpt}(w)]$

(A7) $\forall w \exists Q$ $[Q = \text{fpt}(w)]$

(A8) $\forall w\, w'$ $[\forall Q [Q \iota w \Leftrightarrow Q \iota w'] \Rightarrow$
                              $\forall Q [(Q = \text{stpt}(w) \vee Q = \text{fpt}(w)) \Rightarrow (Q = \text{stpt}(w') \vee Q = \text{fpt}(w'))]]$

(A9) $\forall w\, w'$ $[(\forall Q [Q \iota w \Leftrightarrow Q \iota w'] \wedge \text{stpt}(w) = \text{stpt}(w')) \Rightarrow w = w']$

This description specifies only the relation of precedence with respect to single paths and not the relation between paths in a geometric framework. However, the description is sufficient, because it interrelates the notions necessary for the description of the behavior of paths in the linguistic fragment considered here.

4.2   Geometric Components of Directional Prepositions

Path functions such as TO describe paths with respect to places. As mentioned before, Jackendoff (1990) assumes five such relations. TO expresses that a path leads to the place, FROM that a path starts in the place, VIA that it runs through the place. TOWARDS and AWAY-FROM are imperfect versions of TO and FROM.

All these conditions can be expressed by the paths introduced here and the notions defined on their basis. A path is TO a place, iff the place includes the path's final point but not its starting point (D6). Conversely, a path is FROM a place, iff the place includes the path's starting point but not its final point (D7). A path is VIA a place, if the place includes an inner point of the path but not the starting point or the final point (D8). A path is TOWARDS a place, iff it is part of a path TO the place and they have a common starting point (D9). Correspondingly, a path is AWAY-FROM a place, iff it is part of a path FROM the place and they have a common final point (D10).

(D6) $\text{TO}(w, p)$ $\Leftrightarrow_{\text{def}}$ $\text{fpt}(w) \iota p \wedge \neg(\text{stpt}(w) \iota p)$

(D7) $\text{FROM}(w, p)$ $\Leftrightarrow_{\text{def}}$ $\text{stpt}(w) \iota p \wedge \neg(\text{fpt}(w) \iota p)$

(D8) $\text{VIA}(w, p)$ $\Leftrightarrow_{\text{def}}$ $\exists Q [Q \iota w \wedge Q \iota p] \wedge \neg(\text{stpt}(w) \iota p) \wedge \neg(\text{fpt}(w) \iota p)$

(D9) $\text{TOWARDS}(w, p)$ $\Leftrightarrow_{\text{def}}$ $\exists w' [\text{TO}(w', p) \wedge w \sqsubseteq w' \wedge \text{stpt}(w) = \text{stpt}(w')]$

(D10) $\text{AWAY-FROM}(w, p)$ $\Leftrightarrow_{\text{def}}$ $\exists w' [\text{FROM}(w', p) \wedge w \sqsubseteq w' \wedge \text{fpt}(w) = \text{fpt}(w')]$

These characterizations are very general. They do not require paths to be straight or short. This is appropriate considering that the natural language descriptions examined are very flexible. Complex journeys can be described as 'going to some place', although the concrete path taken can deviate arbitrarily from the direct route. However, the characterization of TOWARDS and AWAY-FROM is critical in this respect because in open space every path can be completed by another path leading to any goal. Thus, further constraints will have to be added based on clarifications of the uses of phrases involving these components.

4.3   Verbs of Motion: Paths as Trajectories

Verbs of motion embed reference to paths via the structure OCCURS-ON(s, GO(x, w)) that signifies an event of x's movement along the path w. To explain our proposal we start with the specification of OCCURS-ON(s, GO(x, w)), assuming that x is a point-like entity. Then we give details on the components used. At the end, we generalize the specification so as to give the conditions of extended objects moving along a path.

(A10) gives the conditions for point-like objects moving along w in situation s.

(A10) $\forall s\, x\, w$

$$[\text{OCCURS-ON}(s, \text{GO}(x, w)) \Leftrightarrow$$
$$\text{OCCURS-ON}(s, \text{CHANGE-PLACE}(x, \text{stpt}(w), \text{fpt}(w))) \wedge$$
$$\forall p\, p'\, [\prec(w, p, p') \Leftrightarrow \text{OCCURS-IN}(s, \text{CHANGE-PLACE}(x, p, p'))]]$$

It says that an object x traverses a path w completely in s, iff the following conditions (a) to (c) are fulfilled: (a) x changes from the starting point of w to the final point of w. (b) For every point p that precedes a point p' on w, x changes from p to p' within s. (c) If x changes from point p to point p' at some time within s, then p precedes p' on w. This characterization guarantees that each point of the path is traversed and the order is determined by the path's direction. The path collects all points the object occupies during its course. Thus, we now have to specify when a change occurs *within* a situation and what it means for an object to change place.

The operators OCCURS-ON, OCCURS-IN, HOLDS-IN used here and in the following are named in accordance with similar operators defined by Galton (1990) relating time entities and situation types. We do not need to restrict the operator OCCURS-ON by general laws but just in interaction with more specific operators such as GO. However, OCCURS-IN can be defined based on OCCURS-ON, if situations can be part of one another ($s' \sqsubseteq s$). This general relation is assumed here without further specification. An event type ε *occurs within* a situation, iff it occurs on a sub-situation (D11).

(D11) OCCURS-IN(s, ε) $\Leftrightarrow_{\text{def}} \exists s'\, [s' \sqsubseteq s \wedge \text{OCCURS-ON}(s', \varepsilon)]$

The specification of an event of changing place is done by (A11). If x changes its place from p to p' in situation s, then p and p' are different and x is in the beginning of s at p and at the end of s at p'.

(A11) $\forall s\, x\, p\, p'$

$$[\text{OCCURS-ON}(s, \text{CHANGE-PLACE}(x, p, p')) \Rightarrow$$
$$p \neq p' \wedge \text{INIT}(s, \text{BE AT}(x, p)) \wedge \text{FIN}(s, \text{BE AT}(x, p'))]$$

The operators INIT, FIN, and BECOME relate a situation with a state type, such that the state type applies at the beginning of, at the end of, or directly after the situation. The additional operator HOLDS-IN is employed in the further specification of these operators. It expresses that the state type holds at least once during the situation. INIT, FIN and HOLDS-IN shall apply, even if the condition specified is valid only for a single moment. Φ is used as the schematic variable for state type descriptions such as BE AT(x, p). State types can be negated using not. However, internal negation and external negation are equivalent in the context of INIT and FIN (A12). We will further assume that situations can directly follow each other ($s < s'$). The final conditions of a situation are the starting conditions for every situation directly following it (A13). s is a situation in which Φ is established (BECOME(s, Φ)), iff Φ does not hold in s and not(Φ) does not hold in a situation directly following s (D12).

(A12) $\forall s$ $[(\neg \text{FIN}(s, \Phi) \Leftrightarrow \text{FIN}(s, \text{not}(\Phi))) \wedge (\neg \text{INIT}(s, \Phi) \Leftrightarrow \text{INIT}(s, \text{not}(\Phi)))]$

(A13) $\forall s\, s'$ $[s < s' \Rightarrow (\text{FIN}(s, \Phi) \Leftrightarrow \text{INIT}(s', \Phi))]$

(D12) BECOME(s, Φ) $\Leftrightarrow_{\text{def}} \neg \text{HOLDS-IN}(s, \Phi) \wedge \exists s'\, [s < s' \wedge \neg \text{HOLDS-IN}(s', \text{not}(\Phi))]$

The axioms (A10), (A11) and (A13) are sufficient to derive that if object x traverses path w in s (OCCURS-ON(s, GO(x, w))) and situation s' directly follows s ($s < s'$), then x is at the final point of w at initially to s' (INIT(s', BE AT(x, fpt(w)))).

Internal double negation of state types cancels in the scope of HOLDS-IN (A14). If Φ does not hold in a situation s, then its negation (not(Φ)) holds in s (A15). If Φ holds in a sub-situation of s, then Φ holds in s (A16). If Φ holds finally (initially) for a situation s', then Φ holds in or finally (initially) for every situation s' is part of (A17), (A18).

(A14) $\forall s$ $[\text{HOLDS-IN}(s, \Phi) \Leftrightarrow \text{HOLDS-IN}(s, \text{not}(\text{not}(\Phi)))]$

(A15) $\forall s$ $[\neg \text{HOLDS-IN}(s, \Phi) \Rightarrow \text{HOLDS-IN}(s, \text{not}(\Phi))]$

(A16) $\forall s\, s'$ $[s' \sqsubseteq s \wedge \text{HOLDS-IN}(s', \Phi) \Rightarrow \text{HOLDS-IN}(s, \Phi)]$

(A17) $\forall s\, s'$ $[s' \sqsubseteq s \wedge \text{FIN}(s', \Phi) \Rightarrow (\text{FIN}(s, \Phi) \vee \text{HOLDS-IN}(s, \Phi))]$

(A18) $\forall s\, s'$ $[s' \sqsubseteq s \wedge \text{INIT}(s', \Phi) \Rightarrow (\text{INIT}(s, \Phi) \vee \text{HOLDS-IN}(s, \Phi))]$

If extended entities shall be considered, the characterization of motion along a path is more complex. The complex condition is given in (A19).

(A19) $\forall s\, x\, w$ $[\text{OCCURS-ON}(s, \text{GO}(x, w)) \Leftrightarrow \exists p\, p'\, [\text{FROM}(w, p) \wedge \text{TO}(w, p') \wedge$
                                      $\text{OCCURS-ON}(s, \text{CHANGE-PLACE}(x, p, p'))]$
             $\forall Q\, [Q \iota w \Rightarrow \exists p\, [\text{HOLDS-IN}(s, \text{BE AT}(x, p)) \wedge Q \iota p]] \wedge$
             $\forall p\, [\text{HOLDS-IN}(s, \text{BE AT}(x, p)) \Rightarrow \exists w'\, [w' = w \sqcap p]] \wedge$
             $\forall p\, p'\, [\text{OCCURS-IN}(s, \text{CHANGE-PLACE}(x, p, p')) \Rightarrow \leqslant_p(w, p, p')]]$

According to (A19), object x traverses path w in situation s, iff x fulfills the following conditions (a) to (d). (a) x changes from a starting position of path w to a final position of path w. (b) Every point of path w is occupied by x in s. (c) x overlaps path w in a convex manner throughout the situation, i.e., the intersection of the object's position and the path is a path. (d) When x changes from position p to position p' during s, then p precedes p' on w.

This characterization uses some additional notions. *A path is the intersection of another path with a region* ($w' = w \sqcap p$), iff for all pairs of points (Q, Q'), Q precedes Q' on w', iff Q precedes Q' on w and both Q and Q' are in p (D13). The relation of precedence is generalized in two ways. First, the relation of *non-strict precedence* ($\preccurlyeq$) is the reflexive hull of the relation of precedence (D14). Second, non-strict precedence is generalized to relate sub-paths of a path ($\preccurlyeq_w(w, w', w'')$). *w' precedes w" with respect to w non-strictly*, iff both are sub-paths of w and the starting point and the final point of w' precede the respective points of w'' on w non-strictly (D15). The third version of non-strict precedence relates positions that intersect a path in a convex manner ($\preccurlyeq_p(w, p, p')$). Consequently *p precedes p' on w*, iff the intersection of w and p precedes the intersection of w and p' on w (D16).

(D13) $w' = w \sqcap p$ $\quad\quad \Leftrightarrow_{def} \forall Q\, Q'\, [\prec(w', Q, Q') \Leftrightarrow \prec(w, Q, Q') \wedge Q\, \iota\, p \wedge Q'\, \iota\, p]$

(D14) $\preccurlyeq(w, Q, Q')$ $\quad\quad \Leftrightarrow_{def} \prec(w, Q, Q') \vee (Q = Q' \wedge Q\, \iota\, w)$

(D15) $\preccurlyeq_w(w, w', w'')$ $\quad\quad \Leftrightarrow_{def} w' \sqsubseteq w \wedge w'' \sqsubseteq w \wedge \preccurlyeq(w, stpt(w'), stpt(w'')) \wedge \preccurlyeq(w, fpt(w'), fpt(w''))$

(D16) $\preccurlyeq_p(w, p', p'')$ $\quad\quad \Leftrightarrow_{def} \exists w'\, w''[w' = w \sqcap p \wedge w'' = w \sqcap p' \wedge \preccurlyeq_w(w, w', w'')]$

5   Conclusion

This article presents an analysis of the lexical structure of four types of German verbs of motion and of directional prepositions. The proposal concerns the argument structure of the verbs and prepositions, the semantic structure, and the geometric specification of the spatial concepts used in these descriptions. The common specification of a path establishes the formal link between verbs of motion and directional prepositional phrases. The path is anchored in the semantic structures of the verbs of motion by a semantic component that is characteristic for these kinds of verbs and assigns the two characteristic roles, the role of bearer of motion and the role of path of motion. Directional prepositional phrases specify geometric characteristics of the path.

The critical point of the analysis of verbs of motion and directional prepositional phrases is to explain restrictions in combinations of verbs of (caused) change of position and directional prepositional phrases. The explanation derived in this article is based on the interpretation of the complex expression on the conceptual level involving spatial background knowledge. Verbs of (caused) change of position refer to both the movement along the path and the arrival at a final position. The final position cannot be directly specified on the linguistic level. Directional prepositional phrases that contribute to the interpretation of the final position on the conceptual level lead to a homogeneous interpretation of the complete phrase. Thus, the preference of verbs of (caused) change of position for (specific) goal phrases can be explained as the effect of the implicit specification of an open parameter of the semantic structure.

References

Bierwisch, M. (1988). On the grammar of local prepositions. In M. Bierwisch, W. Motsch & I. Zimmermann (eds.), *Syntax, Semantik und Lexikon* (pp. 1–65). Berlin: Akademie Verlag.

Bierwisch, M. (1997). Lexical information from a minimalist point of view. In C. Wilder, H.-M. Gärtner & M. Bierwisch (eds.), *The Role of Economy Principles in Linguistic Theory* (pp. 227–266). Berlin: Akademie-Verlag.

Ehrich, V. (1996). Verbbedeutung und Verbgrammatik: Transportverben im Deutschen. In E. Lang & G. Zifonun (eds.), *Deutsch – typologisch* (pp. 229–260). Berlin: de Gruyter.

Eschenbach, C. (1995). *Zählangaben – Maßangaben. Bedeutung und konzeptuelle Interpretation von Numeralia*. Wiesbaden: Deutscher Universitätsverlag.

Eschenbach, C., C. Habel & L. Kulik (1999). Representing simple trajectories as oriented curves. In A.N. Kumar & I. Russell (eds.), *FLAIRS-99, Proceedings of the 12th International Florida AI Research Society Conference* (pp. 431–436). Orlando, Florida.

Eschenbach, C., C. Habel, L. Kulik & A. Leßmöllmann (1998). Shape nouns and shape concepts: A geometry for 'corner'. In C. Freksa, C. Habel & K.F. Wender (eds.), *Spatial Cognition* (pp. 177–201). Berlin: Springer-Verlag.

Galton, A. (1990). A critical examination of Allen's theory of action and time. *Artificial Intelligence, 42*. 159–188.

Habel, C., B. Hildebrandt & R. Moratz (1999). Interactive robot navigation based on qualitative spatial representations. In I. Wachsmuth & B. Jung (eds.), *KogWis99: Proceedings der 4. Fachtagung der Gesellschaft für Kognitionswissenschaft* (pp. 219–224). Sankt Augustin: Infix.

Hilbert, D. (1918). Axiomatisches Denken. *Mathematische Annalen, 78*. 405–415.

Jackendoff, R. (1990). *Semantic Structures*. Cambridge, MA: MIT-Press.

Kamp, H. & U. Reyle (1993). *From Discourse to Logic*. Dordrecht: Kluwer.

Kaufmann, I. (1995). *Konzeptuelle Grundlagen semantischer Dekompositionsstrukturen. Die Kombinatorik lokaler Verben und prädikativer Komplemente*. Tübingen: Niemeyer.

Kulik, L. & C. Eschenbach (1999). A geometry of oriented curves. (Available via ftp://ftp.informatik.uni-hamburg.de/pub/unihh/informatik/WSV/TROrientedCurves.pdf)

Levin, B. (1993). *English Verb Classes and Alternations: A Preliminary Investigation*. Chicago, Il: The University of Chicago Press.

Maienborn, C. (1990). Position und Bewegung: Zur Semantik lokaler Verben. IWBS-Report Nr. 138. Stuttgart: IBM.

Mourelatos, A.P.D. (1978). Events, processes, and states. *Linguistics and Philosophy, 2*. 415–434.

Rauh, G. (1997). Lokale Präpositionen und referentielle Argumente. Linguistische Berichte, 171. 415–442.

Talmy, L. (1975). Semantics and syntax of motion. In J.P. Kimball (eds.), *Syntax and Semantics 4* (pp. 181–238). New York: Academic Press.

Tschander, L.B. (1999). Bewegung und Bewegungsverben. In I. Wachsmuth & B. Jung (eds.), *KogWis99: Proceedings der 4. Fachtagung der Gesellschaft für Kognitionswissenschaft* (pp. 25–30). Sankt Augustin: Infix.

Wunderlich, D. (1991). How do prepositional phrases fit into compositional syntax and semantics. *Linguistic, 29*. 591–621.

Wunderlich, D. & M. Herweg (1991). Lokale und Direktionale. In A. v. Stechow & D. Wunderlich (eds.), *Semantik* (pp. 758–785). Berlin: de Gruyter.

Visual Processing and Representation of Spatio-temporal Patterns

Andreas Eisenkolb[1], Kerstin Schill[1], Florian Röhrbein[1],
Volker Baier[1], Alexandra Musto[2], and Wilfried Brauer[2]

[1] Ludwig-Maximilians-Universität München, Germany
{amadeus,kerstin,florian,volker}@imp.med.uni-muenchen.de
[2] Technische Universität München, Germany
{musto,brauer}@in.tum.de

Abstract. In an ongoing research we address the problem of representation and processing of motion information from an integrated perspective covering the range from early visual processing to higher-level cognitive aspects. Here we present experiments that were conducted to investigate the representation and processing of spatio-temporal information. Whereas research in this field is typically concerned with the formulation and implementation of *visual algorithms* like, e.g., navigation by an analysis of the retinal flow pattern caused by locomotion, we are interested in memory based capabilities, like the recognition of complicated gestures [16].
The result of this array of experiments will deliver a subset of parameters used for the training of an artificial neural network model. Alternatively, these parameters are important for determining the ranges of *symbolic* descriptions like, e.g., the qualitative approach by [11] in order to provide an user interface matched to conditions in human vision. The architecture of the neural net will be briefly sketched. Its output will be used as input for a higher-level stage modelled with qualitative means.

1 Introduction

The recognition of complex motion patterns requires a spatio-temporal memory as basic processing structure. The reason is that dynamic patterns have, in contrast to static patterns, an extension in the time dimension. Thus, for the recognition of a motion gestalt which is defined by its time course rather than a single snapshot, an interval or several snapshots have to be analysed simultaneously. This makes it necessary that spatio-temporal information is stored for some time in order to make it accessible for processes involved in the classification[1] of the spatio-temporal stimulus. A memory structure providing the necessary prerequisites for such an access is proposed in [16]. Application of this structure to transient information storage [1] in the time domain of iconic

---

[1] Not only for *classification*, also for prediction (which is a classification of incomplete data) spatio-temporal information has to be stored for subsequent processes operating on it.

Ch. Freksa et al. (Eds.): Spatial Cognition II, LNAI 1849, pp. 145–156, 2000.

memory [18] predicts basic experimental results on iconic storage and resolves inconsistencies arising with the duration of iconic storage and the temporal course of backward masking. Since it provides necessary structural prerequisites for the processing of motion information it resolves the main critique against the existence of this memory stage (for this discussion see [9]). However, in an analysis of the existing literature on motion processing and motion perception, covering the fields of psychophysics, ecological psychology, neurophysiology and computational vision, it turned out that questions regarding the storage and processing of more complex spatio-temporal information were not taken into account [3]. This is mainly due to the need for a suitable theoretical framework not recognised. This is also reflected in the experimental paradigms where the chosen task to solve did not exhibit the logistic problem of providing a past sequence of the visual input to enable subsequent processes the access to it. This especially holds for the research to the visual low-level functions of motion processing described in [12], like time-to-collision for obstacle avoidance. Whereas in computational vision many tasks can be performed in a memory-less fashion, it can easily be demonstrated that there are capabilities that cannot be realized without a spatio-temporal storage. For instance, with the biological-motion stimuli used by [10] it is obvious that the percept of a motion pattern only rises as a result of processing a sequence of events. The relevance of spatio-temporal gestalt properties, in contrast to simple changes of velocity information, is taken into account by the experiments on representational momentum conducted by [6], but the logical requirements and biological prerequisites for the representation and processing of the respective spatio-temporal patterns are not analysed. In order to gain more understanding about the internal representation of motion information we have conducted several experiments requiring the processing of spatio-temporal information.

The paper is structured as follows: first, we describe the experiments on direction discrimination, followed by the experiments on curvature discrimination. After that we present the experiments on the dynamic Poggendorff stimulus that we conducted in order to investigate in more detail the role of occlusion and the relationship of the processing of static vs. dynamic patterns. In the last part of this paper we briefly scetch our efforts to obtain artificial neural network models to represent spatio-temporal information in a hierarchically layered fashion.

2   Experiments

2.1   Experiments on Direction and Curvature Discrimination

To our knowledge there is no explicit study to the representational prerequisites of spatio-temporal information neither theoretically nor experimentally. Therefore, as a first step we devised a set of experiments with a simple spatio-temporal stimulus configuration. The underlying principle was what we call *comparison of induced internal representations*. Our hope was to explain the results in terms of a memory storehouse conception, i.e., internal representation, capacity, limited availability of an encoded stimulus. In the paper the following terms will frequently be used:

Motion path: A single black dot moving along an invisible straight or curved path on an otherwise white background.

ISI: Elapsed time between offset of the first and onset of the second stimulus (Inter Stimulus Interval).

Induction of internal representations: Subjects (Ss) are presented subsequently with two (or three) visual stimuli and have to indicate the one which conforms to a prespecified criterion. In case the stimuli are no longer physically available Ss have to perform a comparison of the respective *neural encodings*. This was described by [15] as a *side effect* of so called *temporal forced choice* procedures, which can be described as follows: 1) encode and store stimulus 1. 2) encode and store stimulus 2. 3) compare the encodings of stimulus 1 and stimulus 2. We call this scheme the *comparison of induced internal representations*. A spatio-temporal memory is quasi the prerequisite for an internal representation.

Direction Discrimination. The rationale of this experiment was to obtain data to draw conclusions as to the representation of spatio-temporal information. An additional aim was to investigate the processing of static vs. dynamic stimuli. To accomplish this, the dynamic motion paths were presented in a static fashion. We investigated the comparability of two sequentially presented dynamic patterns under conditions of variable temporal (ISI) and spatial separations.

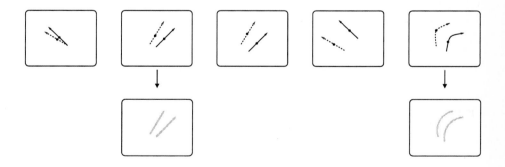

Fig. 1. Dynamic stimuli in experiments 1 to 5 (top, left to right) and corresponding static versions in experiments 6 and 7 (bottom)

Method. We devised seven experimental conditions, five dynamic, two static. For the static conditions, two dynamic conditions were chosen and transformed into respective static versions (see fig. 1). Discrimination performance was operationalized as threshold: per trial two dynamic (motion paths) or static patterns were presented one by one. Ss had to decide in a temporal 2-alternative-forced-choice procedure[2] which of the two patterns was more advanced in time, if considered as pointers of a clock. Temporal variations were introduced by three different ISIs that were tested throughout all conditions, static and dynamic. Spatial variation: in the dynamic conditions we tested the influence of a variable spatial separation on the discriminability of the two motion paths. Four displacements (0, 1.5, 3, 4.5 deg vis) were tested in experiments 1 to 4. In order to introduce more complex dynamic patterns we measured thresholds for orientation discrimination of curved motion paths in experiment 5 (displacement 1.5 deg). In order to compare static and dynamic performances we devised static versions of experiment 2 (experiment 6) and of 5 (7). Presentation time for one pattern was 300 msec in all conditions. Static and dynamic patterns subtended 4.7 deg visual angle. Speed was 15.6 deg/s in all conditions.

Apparatus

Hard- and software: Experiments were conducted on a PC with a graphics board "Matrox Mystique". The monitor used was a Eizo F78 (21"). The experimental program runs under realtime conditions and is based on the public domain graphics package "Allegro 3.0".

Real time conditions: All time critical stimuli were calculated offline and displayed by scrolling the frame buffer of the graphics device. The refresh rate was 135 Hz.

Smooth motion with sub-pixel precision: Moving patterns were interpolated by means of a 8-bit greyscale palette onto the discrete pixel raster of the monitor. Resolution was set to 800x600 pixel.

Illumination and viewing distance: All experiments were conducted in a separate room under a constant dim indirect illumination. Viewing distance was 1 meter.

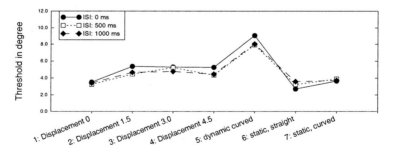

Fig. 2. Results (pooled over Ss)

---

[2] Ss are exposed to sequentially presented dynamic patterns, separated by an ISI, in order to induce corresponding internal representations and be told to choose one of the patterns according to a previously fixed criterion.

Results. Spatial separation leads to a slight increase of threshold, but the threshold elevation seems to be independent from the range of the spatial separation, at least for the three displacements (1.5, 3, 4.5 deg vis) tested in our experiments. Temporal separation of both stimuli to discriminate up to 1000 msec does not influence discrimination thresholds. This holds for all experimental conditions. Thresholds for dynamic stimuli are higher than those of their corresponding static versions.

Discussion. It is a surprising fact that the ISI didn't influence the discrimination threshold throughout all conditions. We had expected that a temporal separation of stimuli to be discriminated would influence the discrimination threshold thus mirroring resource limitations with respect to the storage of spatio-temporal patterns. In experiments with more complex spatio-temporal patterns an influence of a temporal separation should become manifest.

The data suggest that the encoding of direction information is not independent from spatial location. This is in contrast to the results in the time domain. We conclude that on an early level of the visual processing information about direction and position are interweaved.

Our data from the experiments on the dynamic vs. static display mode suggest that static and dynamic spatial information is processed differently: this is in contrast to the often mentioned view that simple dynamic patterns are processed by static mechanisms. Moreover, with static patterns, there is no remarkable increase of threshold as in the dynamic condition when switching from straight motion paths to curved ones.

Curvature Discrimination. The experiments on the direction discrimination didn't exhibit any influence of a temporal separation on the discriminability of simple motion paths. In order to test the ISI invariance we devised another paradigm. Instead of direction discrimination we measured curvature discrimination. The paradigm followed the experiments by [5] to discrete and continuous processing of briefly presented visual stimuli. Now, three internal representations were induced. Moreover the duration of the motion paths was shorter. As a result, a shorter time range with a richer structure could be examined.

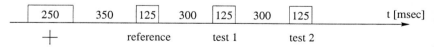

Fig. 3. Temporal order in experiments on curvature discrimination

Method. Ss were presented three temporally separated curved motion paths, *R* (reference path), T_1 (test path 1), T_2 (test path 2), where the order mirrors the order of appearance. The task was to choose the test path that differs in curvature from R by pressing one of two buttons. The temporal separation of R, T_1 and T_2 was defined in three ISI-conditions (in fig. 3 ISI=300 msec).

Motion paths were divided into five curvature classes, each completely determined by a curvature parameter d (see fig. 4). Per trial two neighboured classes were chosen, one for the reference path and one test path, the other one for the remaining test path. According to [5] we assumed that the discrimination was the more difficult the higher the value of d was. The centers of the three motion paths were positioned on the corners of an imaginary rhomb with a diameter of 1 deg vis.

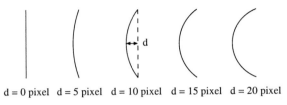

d = 0 pixel   d = 5 pixel   d = 10 pixel   d = 15 pixel   d = 20 pixel

Fig. 4. Curvature classes

The duration of one motion path was set to 125 ms. The moving patterns were well visible black spots with a diameter of 2 arc min. The length of all motion paths was defined as the euclidean distance between start- and endpoint and was set to 57 arc min. As a result the speed of the motion paths increased with curvature. As a counter measure we varied the path length and the presentation time of all motion paths independently from each other on random within carefully determined intervals. Average speed was 7.7 deg vis/s.

Altogether three ISIs were measured for all five discrimination classes, i.e., 15 conditions. For all conditions 60 responses were recorded, all conditions were tested in one session. The duration of one session was 60 minutes. The apparatus was identical to the one described in 2.

Results. As expected, the performance decreases as the degree of curvature is increased, see fig. 5. A temporal separation up to 400 msec does not result in an impairment of inner availability of the motion paths. All Ss showed, partly diverging, strategies to cope with the discrimination task. It turned out that all but one subject had the clear tendency to neglect the first test path in favour of the second.

Discussion. Again, no memory effect due to temporal separation could be observed in the experiments presented above. While in the experiments to direction discrimination the explanation for a lacking memory effect is that processing occurs too fast for a capacity limitation to become manifest here we observe a trade off effect: subjects tend to dedicate more attention to the second path to the disadvantage of the first path. Two Ss show no trade off in the range where the whole stimulus fits into a 500 msec window range which might mean that a strategy cannot be applied. However, this effect has still to be confirmed with a larger number of subjects.

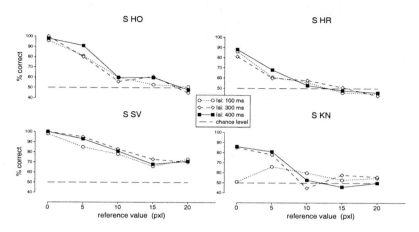

Fig. 5. With the exception of Subject KN no influence of the ISI can be observed

2.2   Processing of Occluded Spatio-Temporal Patterns: The Dynamic Poggendorff Effect

To gain further information about the processing of static vs. dynamic patterns we devised an experimental paradigm we call the "Dynamic Poggendorff Stimulus" (DPS, see fig. 6). A second motivation was that vision often occurs under circumstances of occlusion. In this sense the DPS serves as a spatio-temporal completion task. Spatio-temporal binding is one of the most puzzling problems both in psychophysical research and robot vision.

Ss saw a single black dot moving along a (invisible) straight oblique line. The middle segment of the motion path was occluded by a vertical bar so as to divide it in two visible segments. The stimulus can be thought of as a dynamic version of the "Poggendorff" stimulus. Ss had to adjust the position where the moving dot reappeared in order to bring both visible segments into alignment. We varied the dot velocity (5/10/15 deg/s), the width of the occluding bar (0.8/2.4/4.2 deg) and the filling mode of the occluding bar (white/black-white/black). 27 conditions were obtained by permuting all conditions. As dependent variable the alignment error (AE) was measured. A positive AE means that the adjusted segment was lower than the correct one. Apparatus was identical to the one described in 2.

Results. The filling mode has no influence on the AE (no figure). An occluding bar leads to a positive AE, in analogy to the static Poggendorff effect. Increasing

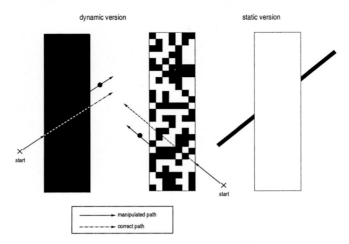

Fig. 6. The dynamic Poggendorff stimulus

the bar width leads to an increase of the AE by a factor of 2.3. Increasing the dot velocity leads to a decrease of the AE (factor 0.9).

Discussion. It is an interesting result that the presence of an occluding bar leads to a systematic (positive) AE. The dynamic Poggendorff effect (DPE) thus clearly suggests a qualitative analogy between static and dynamic processing modes. Research reported in [4] on the other hand suggests a drastical quantitative difference of the processing of static and dynamic stimuli. It is also probable that the task employed involves memory based mechanisms important for prediction of courses of motion. This on the other hand does by no means explain why there is a DPE.

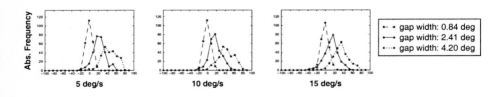

Fig. 7. Influence of gap width (top) and velocity (bottom) on the alignment error

3 Modelling the Processing of Spatio-Temporal Information with Artificial Neural Networks

Our overall aim is the modelling of spatio-temporal processing within a hierarchical system architecture. The structural constraints on the storage of spatio-temporal information have been gained by modelling motion information [17] and have been applied successfully e.g. to the task of generalization [14]. This memory structure will be the basic design principle for the different layers of the neural network. For the training of the system results gained in the experiments described in the former chapters and in [3] will be used.

The task that our model should be able to solve is the representation of spatio-temporal sequences on different time scales, their categorization and prediction. The output could consist for example of a set of motor commands that can be seen as a higher order input for a symbolic qualitative stage, which uses loops, corners or even higher abstractions of shapes as described in [11]. Through the possibility of representing multiple scales the system encloses different granularities of abstraction. In the first computing level there might be a representation of orientation selective filters. The next level combines the output of filters to pairs or longer sequences. In the highest processing level we should have a representation of higher order shapes. The system has the ability to classify given spatio-temporal signals into higher abstractions of motor command sequences. These sequences could be associated with a symbolic representation of shapes. The overall aim is the development of an integrated model [17] ranging from early vision processing as modelled by an artificial neural network (ANN) model to a symbolic description as modelled by the qualitative approach from [11].

Several models were developed for representing temporal signals, for example for speech recognition. The idea was to take these linguistically inspired models, which are capable of representing time structures, and extend them to represent spatial knowledge, too. Self organizing structures seem to be tailor-made for the task of representing spatial knowledge, so we decided to set our focus on these methods. There exist several models of ANN for representing information over time sequences, like *Backpropagation Through Time* [19], *Real Time Recurrent Neural Networks*, *Adaptive Resonance Theory* [8] and *Dynamic Self Organizing Maps* [13], but the combination of spatial and temporal information is, in our opinion, easiest done with *Learning And Processing Sequences* by [2]. As mentioned above, we have to construct a multi layered structure where each level represents a single scale of abstraction. We set our focus on self organizing systems because of their specifications and abilities in recognizing structures automatically, and the unsupervised manner of adaptation. [13] published a model consisting of three self-organizing maps connected recursively to each other. His model has the ability to classify a spatio-temporal signal but no ability to generate and predict longer sequences. *Learning And Processing Sequences* closes this gap of generating sequences. This model, however, is not able to represent multiple scales. So we expanded it by several maps to get the ability of representing different steps in abstraction of shapes. A simplified sketch is shown in fig. 8.

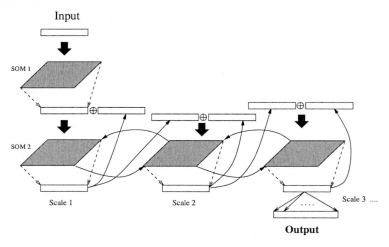

Fig. 8. LAPS model with self organizing maps (SOM) on multiple scales

Intermodal connections are also possible. For example a connection of visual and motor system. Through the possibility of online learning, the system will be able to adapt new environmental conditions whenever they occur. Sensory inputs that do not often occur are only represented weak, and get even weaker as more other inputs follow: the entire system reacts as if it *forgot* less relevant information with time. This information handling can be modelled through parameters and be used to tune how learned information is weakened when underrepresented with respect to the input signal. In navigation tasks, this would be a step from the currently often used closed world assumption to an open world assumption. The system can be defined even more generally, if *Growing Self Organizing Networks* as invented by [7] would be used in each computational layer. So the structure adapts itself to new circumstances in the environment. These structures would entirely adapt to the given task. With such a model it will be possible to plan routes in dynamic environments with collision avoidance through collision forecasting.

4   General Discussion

Three experimental paradigms were presented: direction discrimination, curvature discrimination, dynamic Poggendorff adjustment. In the first paradigm we investigated the influence of spatial and temporal factors on the access to internal representations of dynamic patterns. The access performance was operationalized as discrimination threshold. Dynamic patterns were well visible single black dots moving along a (invisible) straight or curved motion path. It turned out that the access to dynamic patterns remained invariant under conditions of a temporal separation (ISI) but was impaired by a spatial separation of the motion patterns to be discriminated. Experiments on curvature discrimination were conducted, which involved a different task on different degrees of difficulty. Moreover, a time range other than in the experiments on direction discrimination was examined. Again, a temporal separation of the patterns to be discriminated had no influence on the performance. The results confirmed the ISI-invariance. We had expected that a temporal separation would influence the discrimination and curvature threshold thus mirroring resource limitations with respect to the storage of spatio-temporal patterns. Surprisingly, this is not the case. A possible explanation is that the stimuli can transit the "iconic bottleneck" without bigger damage so that a comparison as necessary in the presented discrimination task would not be impaired by a temporal separation. The question remains why a *spatial* separation, irrespective from its extent, leads to threshold elevation. In a future study parameters, e.g., eye movements causing this effect should be examined in more detail. To gain further information about the processing of static vs. dynamic patterns we devised an experimental paradigm we call the "Dynamic Poggendorff Stimulus". The result was that an occluding bar leads to a positive alignment error, in analogy to the static Poggendorff effect.

Using very simple stimuli like the motion paths presented in this paper often leaves the experimenter in ambiguity caused by the discrepancy of claim and means. Humans can extract effortlessly "rich" biological motion patterns whereas "poor" stimuli used in threshold experiments can be very demanding. The strategy in the experimental series presented here was to start with simple spatio-temporal configurations, straight motion paths, separated by a variable ISI and to test spatial variations. As a next step we introduced curved motion paths which lead to an increase of threshold. As a third step we packed three curved motion paths in time windows ranging from 500 msec to 1200 msec and tested, like in the experiments before, the comparability. In future experiments more complex stimuli will be employed to investigate the spatio-temporal memory, like, a.o., displays with multiple points that we call here *events*. Johanssons biological motion patterns could be called events in this sense. An event, however, can be formed by one single point also. The interesting fact about multiple points is that under some conditions (like in the case of cycloidal wheel motion used by the Gestalt psychologists) a percept of moving units may arise. Event perception, thus, involves higher processing stages, requiring up to 500 msec processing time and stimuli may possibly be allocating a huge portion of the visual field which in turn may mean that higher visual areas are involved. At this point a pure bottom-up approach cannot account for the perceptual phenomena mentioned above. This in turn makes it necessary to integrate top-down mechanisms explicitly.

References

1. E. L. Bjork and R. A. Bjork, editors. *Memory, Handbook of Perception and Cognition.* Academic Press, 2nd edition, 1996.
2. G. Briscoe. *Adaptive Behavioural Cognition.* PhD thesis, Curtin University of Technology, School of Computing, September 1997.

3. A. Eisenkolb, A. Musto, K. Schill, D. Hernandez, and W. Brauer. Representational Levels for the Perception of the Courses of Motion. In C. Freksa, C. Habel, and K. F. Wender, editors, *Spatial Cognition*, volume I of *Lecture Notes in Artificial Intelligence*, pages 129–155. Springer Verlag, 1998.
4. A. Eisenkolb, C. Zetzsche, A. Musto, W. Brauer, and K. Schill. Analysis and modeling of the visual processing of spatiotemporal information: Influence of temporal and spatial separation on the discrimination of trajectories. In *Perception*, volume 27, page 168, 1998.
5. M. Ferraro and D. H. Foster. Discrete and continuous modes of curved-line discrimination controlled by effective stimulus duration. *Spatial Vision*, 1(3):219–230, 1986.
6. J. J. Freyd. Dynamic mental representations. *Psychological Review*, 94(4):427–438, 1987.
7. B. Fritzke. Growing self-organizing networks - why? In M. Verleysen, editor, *ESANN'96*, pages 61–72. D-Facto Publishers, 1996. http://www.neuroinformatik.ruhr-uni-bochum.de.
8. S. Grossberg. Competitive learning: From interactive activation to adaptive resonance. In S. Grossberg, editor, *Neural Networks and Natural Intelligence*, pages 213–250. MIT - Press, 1988.
9. R. N. Haber. The impending demise of the icon: A critique of the concept of iconic storage in visual information processing. *The Behavioral and Brain Sciences*, 6:1–54, 1983.
10. G. Johansson. Visual perception of biological motion and a model for its analysis. *Perception & Psychophysics*, 14:201–211, 1973.
11. A. Musto, K. Stein, A. Eisenkolb, K. Schill, and W. Brauer. Generalization, segmentation and classification of qualitative motion data. In Henri Prade, editor, *Proc. of the 13th European Conference on Artificial Intelligence (ECAI-98)*. John Wiley & Sons, 1998.
12. K. Nakayama. Biological Image Motion Processing: A Review. *Vision Research*, 25(5):625–660, 1985.
13. C. M. Privitera and L. Shastri. Temporal compositional processing by a dsom hierarchical model. Technical Report 94704-1198, Int. Computer Science Inst. Berkeley, California, 1947 Center St. - Suite 600 - Berkeley, 1996.
14. H. Raith. Methoden zur Analyse von Bewegungsinformationen in technischen Systemen und Implementierung eines biologischen Modells zur Repräsentation spatiotemporaler Informationen. Master's thesis, Institut für Informatik, Technische Universität München, 1999.
15. D. Regan. Storage of spatial-frequency information and spatial-frequency discrimination. *Journal of the Optical Society of America*, 2(4):619–621, 1985.
16. K. Schill and C. Zetzsche. A model of visual spatio-temporal memory: The icon revisited. *Psychological Research*, 57:88–102, 1995.
17. K. Schill, C. Zetzsche, W. Brauer, A. Eisenkolb, and A. Musto. Visual representation of spatio-temporal structure. In B.E. Rogowitz and T.N. Papas, editors, *Human Vision and Electronic Imaging*, pages 128–138. Proceedings of SPIE, 3299, 1998.
18. G. Sperling. The information available in a brief visual presentation. *Psychol. Monogr.: Gen. Applied*, 74:1–29, 1960.
19. J. Tani. Model-based learning for mobile robot navigation from the dynamical systems perspective. *IEEE Trans. System, Man and Cybernetics (Part B)*, 26(3):421–436, 1996.

Orienting and Reorienting in Egocentric Mental Models[*]

Robin Hörnig, Klaus Eyferth, and Holger Gärtner

Technical University of Berlin
FR 5–8, Franklinstraße 28/29, D 10587 Berlin
sppraum@cs.tu-berlin.de
http://ki.cs.tu-berlin.de/~sppraum

Abstract. In a text, several objects may linguistically be localized relative to a protagonist, the orientation of which remains unchanged. Such a *static description* yields a distinctive *description perspective*. On the other hand, the same layout can be described with the protagoist continually reorienting, while only objects are described which he is currently facing. Such a *dynamic description* does not yield an invariant description perspective. The special character of the description perspective in egocentric mental models was experimentally confirmed. Two accounts are discussed to explain it. The functional account considers a difference in salience of egocentric directions before and after a first imagined reorientation. The representational account suggests that only an invariant orientation allows to maintain a *mental image* in addition to the *mental model*.

1 Imagined Orientation and Object Localization

Repeatedly, we asked subjects in experiments to comprehend texts describing spatial relations. Having asked her for instance to imagine stepping into the middle of a yet unspecified office, we may have told her: *In front of you is the desk*. What does this message mediate? It is commonly suggested that by uttering this sentence, we linguistically localize the desk relative to her as a relatum. Interpreting the localization, she will construct an egocentric mental model of the office with the desk in the frontal region of her egocentric reference frame.[1] Now, suppose that the text had already depicted this office and she knows the desk's location. In this case, telling her to imagine being in the middle of the office with the desk in front of her, would mean to tell her how she should imagine to be oriented within the office rather than to tell her where the desk is. But even with familiar environments, the situation is ambiguous. It could be

---

[*] This research was supported by the Deutsche Forschungsgemeinschaft (DFG) in the project "Modelling Inferences in Mental Models" (Wy 20/2–2) within the priority program on spatial cognition ("Raumkognition").

[1] Our formal approach to spatial mental models is described in Claus, Eyferth, Gips, Hörnig, Schmid, Wiebrock, and Wysotzki (1998); see also Wiebrock, Wittenburg, Schmid, and Wysotzki, this volume.

Ch. Freksa et al. (Eds.): Spatial Cognition II, LNAI 1849, pp. 157–168, 2000.
© Springer-Verlag Berlin Heidelberg 2000

that we suppose her to face the desk after entering the office from the door. Then we could have told her that the desk had been moved away from its expected location, e.g., to her left as she enters the office. Linguistic localizations of objects in familiar settings may function in either of two ways. Taking for granted where objects are, we can tell her how she is oriented within the setting. Or we can inform her about where objects are, implying her orientation. Hence, describing an unfamiliar spatial layout in the way sketched above, seems to do two jobs at a time. We do not only tell her where things are, but we also establish her imagined orientation within that layout. Orientation becomes determined by localizing objects. In this text, we will treat only orientation and reorientation in unfamiliar but stable object constellations.

Suppose we had described already the whole spatial layout – the objects in front, behind, to the left, and to the right of the subject – and then asked her to imagine turning to her left. An imagined reorientation will force her to update her egocentric mental model accordingly. All objects surrounding her will be displaced after turning, e.g., the desk previously in front of her will now be to her right. After this reorientation, rather the locations of the objects seem to be determined by the new orientation than the orientation designated by the location of objects. Performing a mental reorientation in an egocentric mental model disentangles imagined orientation from object localizations. Thus, if later on we asked her to remember the desk's place, it could make a difference, whether she has imagined previously a turning or not.

2   Reorientation and Updating

According to Barbara Tversky (1996, p. 472), people "construct a mental spatial framework, consisting of extensions of the three body axes, and associate objects to the appropriate direction." Since the egocentric reference frame, i.e. the *spatial framework*, is the only one relevant for the mental model, a reorientation has to be simulated by *re*-locating objects: dissociating and associating them anew according to the mentioned reorientation: "The mental framework preserves the relative locations of the objects as the subject mentally turns to face a new object, allowing rapid updating." (ibid). According to this, imagined orientation is not explicitly represented. Since we take *orientation* (i.e. *heading* in the terminology of Klatzky 1998) to be a relational concept (see Hörnig, Claus, and Eyferth 1997; see also Habel 1999), orientation is not addressable as such but only indirectly by relating two successive states of the mental model to each other. If these are the same, orientation is unchanged. If they are different, either a reorientation has taken place or the objects have moved (or a mixture of both). Orientation as a relational notion cannot sensibly be applied to a single state of the mental model. We regard imagined orientation as confounded with object localization at any one time within such an approach. Remembering an object's place will be the same, regardless of whether a reorientation has been previously imagined or not.

A self-reorientation can be explicitly represented by relating the egocentric reference frame to a reference frame external to the egocentric one: to an allocentric reference frame. As Klatzky (1998, p. 10) suggested, an allocentric reference direction "could be defined by the geometry of a room". Accordingly, we propose that in our case the office provides an allocentric reference frame. It is allocentric, because identical locations of objects before and after a self-reorientation are covered by an invariant mental model related to the reference frame of the office. Then, when we ask her to imagine herself in the middle of the office, her mental model will provide its allocentric reference frame. Imagined orientation can now be expressed by relating the egocentric reference frame to the allocentric one. As soon as she is told that the desk is in front of her, her egocentric reference frame has to be oriented towards the desk's place, which is specified with respect to the allocentric reference frame. This requires that (the allocentric reference frame of) the office bears an orientation in its own right.

For unfamiliar spatial layouts imagined orientation and object localizations are not distinguished. So, we assume the allocentric reference frame to be initially co-aligned with the egocentric one, to cover the fact that imagined orientation is implied rather than expressed. As long as the subject does not reorient within her egocentric mental model, object localizations within the allocentric reference frame do not give any direction information beyond the localizations in the egocentric one and might be dispensable. But as soon as the subject imagines a self-reorientation, the allocentric reference frame becomes crucial for an explicit representation of self-movements. Any subjective reorientation is explicitly represented as a change in orientation of the egocentric reference frame relative to the allocentric one. We assume that at least after the first reorientation, the recipient will build up a spatial reference of the egocentric conception to the allocentric one.

3   Static and Dynamic Descriptions

Generalizing the work of Shepard on mental rotation of objects, Franklin and Tversky (1990) proposed the *mental transformation model* for object access in egocentric mental models that roughly corresponds to visually searching the surrounding environment. Only objects in front are immediately accessible. Access to objects located elsewhere requires the ego to reorient until the object is faced and becomes accessible. Since access latencies are assumed to monotonically increase with the required degree of rotation, this model predicts fastest access to objects in front and slowest access to objects behind, while access latencies for objects on either side should lie in between. In contrast, Franklin and Tversky suggested the *spatial framework model* that postulates that objects in any direction are immediately accessible. Access latencies are assumed to be a function of the properties of the egocentric reference frame. The front-back axis dominates the left-right axis, which is derived from the former. Therefore, objects on either side are less accessible than objects in front or behind. Furthermore, the asymmetry of the front-back axis (greater visual and functional salience of front as compared with back) leads to faster access for front. Using second person narratives as a testing paradigm, where subjects are given directions and have to respond with the objects located there, Franklin and Tversky demonstrated that latency patterns agreed with the spatial framework model but not with the mental transformation model. In the horizontal plane, objects in front were accessed fastest and objects behind were judged faster than objects on either side. Bryant and Tversky (1992) replicated these findings with object probes: subjects were presented object names and responded with the corresponding directions. Again, objects in front were judged faster than objects behind, and responding to objects behind was faster than to objects to the left or to the right.

In all those experiments, static descriptions were used as narratives. We refer to a *static description* if objects surrounding the ego are localized by different projective prepositions (e.g., *in front of, to the left of*), while orientation remains invariant. With a static description we might begin to describe the office as follows:

> *In front of you is the desk. On the desk is the hand calculator. To your left is the shelf.* etc.

With such a static description, we gain a distinctive orientation as outlined above. This special orientation may be called *description perspective*, since the same layout can be described from a different perspective gained by a reorientation within the layout:

> *To your right is the desk. On the desk is the hand calculator. In front of you is the shelf.* etc.

Neither Franklin and Tversky (1990) nor Bryant and Tversky (1992) tested their subjects with the description perspective left unchanged. Subjects were always oriented anew before the testing phase began.

While a static description establishes a description perspective, i.e. a distinctive orientation, a dynamic description does it not. The same layout can be described in a dynamic way by continually reorienting the recipient within the room and describing solely objects she is currently facing:

> *In front of you is the desk. On the desk is the hand calculator. Now you turn to your left. In front of you is the shelf.* etc.

Since with a dynamic description, the recipient continually changes her orientation already during acquisition, there is no distinctive orientation. An allocentric reference frame, against which the egocentric frame might be rotated, is required to represent explicitly the self-reorientations during acquisition.

Static descriptions should not only differ from dynamic ones by yielding a description perspective. Moreover, they should be easier to comprehend than dynamic ones, as claimed by Tversky (1996, p. 469): "Even if the possibility of different perspectives is recognized, consistency of perspective within a discourse can provide coherence to a discourse, rendering it more comprehensible. Switching perspective carries cognitive costs, at least for comprehension." Although Tversky addresses perspective switches other than simple reorientations within a layout, we assume that due to the continuing need for updating the egocentric mental model, processing a dynamic description should be cognitively more demanding than a static one.

People do not only construct egocentric mental models when they get second person narratives (Franklin and Tversky 1990). Even a third person narrative may mediate such a model by taking the perspective of the protagonist, as reported by Tversky (1996, p. 476): "In any case, readers readily take the perspective of either a character or an object central in a scene, even when the character or object is described in the third person." Such perspective taking was observed in investigations that use static descriptions, too (Franklin, Tversky, and Coon 1992; Bryant and Tversky 1992) (Hörnig, Eyferth, and Claus 1999). Since we assume dynamic descriptions to be more difficult to comprehend than static ones, subjects possibly avoid to take the protagonist's perspective when processing dynamic descriptions.

4   Remembering Egocentric Locations of Objects before and after a Reorientation

In this section we will report on four experiments on mental models constructed by subjects during listening to third person narratives.[2] Subjects were expected to take the protagonist's perspective, and hence, to construct egocentric mental models. Eight narratives told about a single protagonist standing in the middle of a room (e.g., an office) and being surrounded by four pairs of objects (e.g., the desk with the hand calculator on it) in any of the four directions in the horizontal plane: in front of the protagonist and to his right, back, and left.

Acquisition conditions were varied along two dimensions between the experiments. The variation of main interest was the *description condition*. Descriptions in the first two of the experiments were static and established a description perspective, which was systematically varied between subjects by rotating the layout. In the other two experiments, descriptions were dynamic, establishing no distinctive orientation. In both cases, every time subjects pressed a key, one pair of objects was described. Subjects could press these keys as often as they wished until they thought they had learned the layouts well enough to be able to start the testing phase. Learning rates, i.e. number of key-pressings, were recorded. The variation of minor importance was that acquisition was either *self-determined* or *predetermined*. In the self-determined condition, subjects could indicate, which direction they wished to be described next, by pressing a corresponding direction key. For example, if in the static description condition, a subject pressed the *right*-key, the objects to the right of the protagonist was described next. If a subject in the dynamic description condition pressed the *right*-key, the protagonist

---

[2] We are trying to publish soon a detailed report on the experiments.

turned right, and then the objects in front were described. So, if the *right*-key was pressed twice, the same object pair was described two times in the static description condition, but two different object pairs were described in the dynamic description condition, since the protagonist turned right two times. In the predetermined acquisition condition, subjects pressed the space bar to get the next object pair presented and did not know what direction would be described next. The acquisition condition (*self-determined* versus *predetermined*) was introduced to check if difficulties in comprehending dynamic descriptions arise only if self-determined acquisition is ruled out. No notable difference between conditions was expected for static descriptions.

As testing procedure we adopted the object probes paradigm of Bryant and Tversky (1992). In the testing phase for each of the eight narratives, subjects heard an object name and had to remember the corresponding direction. As soon as subjects knew the direction, they pressed a button and the latency for this response was recorded. Then they indicated the direction by pressing a corresponding direction key, so that the correctness of the response could be judged. This procedure was then repeated two times for further objects, named always after a new orientation of the protagonist was indicated. Each reorientation demanded an updating of the mental model.

Beyond the direction dependent availabilities of objects, we wanted to know whether the first reorientation in the testing phase has an impact on response latencies. The first object probe for each narrative was always given before any reorientation was mentioned in the testing phase. Thus, no reorientation at all had been induced under the static condition. Under the dynamic condition reorientations were repeatedly induced during the acquisition phase, but also here started the testing phase without reorientation. The testing procedure was identical in all four experiments.

4.1   Static Descriptions

For static descriptions, latencies for correct responses of the self-determined condition (25 subjects) and the predetermined condition (24 subjects) did not indicate any differential effect of these acquisition conditions. Therefore, these data were analysed together. Mean response latencies corresponded to the spatial framework pattern, i.e. objects in front were judged faster than objects behind, and those were judged faster than objects on either side. Both differences are statistically significant. No difference was found between objects to the left or right. Mean latencies increased after the first reorientation was mentioned in the testing phase. But, more interestingly, the strongest effect was found for the interaction of direction with reorientation (*before* versus *after reorientation*). A closer inspection revealed that only after the first reorientation response times agreed with the spatial framework pattern, except for the asymmetry of the front-back-axis. After the first reorientation, objects in front were judged as fast as objects behind, and both were judged faster than objects on either side. Again, this difference was statistically confirmed. That the asymmetry of the front-back-axis showed up in the data collapsed over the reorientation condition resulted from a marked change in latency patterns before and after reorientation. Previous to any reorientation, latencies clearly deviate from the spatial framework pattern. Objects behind were judged significantly slower than any other objects. Judgments for objects in front were about 300 ms faster than for objects to the right and about 230 ms faster than for objects to the left, but neither difference was statistically reliable, as were the differences between behind and right (550 ms) and between behind and left (610 ms). The first reorientation caused a significant decrease in response times for objects behind, and at the same time a significant increase in response times for objects to the left or right. This explains why mean latencies before reorientation are shorter than after reorientation: the slow responses to lateral objects (left and right) are counted in this average twice as often as the response times to objects behind.

Thus, the data of the static description condition clearly confirm the special character of the description perspective. Only after reorientation, the spatial framework pattern was observed. That the spatial framework pattern of our data lacks the expected asymmetry of the front-back axis does not argue against an internal framework pattern.[3] In the narratives of Tversky and colleagues reorientation was expressed by naming the object faced subsequent to the reorientation (e.g., Franklin and Tversky 1990, p. 66: ... *and you now face the lamp.*), whereas in our experiments, a reorientation was induced in terms of directions (e. g., ... *turns to the left.*) Obviously, presenting a filler sentence after a reorientation in Franklin and Tversky's experiments was not sufficient to cancel the advantage in accessing objects in front, mentioned already in the reorientation sentence, over objects behind, which had not been mentioned before.

Given the description perspective before a reorientation, response latencies agree with the mental transformation model. Franklin, Henkel, and Zangas (1995, p. 399) report that the conceptual frontal regions of their subjects comprise 124°. Accordingly, the left and right boundary of the frontal region end at 298° and 62°, respectively. If we take the conceptual frontal region as the one where objects become directly accessible according to the mental transformation model, then a reorientation of only 28° to the left or to the right would suffice for an object located exactly to the left (270°) or to the right (90°) to reach the boundary of the frontal region. But it would need another 90° of rotation for an object located exactly behind (180°) to reach the boundary of the frontal region, and thus 118° of rotation altogether. So, a rotation of only 28° might occur so fast, in about 300 ms in our data, that it could not be validated statistically.

4.2   Dynamic Descriptions

Unexpectedly, learning rates of subjects receiving dynamic descriptions were not higher than those of subjects receiving static descriptions. However, error rates were considerably higher for dynamic descriptions, but only if acquisition was

---

[3] See the claim of Tversky (1996) that lacking of asymmetry is indicative of an external framework pattern. In a personal conversation, Barbara Tversky agreed that her argument does not apply to our findings.

predetermined. Learning rates and error rates of the self-determined condition did not differ from those for static descriptions. Although subjects in the predetermined acquisition condition occasionally remarked that the task was difficult, learning was rather worse than longer. That for dynamic descriptions acquisition conditions caused differences was further supported by a significant interaction with direction in a conjoint analysis of response latencies (20 subjects in each acquisition condition). Therefore, the data for both conditions were analysed separately in this case.

With self-determined acquisition, only direction had an impact on latencies, whereas the factor *before* versus *after reorientation* showed no effect. Mean latencies agree with the spatial framework pattern without asymmetry on the front-back-axis, as was the case for static descriptions subsequent to the first reorientation. Objects in front and behind were judged reliably faster than objects on either side. Neither latencies for objects in front or behind, nor for objects to the right or to the left differed. In contrast to static descriptions, the spatial framework pattern is observable already before the first reorientation in the testing phase. This is exactly what we expected since the first reorientation in the testing phase was not the first reorientation altogether. Reorientations had to be performed already during acquisition.

For the twenty subjects in the predetermined acquisition condition, direction as well as reorientation had a substantial effect on latencies, but both did not interact. As the interaction of the conjoint analysis indicated, latencies with predetermined acquisition did not agree with the spatial framework pattern. Instead, objects in front and to the left were judged faster than objects behind. Moreover, judgments were faster after the first reorientation in the testing phase. At present, we doubt that our findings for predetermined acquisition of dynamic descriptions are decisive. Contrary to what we expected, the greater difficulty of predetermined acquisition of dynamic descriptions did not motivate subjects to enhance their effort in learning, but instead, they lowered their learning criterion. The consequence were much higher error rates leading to a lower level of statistical confidence than in the other three experiments. Under the assumption that our subjects in this fourth experiment could not cope with predetermined acquisition adequately, and under preservation of further analysis of the data, the following discussion will be primarily based on the first three experiments that yield a consistent picture, in which the fourth experiment does not fit.

5   What Makes the Distinctive Orientation Special?

We could demonstrate that static descriptions, in contrast to dynamic ones, result in a distinctive orientation. The description perspective reliably yields response latency patterns different from those obtained after a first reorientation. Since we take response latencies to reflect object access, we conclude that objects are accessed differently within a static mental model as compared with one that already had to be updated previously, according to the required reorientations, suggesting an allocentric perspective.

5.1   Salience of Egocentric Directions

Reorientation in an egocentric mental model can be explicitly represented only if an allocentric reference frame is available. Accordingly, we argue that the spatial framework pattern is not indicative of a purely egocentric representation, in which objects are associated to egocentrically defined directions that have to be reallocated in case of a reorientation. Instead, we take the spatial framework pattern to reflect an interaction of the egocentric reference frame with an allocentric one.[4] Because an allocentric reference frame is necessary to explicitly represent self-movement, a purely egocentric model would be sufficient only if no self-movement has to be represented. This holds only for static descriptions before the first reorientation is mentioned. Therefore, objects located in a purely egocentric mental model could be accessed differently than those located in an egocentric mental model connected to an allocentric reference frame.

If we explain the lesser accessibility for objects behind before the first reorientation in terms of the mental transformation model, a purely egocentric mental model is not very convincing on a priori arguments. The mental transformation model explains object access as mediated by a required transformation, i.e. reorientation. Obviously, it presumes that objects are not immediately located and accessible within the egocentric reference frame. Rather, objects are localized allocentrically but are accessed egocentrically.

Even to consider a static description to establish a distinct orientation within an egocentric mental model calls for an allocentric reference frame. Although we said above that an allocentric reference frame does not yield any information beyond a purely egocentric mental model for an invariant description perspective, an allocentric reference frame seems necessary to be able to speak of imagined orientation at all. Only then, we can conceive of establishing the imagined orientation as a concurrent egocentric and allocentric localization of objects (see our suggestions for the function of so-called *anchoring objects* in Hörnig, Claus, and Eyferth 1999). Localizing objects egocentrically and allocentrically at the same time connects both reference frames and can then be said to establish the imagined orientation within an allocentrically defined environment. With an invariant description perspective, objects on both sides might be more important than objects behind in order to establish such a distinct orientation. Such a claim is consistent with the order of self-determined acquisition observed in our experiments. Beginning with objects in front, subjects did not continue with objects behind, as one would expect if those two directions were of highest importance for gaining orientation. Instead, they explored the layouts clockwise or counter-clockwise. Since the spatial framework model relies on the greater dominance or salience of the front-back dimension as compared with the left-right dimension, it would not correctly predict the order of acquisition, which in turn might reflect the importance of egocentric directions in establishing orientation.

---

[4] See also Werner, Saade, and Lüer (1998) who demonstrated an influence of the geometry of a room-based reference frame on egocentric object access.

5.2   Mental Models and Mental Images

If we take response latencies before the first reorientation with static descriptions to indicate a mental transformation, it is not very likely that objects in one and the same representation are accessed differently before and after reorientation. Why should object access require a transformation before reorientation but be no longer necessary thereafter? But different access operations are plausible if more than one representation is in question. Johnson-Laird (1983, p. 165) distinguishes *mental models*, "which are structural analogues of the world", from *mental images*, "which are perceptual correlates of models from a particular point of view." We adopt here a proposal of a minimal architecture by Schlieder and Behrendt (1998) that relies on the work of Kosslyn (1994). Schlieder and Behrendt conceive of a *mental model* as an ordinal spatial relational representation and of a *mental image* as a metric visual representation. Both kinds of representations are assumed to support each other in a stabilizing feedback loop. In our view, the spatial framework model account for mental models – supplemented by an allocentric reference frame – is relational in nature (see Tversky, 1991, p. 141). However, the mental transformation model was formulated in analogy to visual search in the environment and might adequately account for object access in a visual representation rather than a spatial relational one. If subjects construct a *mental model* as well as a *mental image* while listening to a description of a spatial layout, they may access objects later on within either of the two representations (on the distinction of two spatial representations, see also Wiebrock et al., this volume). But if updating of both representations is cognitively too demanding, only one of the representations is updated while the other one has to be abandoned. Since the task at hand is relational in nature, it is plausible to assume that updating the relational *mental model* is unavoidable, but that the *mental image* is dismissed as soon as updating becomes necessary. The drastic change in response latencies before and after reorientation then could be explained by object access in the *mental image* before updating is required. After a reorientation, objects have to be accessed in the updated *mental model*, since it has become the only representation available. If access in a *mental image* requires a mental transformation but in a *mental model* does not, then the mental transformation model correctly predicts access latencies with the description perspective left unchanged, but the spatial framework model correctly predicts access latencies afterwards.

If we strictly interpret the statistical analysis, objects in front are as well accessible as objects on either side. Only access of objects behind is worse. This would be well compatible with a mental image tied to a distinctive egocentric view. Because such a mental image could only entail potentially visible objects, it would definitely not entail objects behind, since these are not visible. As a consequence, objects behind would be the only ones that cannot be accessed within the mental image but have to be accessed within the mental model. If people begin searching for objects within the mental image as long as both representations are available, access within the mental model would be initiated only after the mental image had been unavailingly searched. This would explain the disadvantage of objects behind.

We found evidence quite consistent with an explanation in terms of two distinguishable spatial representations in two previous experiments. In those experiments, subjects read static descriptions instead of hearing them (see Hörnig, Eyferth, and Claus 1999). The two reading experiments differed with respect to the modality of presentation in the testing phase. When probes were presented visually (*pure reading condition*), we obtained the spatial framework pattern without asymmetry on the front-back-axis even before the first reorientation occurred. This difference to auditive presentation is explainable if we assume that reading interferes with the construction of a visual representation, but allows to construct a mental model. Then readers would have to access objects within the mental model from the very beginning of the testing phase, since they form no mental image. However, in the second experiment (*mixed modality condition*) subjects read the narratives, too, but were tested auditively. Under this condition, a decisive impact of the first reorientation was observed. As in the present experiments, objects behind were judged reliably slower than all other objects before a reorientation was mentioned. But after the first reorientation, latencies agreed with the spatial framework pattern, again, without asymmetry on the front-back-axis. This last result contradicts the claim that reading hinders subjects from forming a mental image. However, in the acquisition phase, subjects in the mixed modality condition studied the narratives significantly longer than in the pure reading condition. So, we might argue that this additional study time in the mixed modality condition reflects the cognitive demand of forming a mental image although reading made it difficult. Subjects in the pure reading condition might avoid these extra costs since interference would be expected to occur also during testing.

5.3   Summary and Outlook

We have discussed rather different accounts of the special character of a static description perspective. One of them explains it in terms of a shift in egocentric direction salience once imagined orientation becomes disentangled from object localization. This happens if the description perspective has to be abandoned. While *left* and *right* play a more important role in establishing orientation in a spatial layout, *behind* becomes more salient once one has begun to navigate within the layout. The second account assumes that two different spatial representations are involved: a mental model and a mental image, with the latter only being available if no updating is necessary. There are several possibilities to test for a visual representation (e.g, distractor tasks), which we will not consider here. At present, it seems especially interesting to us to investigate what happens if the distinctive orientation corresponding to the description perspective is re-established by a second reorientation (e.g., turning back twice, or a left turn followed by a right turn). If it turns out that regaining the description perspective amounts to a latency pattern corresponding to the one obtained before any reorientation occurred, the functional account seems more plausible than the representational one, since it is not likely that a mental image once discarded becomes available again later.

References

Bryant, D. J. and B. Tversky (1992). Assessing spatial frameworks with object and direction probes. *Bulletin of the Psychonomic Society 30*, 29–32.

Claus, B., K. Eyferth, C. Gips, R. Hörnig, U. Schmid, S. Wiebrock, and F. Wysotzki (1998). Reference frames for spatial inferences in text comprehension. In C. Freksa, C. Habel, and K. F. Wender (Eds.), *Spatial Cognition*. Berlin: Springer Verlag.

Franklin, N., L. A. Henkel, and T. Zangas (1995). Parsing surrounding space into regions. *Memory and Cognition 23*, 397–407.

Franklin, N. and B. Tversky (1990). Searching imagined environments. *Journal of Experimental Psychology: General 119*, 63–76.

Franklin, N., B. Tversky, and V. Coon (1992). Switching points of view in spatial mental models. *Memory and Cognition 20*, 507–518.

Habel, C. (1999). Drehsinn und Reorientierung – Modus und Richtung beim Bewegungsverb *drehen*. In G. Rickheit (Ed.), *Richtung im Raum*. Opladen: Westdeutscher Verlag.

Hörnig, R., B. Claus, and K. Eyferth (1997). Objektzugriff in Mentalen Modellen: Eine Frage der Perspektive. In C. Umbach, M. Grabski, and R. Hörnig (Eds.), *Perspektive in Sprache und Raum*. Wiesbaden: Deutscher Universitätsverlag.

Hörnig, R., B. Claus, and K. Eyferth (2000). In search for an overall organizing principle in spatial mental models: a question of inference. In S. O'Nuallain and M. Hagerty (Eds.), *Spatial Cognition; Foundations and Applications*. Amsterdam: John Benjamins. Forthcoming.

Hörnig, R., K. Eyferth, and B. Claus (1999). Egozentrische Inferenz von Objektpositionen beim Lesen und Hören. *Zeitschrift für Experimentelle Psychologie 46*, 140–151.

Johnson-Laird, P. N. (1983). *Mental Models*. Cambridge: Cambridge University Press.

Klatzky, R. L. (1998). Allocentric and Egocentric Spatial Representations: Definitions, Distinctions, and Interconnections. In C. Freksa, C. Habel, and K. F. Wender (Eds.), *Spatial Cognition*. Berlin: Springer Verlag.

Kosslyn, S. (1994). *Image and Brain*. Cambridge, MA: MIT Press.

Schlieder, C. and B. Behrendt (1998). Mental model construction in spatial reasoning: a comparison of two computational theories. In U. Schmid, J. Krems, and F. Wysotzki (Eds.), *Mind modelling*. Lengerich: Pabst Science Publishers.

Tversky, B. (1991). Spatial mental models. *The Psychology of Learning and Motivation 27*, 109–145.

Tversky, B. (1996). Spatial perspective in descriptions. In P. Bloom, M. A. Peterson, L. Nadel, and M. F. Garrett (Eds.), *Language and space*. Cambridge, MA: MIT Press.

Werner, S., C. Saade, and G. Lüer (1998). Relations between the mental representation of extrapersonal space and spatial behavior. In C. Freksa, C. Habel, and K. F. Wender (Eds.), *Spatial Cognition*. Berlin: Springer Verlag.

Wiebrock, S., L. Wittenburg, U. Schmid, and F. Wysotzki (2000). Inference and Visualization of Spatial Relations. This volume.

Investigating Spatial Reference Systems through Distortions in Visual Memory

Steffen Werner and Thomas Schmidt

Institute of Psychology, University of Göttingen, D-37073 Göttingen, Germany
{swerner,tschmid8}@uni-goettingen.de

Abstract. Memory representations of spatial information require the choice of one or more reference systems to specify spatial relations. In two experiments we investigated the role of different reference systems for the encoding of spatial information in human memory. In Experiment 1, participants had to reproduce the location of a previously seen dot in relation to two landmarks on a computer screen. The placement of the two landmarks was varied so that they were horizontally or vertically aligned in half of the trials, and diagonally aligned in the other half of the trials. Reproductions showed a similar pattern of distortions for all four different orientations of the landmarks, indicating the use of the landmarks as an allocentric reference system. In Experiment 2, the influence of this allocentric reference system for very brief retention intervals (100 and 400 ms) was demonstrated in a visual discrimination task, extending previous work. The results suggest that landmark-based spatial reference systems are functional within 100 ms of stimulus presentation for most of the observers. Allocentric reference sytems therefore are an essential part even of early mental representations of spatial information.

1 Introduction

Whenever we try to communicate the location of an object to another person, we must present a reference system in which the location is specified. Different types of such reference systems exist, as many authors have pointed out (e.g., Klatzky, 1998; Levinson, 1996). For example, we can use ourself or another external object as the referent. This basic distinction pits an egocentric, or body-centered reference system against an allocentric, or environment-centered reference system (Klatzky, 1998). It is often important to make clear which reference system is used in the current course of communication. During a medical examination, for example, the phrase "raise your left foot" might lead to confusion because it is not clear whether it is meant from the doctor's or the patient's point of view (i.e., a conflict of different egocentric reference systems). Although the use of different reference systems might be most obvious in the course of communication, reference systems are as much required in nonverbal situations. To remember the location of an object, for example, we have to know the reference system used at encoding, just as partners in communication must know which reference system the other one is currently using.

Ch. Freksa et al. (Eds.): Spatial Cognition II, LNAI 1849, pp. 169-183, 2000.

In many perceptual and memory systems, spatial reference systems are hard-wired and cannot be easily changed. In visual perception, for example, different reference systems are used at different stages of processing. Whereas all information is initially encoded in a retinocentric reference system, the information is later represented in head-centered coordinates and eventually transformed into effector-specific coordinates, e.g., the location of an object relative to one's hand if one plans to grasp it (e.g., Berthoz, 1991; Soechting & Flanders, 1992). Of course many of these representations may exist in parallel. In contrast, the reference systems used in spatial memory and verbal communication are much more varied and flexible (see, for example, the overview in Levinson, 1996).

1.1   Investigating Spatial Reference Systems

There are a number of different methods used to probe the spatial reference systems used in human spatial memory. Linguistic analyses of spatial descriptions are one way to assess the reference system used to describe spatial relations. To increase the amount of control, linguistic acceptability ratings are often used. Observers in such studies are asked to rate the appropriateness of a verbal description of a real or depicted situation (see Levelt, 1996, for some examples). If the spatial configurations and the verbal descriptions are carefully chosen, the ratings will indicate which reference system the observers preferred. A different attempt to investigate the role of spatial reference systems in linguistic studies relies on the analysis of reaction times and relative error rates when judging the correctness of verbally presented spatial expressions. In one such study, Carlson-Radvansky and Jiang (1998) used the effect of negative priming to study the activation of multiple reference systems when judging spatial relations. They were able to show that a reference system available but not used in a prior task was more difficult to use in a following task than when the reference system was not available in the prior task. This suggests that the reference system not used in the prior task was nevertheless activated, even though it was of no relevance to the task at hand.

When investigating the spatial reference systems used in human memory, however, linguistic studies have one critical drawback. Spatial memory, in these cases, is accessed not directly but mediated by verbal processing at some stage. This leaves open the possibility that the reference systems observed might not reflect any of those used in spatial memory, but rather the reference systems used in language to express spatial relations. A number of non-verbal methods have therefore tried to investigate spatial reference systems without intervening verbal processing.

Several recognition or reproduction tasks have been used for this purpose. Pederson (1993), for example, had observers look at a configuration of objects on a table. They then had to turn around to face another table, which had been standing directly behind them. Their task was to reconstruct the configuration as they remembered it. Most people who were brought up in a western culture reconstructed the layout so that it matched their remembered, egocentric image (e.g., the object on the left in egocentric terms was again the object on the left in egocentric terms). Participants from cultures using mainly cardinal directions to indicate spatial relationships, however, tended to place the objects so that their cardinal relations stayed the same (e.g., the northern object was again placed on the northern end of the new table, the eastern object to the East, etc.), thus mirror-reversing the egocentric relations. Although verbal coding of the spatial relations might still be a factor in this type of study, it was not explicitly required as in the other studies mentioned above.

A different non-verbal approach to probing the reference systems used in spatial memory relies on the effect of orientation-specificity (Presson & Hazelrigg, 1984). A spatial behavior is termed orientation-specific whenever some part of it critically depends on the real or imagined orientation of the agent. A simple example of this is the use of map knowledge. It is commonly assumed that the orientation of a map, which is usually North-up, corresponds to the main reference-axis used in spatial memory for large-scale spaces (e.g., Sholl, 1987). Therefore, if a person is asked to imagine standing in Rome, facing Oslo, pointing in the direction of Madrid, this is usually an easier task than imagining standing in Madrid, facing Rome, pointing to Oslo. The reason for this is that the Rome-Oslo axis coincides with the North-South reference axis in spatial memory and is therefore easier to imagine. In a number of recent studies, this effect has been used to investigate the role of different kinds of reference systems (Roskos-Ewoldsen, McNamara, Shelton & Carr, 1998; Shelton & McNamara, 1997; Sholl, 1987; Werner, in preparation; Werner, Saade & Lüer, 1998; Werner & Schmidt, in press).

The approach which we will focus on in this paper relies on distortions in spatial memory as an indicator of the reference system used (e.g., Huttenlocher, Hedges, & Duncan, 1991; Laeng, Peters, & McCabe, 1998). A large body of psychological research shows that people often err systematically when remembering spatial locations. These distortions in spatial memory are commonly seen as evidence that perceived space is structured in some way, e.g. by different regions, groupings etc., which biases the way in which locations are remembered. The perceived structure partly determines the available reference systems. A simple example will illustrate this point. When observers see a dot together with two landmarks, as depicted in Figure 1, they show a systematic pattern of errors (Werner & Diedrichsen, submitted). Dot locations close to the two landmarks and the midpoint between the landmarks are reproduced further away from these points than they really are, exaggerating small deviations. Similarly, locations directly above or below the horizontal line connecting the two landmarks are usually reproduced further above or below. Other dot locations, such as the one right on the landmarks or the midpoint, are reproduced without bias. Memory for the dot location is thus tied to the two landmarks as the basic elements of the reference system.

There are a number of advantages of using the analysis of spatial memory distortions as an indicator of spatial reference systems. The non-verbal character eliminates the potential problems associated with verbal tasks or responses as was mentioned above. It also makes this procedure viable for studies with pre-verbal children (e.g., Huttenlocher, Newcombe & Sandberg, 1994) or even animals. A third advantage lies in its potential for comparing different actions or effectors on similar spatial tasks, such as reproducing a location on a small piece of paper, pointing to it, or walking to it on a field (Werner, Saade, & Lüer, 1998).

Finally, the time course of spatial distortions in memory can be traced by using a visual discrimination method instead of a location production method. In a previous experiment in our laboratory (Werner & Diedrichsen, submitted), observers first viewed the two landmarks together with the target dot on a computer screen. After 200 ms the two landmarks and the dot disappeared and were masked by random line patterns for a variable masking interval of 100 ms to 800 ms. After this time, the two landmarks and the dot reappeared and the observers had to judge whether or not the dot had moved. On some trials the dot remained in the original location, on some others it changed its location in one of two directions. The main result was a bias of the observers towards reporting no changes when the dot had changed its location in the direction of the memory distortion, whereas more changes were reported when it had changed in the opposite direction.

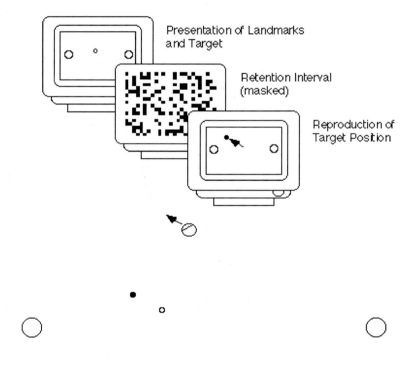

Fig. 1. *Top:* The two-landmarks reproduction paradigm. The observer has to reproduce the presented location after a retention interval with a mouse (the presented location is depicted as an open circle, the reproduction as a filled circle). *Bottom:* Schematic depiction of the observed distortion pattern. The arrows originate at the presented location and end at the reproduced location (based on data from Werner & Diedrichsen, submitted)

This asymmetry in responses was observable after only 100 ms, and did not increase any further after 400 ms. These results suggest that spatial memory is already distorted after retention intervals of only 100 ms and that the distortion reaches an asymptotic maximum level after less than half a second. This also indicates that spatial relations are encoded in allocentric reference systems at early stages of processing.

In this paper, we will extend the findings of Werner and Diedrichsen (submitted) by investigating the landmark-based reference system involved in the course of spatial information processing for a simple spatial arrangement, namely the location of a dot in relation to two horizontally aligned landmarks as depicted in Figure 1. In Experiment 1 we analysed the pattern of spatial distortions by asking participants to reproduce the location of the dot in relation to the two landmarks after a 400 ms retention interval. By changing the orientation of the stimulus configuration we controlled for effects of other potential reference systems. In Experiment 2, we used several of the previous targets to study the time course of their distortion in a visual discrimination task.

2   Experiment 1

In previous experiments, the two-landmarks configuration had only been used in horizontal or vertical orientations. This intrinsic stimulus orientation coincides with other reference systems that might possibly be used by the participants (e.g., the edges of the monitor, the direction of gravity, or the vertical and horizontal retinal axes). Experiment 1 was designed to assess whether misalignment of the intrinsic stimulus orientation with these frames of reference would change the distortion pattern. We used a total of 13 target positions between the two landmarks where we expected especially salient effects of distortion due to previous results (Werner & Diedrichsen, submitted). In one condition, the two landmarks were horizontally or vertically aligned, whereas in a second condition the whole configuration of landmarks and targets was rotated by 45°, resulting in two diagonally aligned landmark configurations. All stimuli were presented within a rectangular frame that was aligned with the monitor's sides (and thus also the gravitational axis) regardless of the condition. The experimental task was simply to reproduce the exact location of a briefly presented target with respect to the two landmarks.

2.1 Method

Participants. Six undergraduate students (age 23 to 31, all female, all right-handed) of the Institute of Psychology at the University of Göttingen participated for course credits or for a payment of 15,- DM per hour. Their vision was normal or corrected-to-normal.

Apparatus. The experiment was controlled by a Personal Computer with an AMD K-2 processor (300 MHz). Stimuli were presented on a 14" VGA color monitor (640 by 480 pixel [px]), synchronized with the monitor retrace rate of 60 Hz. Participants were seated on a height-adjustable chair at a distance of approximately 100 cm.

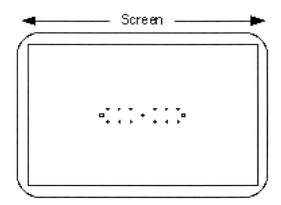

Fig. 2. Display configuration in Experiment 1 in the horizontally aligned condition. Stimuli are drawn to scale and were presented bright against a dark background. All thirteen possible target locations are depicted, but only one was presented at any given trial

Stimuli. All stimuli were presented within a white rectangular frame (600 x 380 px, 17.16° x 10.86°) at the center of the screen against a black background (Fig. 2). Landmarks were two green unfilled circles, 9 px (.26°) in diameter and 100 px (2.86°) away from the center of the screen. The target was a small white dot with a diameter of 3 px, presented at one of 13 possible locations. Landmarks and target could appear in any of four orientations. In the situation where the two landmarks were horizontally aligned (0° orientation), one of the possible locations was at screen center, the other 12 at y-coordinates of ±15 px and x-coordinates of ±90, ±60, and ±30 px. For the 90° orientation, the x- and y-coordinates were exchanged. The diagonally aligned condition (45° and 135° orientations counterclockwise) resulted from rotating the two landmarks and the corresponding targets around the center of the display. The frame and the mask remained unrotated. A dynamic pattern similar to static interference on a television screen that filled the rectangular frame was used. It consisted of randomly chosen black and white elements (2 x 2 px), with one quarter of the elements white at any given time. Four different random patterns were presented in succession for 33 ms each, after which the sequence repeated itself.

Procedure. Each trial began with the presentation of the two landmarks within the white frame. After 500 ms, the target appeared at one of the five possible locations and remained on the screen for 500 ms before it was replaced by the dynamic mask for 400 ms. The landmarks remained visible until the participant responded and were also visible during the masking interval. The participants' task was to use the mouse cursor (which looked exactly like the target) to reproduce the target's location as exactly as possible and to press the left mouse button when finished. The mouse cursor appeared randomly in the center of one of the landmarks to prevent the use of the initial cursor position as an additional spatial referent. The button press elicited a 1000 Hz, 100 ms tone. After an intertrial interval of 500 ms, a new trial began. The instruction emphasized accuracy rather than speed.

Stimulus conditions were counterbalanced such that each combination of target position and stimulus orientation occured randomly and equiprobably, with each combination appearing once every four blocks. The center target position appeared twice as often as any other target position to yield equal numbers of observations for all x and y coordinates.

Each participant performed one session of 20 blocks with 28 trials each. The session started with an additional practice block of 28 trials. After each block, participants received summary feedback of their average euclidean deviation from the target. After the session, participants were debriefed and received an explanation of the purpose of the experiment.

2.2   Results

For the following analyses, response times shorter than 100 ms and longer than 6000 ms were excluded (2.14 %). We also excluded all trials where responses were more than 30 px (0.86°) away from the original target (0.24 %). Additionally, we excluded all trials where the deviation from the true target was more than three standard deviations larger or smaller than the individual average target deviation (0.95 %). Practice blocks were not analysed.

Deviations of the participants' responses from the original targets were analysed with three-factorial ANOVAs (Target Position × Display Orientation × Participant). Distortions along the x- and y-axes were analysed seperately. There were no significant differences or interactions between the two different orientations in each condition. The data were therefore collapsed over these pairs of orientations. For simplicity, we will generally not report effects associated with the participant factor.

Results are shown in Fig. 3. As expected, there was a distortion of responses away from the nearest landmark and away from the midpoint between the landmarks, with smaller distortions at the midpoint itself or distant from the landmarks and midpoint. The main effect of target position on spatial distortion was significant for the x coordinate, $F(6, 30) = 3.96$, $p = .005$, as well as for the y coordinate, $F(2, 10) = 9.65$, $p = .005$. There was no main effect of display orientation, i.e. the two diagonal display orientations showed the same average distortion as the two other orientations, $F(1, 5) < 1$ for the x coordinates, $F(1, 5) = 1.71$, $p > .200$ for the y coordinates. However, patterns of distortion differed slightly across display orientations: There was an interaction of target position and orientation for the x coordinates, $F(6, 30) = 2.45$, $p < .050$, but it was not significant for the y coordinates, $F(18, 90) = 2.12$, $p = .100$. The most salient difference between display orientations is the reduced distortion for the diagonal orientations at the targets closest to the landmarks (both $p < .005$ in *post hoc* Tukey tests).

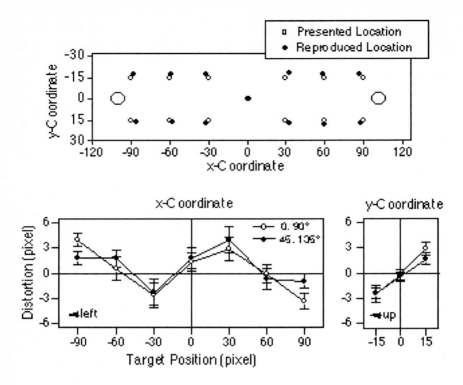

Fig. 3. *Top:* Presented and reproduced target locations in Experiment 1 with respect to the two landmarks (large circles). The rectangular frame does not correspond to the frame presented during the experiment. *Bottom:* Amount of distortion for x- and y-dimension separately

2.3 Discussion

Experiment 1 shows that visual memory is spatially distorted (Huttenlocher, Hedges, & Duncan, 1991; Laeng, Peters, & McCabe, 1998; Nelson & Chaiklin, 1980) and the results replicate the typical pattern in the two-landmarks task (Werner & Diedrichsen, submitted). The reproduced locations are distorted away from the landmarks and from the midpoint between the landmarks. This pattern is similar for different display orientations, indicating that distortions develop within an allocentric reference system defined by the two landmarks. By changing the location or alignment of different reference systems, as was done in this experiment, spatial distortion patterns can thus nicely identify the dominant reference system used to encode spatial relations in memory.

One possible reason why the distortions are smaller than expected for targets close to the landmarks at diagonal display orientations might lie in the use of one of the alternative reference systems mentioned above. These targets are so close to either a horizontal or a vertical alignment with one landmark that the use of these reference systems might be very helpful. Participants might thus adopt a strategy of switching reference systems if it allows them to encode the spatial relations more efficiently or more accurately. It would be interesting to test, for example, if a similar effect would occur in situations where the landmark configuration might change between the first and the second presentation (e.g., from horizontal to diagonal). In this case, reference systems other than the two landmarks would be of only little use in situations where the configuration changed.

Most importantly, however, the results show that even such a simple pattern of landmarks as the one used in this experiment suffices to establish an allocentric spatial organization that influences the memory of spatial location. The results do not tell us, however, at which point of processing the allocentric reference system induced by the landmarks is established and used to encode the necessary spatial relations. Like most studies using the distortions of spatial locations as a means to identify the dominant reference system, the time it takes to reproduce a location limits the temporal sensitivity of the measure. Other methods, such as the priming paradigm employed by Carlson-Radvansky and Jiang (1998), can identify effects of reference systems at a much higher temporal resolution. While the results of Experiment 1 show that a few seconds, consisting of retention interval and spatial reproduction time, are sufficient to produce spatial distortions, it would be interesting to trace the role of the landmark-based reference system over brief periods of time.

The solution to this problem lies in using a visual discrimination task to investigate spatial memory distortions (Freyd & Johnson, 1987; Werner & Diedrichsen, submitted). In this case, the observer does not have to physically reproduce a spatial location, but merely judge whether the location of a target has changed between two presentations. This allows for a tight control of presentation times and very short retention intervals. As Werner and Diedrichsen were able to show, spatial distortions due to a landmark-based, allocentric reference system were observable at retention intervals of only 50 ms, suggesting that allocentric coding is already used at early stages of spatial processing. The following experiment employs a similar strategy to probe the time course of allocentric spatial coding.

3   Experiment 2

The purpose of this experiment was to investigate the time course of the distortion effect in the two-landmarks task by using a visual discrimination paradigm. Instead of asking participants to reproduce a target location, they now had to tell whether a target had been displaced to the left or to the right during the masking interval. In addition to the 400-ms mask used in experiment 1, we used a 100-ms mask to see whether distortion effects were already present at this early stage.

Unlike the method employed by Werner and Diedrichsen (submitted), where the participants had to judge whether or not a dot had moved between two presentations, the participants' task in this experiment was to report the *direction* of target displacement rather than to simply detect its presence. This allowed us to sample psychometric functions that could be readily analysed by statistical standard procedures, yielding estimates for the bias and the sensitivity of individual participants.

More specifically, assume that a target is in a region of the display where the landmark induces a *rightward* bias in the memory representation of the target. Now consider that the target is physically displaced to the right during the masking interval. Because the memory representation has also drifted to the right, the apparent displacement should be small, and the likelihood of participants reporting a rightward displacement should also be small. Compare this with the situation in another region of the display where a *leftward* bias of the memory representation is induced. If the target is still physically displaced to the right, the apparent displacement should appear large, and the likelihood of reporting a rightward displacement should be large, too.

3.1   Method

Participants. Six undergraduate students (age 22 to 29, all female, two of them left-handed) of the Institute of Psychology at the University of Göttingen participated for course credits or for a payment of 12,- DM per hour. Their vision was normal or corrected-to-normal.

Apparatus. The setup was the same as in experiment 1, only that the viewing distance was 80 cm.

Stimuli. To allow for direct comparisons between experiments, stimuli were the same as in experiment 1, with the following exceptions. We used only one stimulus orientation, so that the two landmarks were always horizontally aligned. The landmarks were filled in this experiment. There were only five target positions: one at the center of the screen, the other four positions at y-coordinates of ±15 px (±.54°) and x-coordinates of ±30 px (±1.07°). These targets corresponded to the five innermost target positions from Experiment 1. They were chosen because they had shown strong distortions along the x coordinate in experiment 1, and this distortion had been independent of display orientation. Furthermore, they were all near the center of the display which reduces possible effects of stimulus eccentricity when fixating on the center.

Procedure. A trial began with the presentation of the two landmarks within the white rectangular frame as in Experiment 1 (Fig. 4). After 500 ms, the target appeared at one of the five possible locations and remained on the screen for 500 ms before it was replaced by a 320 x 200 px dynamic mask for either 100 or 400 ms. Immediately following mask presentation, the target was presented again, but this time with a displacement of 0, 1, 2, or 3 px to the left or right of its original position. Participants were not informed that in some cases no displacement occured. The landmarks remained visible until the participant responded and were also visible during the masking interval. The participants' task was to indicate whether the target had been displaced to the left or right by pressing the appropriate key ("4" for "left", "6" for "right" on the numerical pad of the computer keyboard) with the index or ring finger of their right hand, respectively. The instruction emphasized accuracy rather than speed and encouraged participants to guess when they were not sure about the direction of target displacement. The keypress response elicited a 2000 Hz, 100 ms tone, and a warning tone (100 Hz, 500 ms) was sounded when a key other than the two permitted was used. After an intertrial interval of 500 ms, the next trial began. Throughout the experiment, no feedback concerning the level of performance was given.

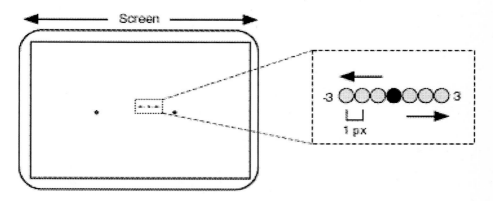

Fig. 4. Display configuration in experiment 2. Stimuli were presented bright against a dark background. The panel on the right represents an enlarged section of the display, showing schematically the 7 possible target displacements for one target position

Stimulus conditions were counterbalanced such that each combination of target position, mask duration, and target displacement occured randomly and equiprobably, with each combination appearing once every two blocks. The center target position appeared twice as often as any other target position to yield equal numbers of observations for x and y coordinates.

Each participant took part in four sessions of 16 blocks with 42 trials each, resulting in a total of 2,688 trials. Each session started with an additional practice block of 42 trials. At the beginning of the first session, participants trained the discrimination task in a short demonstration program using larger target displacements. After the last session, they were debriefed and received an explanation of the purpose of the experiment.

3.2 Results

For the following analyses, response times shorter than 100 ms and longer than 999 ms were excluded (0.92 %). Additionally, we excluded all trials where a participants' response times were more than three standard deviations above or below her average response time (2.37 %). Practice blocks were not analysed.

Results are shown in Figure 5. Psychometric functions were analysed by multiple logistic regression (Agresti, 1996). We used a Wald statistic, reported here as $W(df)$, where df denotes the degrees of freedom. The regression model contained the Target Displacement, Target Position, and Mask Duration main effects, the Displacement x Mask Duration interaction, and the Position x Mask Duration interaction. Model fit was excellent[1], with a high correlation of the observed and predicted means, $r = .996$.

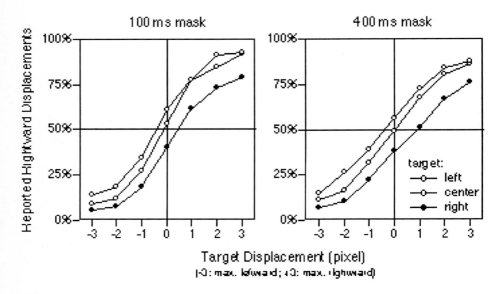

Fig. 5. Psychometric functions from experiment 2, separately for the two masking intervals. Spatial memory distortions in one direction lead to an increase of displacement judgements in the opposite direction (see text)

Not surprisingly, there was a main effect of Target Displacements, indicating that participants were able to discriminate leftward and rightward displacements, $W(1) = 3907.66$, $p < .001$. Importantly, there was a clear effect of Target Position in the predicted direction, $W(2) = 352.78$, $p < .001$, indicating that compared with targets at the midpoint between the two landmarks, participants were more likely to report a rightward displacement for targets on the left, $W(1) = 40.82$, $p \leq .001$, and a leftward displacement for targets on the right, $W(1) = 153.06$, $p < .001$.

There was a small main effect of Mask Duration, reflecting a tendency to report leftward rather than rightward displaments, which was more pronounced at the longer mask duration, $W(1) = 5.92$, $p < .050$. More importantly, psychometric functions were steeper at the shorter masking interval (i.e., a Displacement x Mask Duration interaction), indicating a loss of sensitivity with increasing mask duration, $W(1) = 50.13$,

---

[1] The same model was used for the data of individual participants, also with excellent fit, $.967 \leq r \leq .996$.

$p < .001$. There was no evidence of a Position x Mask Duration interaction, suggesting that the distortion effect did not vary with mask duration, $W(2) = .92, p = .632$.

Participants differed widely in sensitivity and overall response bias. The predicted distortion effect occured in four out of six participants, causing responses to be biased towards the side opposite to the target position, $46.77 \leq W(2) \leq 186.14$, all $p < .001$. One participant, however, showed the reverse effect, $W(2) = 18.76, p = .001$.

A significant increase of distortion with increasing masking interval occured in only one participant, $W(2) = 22.63, p < .001$. In two other participants, however, the distortion effect *decreased* with increasing mask duration, $W(2) = 23.98, p < .001$, and $W(2) = 10.25, p = .006$, respectively. In the remaining participants, this effect was not significant, $2.37 \leq W(2) \leq 6.79, .034 \leq p \leq .306$.

3.3 Discussion

Experiment 2 examined the distortions in spatial memory observed for the five innermost target positions in Experiment 1 in more detail. The results are consistent with the assumption that memory representations are biased away from the midpoint between the two landmarks, thereby increasing the probability of reporting a physical target displacement in the opposite direction. Judgements for targets presented directly at the midpoint, however, remain essentially undistorted.

The significant bias in judgments with the short masking interval indicates that even for very brief retention intervals of only 100 ms the landmark-based, allocentric reference system affects the memory for spatial locations. This finding replicates the results of Werner and Diedrichsen (submitted) who found evidence for spatial memory distortions after only 50 ms. However, in the experiments reported here there was no effect of mask duration on the size of the distortion. Unlike previous findings, the distortion seems to be as strong after 100 ms as after a 400 ms retention interval. This suggests that spatial memory distortions can be fully developed after only 100 ms, which is much shorter than the asymptotic 400 ms reported by Werner and Diedrichsen (submitted). It thus seems clear from the results that spatial memory distortions can develop over brief time spans, suggesting that memory representations are involved quite early in the course of visual processing.

As expected, the sensitivity for target displacements diminishes for the longer masking interval. This does not, however, coincide with an increase in spatial distortions, indicating a dissociation between the size of a distortion and an observer's sensitivity.

4   Conclusion

The goal of this paper was to introduce the analysis of spatial distortions as a useful tool to identify the role of different spatial reference systems in human memory. In the first experiment, systematic biases were found for observers' reproductions of dot locations in a simple two-landmarks situation. When changing the orientation of the two landmarks, the distortion pattern followed the new orientation, establishing the dependence of the distortion on the allocentric reference system induced by the landmarks. The distortion pattern of remembered locations thus can identify the dominant spatial reference system used to encode location information within a particular task (see also Huttenlocher, Hedges, & Duncan, 1991; Nelson & Chaiklin, 1980).

By changing the experimental procedure in the second experiment, we were able to demonstrate the presence of spatial memory distortions even for very brief retention intervals of 100 ms. Although this does not necessarily imply that the spatial memory distortion has fully developed 100 ms after the offset of the stimulus, it clearly marks an upper limit for the relative processing lag that is needed between two stimuli, so that one is already showing the biasing effect of the allocentric reference system, while the other is not yet affected by it. Moreover, the results have strong implications for the comparison process between the two mental representations involved, the representations of the remembered and the presented location. One explanation could be that the biasing effect of the allocentric reference system develops gradually as soon as the visual representation of a stimulus starts to decay, so that a comparison process has to match a biased memory representation with a yet undistorted visual representation of a stimulus (also compare Werner & Diedrichsen, submitted). Simulations of early visual cortical areas suggest that topographic representations of spatially extended stimuli decay in a gradual fashion, with interactions between stimulus representations leading to nonhomogenous rates of decay across the visual field (Francis, 1997; Francis & Grossberg, 1994). Because the distortion effects described here arise quite early in visual processing, one might speculate whether a landmark-induced allocentric reference system could lead to spatial distortions in topographically organized visual areas such as V1 to V5.

By using a combination of spatial reproduction and spatial discrimination tasks, as demonstrated in this paper, the role of different reference systems can be investigated at a high spatial and temporal resolution (Werner & Diedrichsen, submitted), allowing researchers to investigate the time course of spatial reference systems at a behavioral and neuropsychological level. The exact description of spatial distortion allows more than just a classification of the general type of reference system used to encode a location (e.g., allocentric vs. egocentric). It additionally can shed light on the particular way space is structured by a reference system. Eventually, it might even lead to general theories of how spatial and geometrical relations are perceived and encoded (e.g., Huttenlocher et al., 1991). This might help to identify the computational dynamics and brain structures associated with particular reference systems.

References

Agresti, A. (1996). *An introduction to categorical data analysis.* New York: Wiley.

Berthoz, A. (1991). Reference frames for the perception and control of movement. In J. Paillard (Ed.), *Brain and space* (pp. 81-111). Oxford: Oxford University Press.

Carlson-Radvansky, L.A. & Jiang, Y. (1998). Inhibition accompanies reference-frame selection. *Psychological Science, 9*, 386-391.

Francis, G. (1997). Cortical dynamics of lateral inhibition: Metacontrast masking. *Psychological Review, 104*, 572-594.

Francis, G. & Grossberg, S. (1994). Cortical dynamics of form and motion integration: Persistence, apparent motion, and illusory contours. *Vision Research, 36*, 149-173.

Freyd, J. J. & Johnson, J. Q. (1987). Probing the time course of representational momentum. *Journal of Experimental Psychology: Learning, Memory, and Cognition, 13*, 259-268.

Huttenlocher, J., Newcombe, N., & Sandberg, E. H. (1994). The coding of spatial location in young children. *Cognitive Psychology, 27*, 115-147.

Huttenlocher, J., Hedges, L.V., & Duncan, S. (1991). Categories and particulars: Prototype effects in estimating spatial location. *Psychological Review, 98*, 352-376.

Klatzky, R.L. (1998). Allocentric and egocentric spatial representation: Definitions, distinctions, and interconnections. In K.-F. Wender, C. Freksa & C. Habel (Eds.), *Spatial cognition. An interdisciplinary approach to representing and processing spatial knowledge* (pp. 107-127). Berlin: Springer.

Laeng, B., Peters, M., & McCabe, B. (1998). Memory for locations within regions: Spatial biases and visual hemifield differences. *Memory & Cognition, 26*, 97-107.

Levelt, W.J.M. (1996). Perspective taking and ellipsis in spatial descriptions. In P. Bloom, M.A. Peterson, L. Nadel, & M. Garrett (Eds.), *Language and Space* (pp. 77-107). Cambridge. MIT-Press.

Levinson, S. (1996). Frames of reference and Molyneux's questions: Cross-linguistic evidence. In P. Bloom, M.A. Peterson, L. Nadel, & M. Garrett (Eds.), *Language and space* (pp. 109-169). Cambridge, MA: MIT Press.

Nelson, T. O. & Chaiklin, S. (1980). Immediate memory for spatial location. *Journal of Experimental Psychology: Human Learning and Memory, 6*, 529-545.

Pederson, E. (1993). Geographic and manipulable space in two Tamil linguistic systems. In A.U. Frank & I. Campari (Eds.), *Spatial information theory* (pp. 294-311). Berlin: Springer.

Presson, C.C. & Hazelrigg, M.D. (1984). Building spatial representations through primary and secondary learning. *Journal of Experimental Psychology: Learning, Memory, and Cognition, 10*, 716-722.

Roskos-Ewoldsen, B., McNamara, T.P., Shelton, A.L., & Carr, W.S. (1998). Mental representations of large and small spatial layouts are orientation-dependent. *Journal of Experimental Psychology: Learning, Memory, and Cognition, 24*, 215-26.

Shelton, A. L. & McNamara, T. P. (1997). Multiple views of spatial memory. *Psychonomic Bulletin & Review, 4*, 102-106.

Sholl, M.J. (1987). Cognitive maps as orienting schemata. *Journal of Experimental Psychology: Learning, Memory, and Cognition, 13*, 615-628.

Soechting, J.F. & Flanders, M. (1992). Moving in three-dimensional space: Frames of reference, vectors, and coordinate systems. *Annual Review of Neuroscience, 15*, 167-191.

Werner, S. (in preparation). The effect of egocentric and allocentric frames of reference on the mental representation of extrapersonal space.

Werner, S. & Diedrichsen, J. (submitted). The time course of spatial memory distortions.

Werner, S. & Schmidt, K. (in press). Environmental reference systems for large-scale spaces. *Spatial Cognition and Computation.*

Werner, S., Saade, C., & Lüer, G. (1998). Relations between the mental representation of extrapersonal space and spatial behavior. In K.-F. Wender, C. Freksa & C. Habel (Eds.), *Spatial cognition. An interdisciplinary approach to representing and processing spatial knowledge* (pp. 107-127). Berlin: Springer.

Towards Cognitive Adequacy of Topological Spatial Relations

Jochen Renz[1], Reinhold Rauh[2], and Markus Knauff[2]

[1] Institut für Informatik, Albert-Ludwigs-Universität
Am Flughafen 17, 79110 Freiburg, Germany
renz@informatik.uni-freiburg.de
[2] Abteilung Kognitionswissenschaft, Institut für Informatik und Gesellschaft,
Albert-Ludwigs-Universität
Friedrichstr. 50, 79098 Freiburg, Germany
{reinhold,knauff}@cognition.iig.uni-freiburg.de

Abstract. Qualitative spatial reasoning is often considered to be akin to human reasoning. This, however, is mostly based on the intuition of researchers rather than on empirical data. In this paper we continue our effort in empirically studying the cognitive adequacy of systems of topological relations. As compared to our previous empirical investigation [7], we partially lifted constraints on the shape of regions in configurations that we presented subjects in a grouping task. With a high level of agreement, subjects distinguished between different possibilities of how spatial regions can touch each other. Based on the results of our investigation, we propose to develop a new system of topological relations on a finer level of granularity than previously considered.

1 Introduction

Reasoning about spatial information is an important part not only of human cognition but also of industrial applications such as geographical information systems (GIS). While in most existing GIS spatial information is represented in a quantitative way, which is certainly useful from an engineering point of view, new approaches to GIS try to come closer to the way spatial information is communicated by natural language and, thus, to the way human cognition is considered to represent spatial information, namely, by representing the relationships between spatial entities. This has the advantage of being more user-friendly—queries to a spatial information system or space related instructions to an autonomous agent, for instance, can be formulated in natural language and immediately be translated into the internal representation.

Qualitative spatial reasoning (QSR), a sub-discipline of artificial intelligence, is devoted to this way of representing and reasoning about spatial information. Many different aspects of space can be treated in a qualitative way, e.g., distance, size, direction, shape, or topology. But, as Cohn [1, page 22] wrote in his overview article on qualitative spatial reasoning, "An issue which has not been much addressed yet in the QSR literature is the issue of cognitive validity – claims are

Ch. Freksa et al. (Eds.): Spatial Cognition II, LNAI 1849, pp. 184–197, 2000.

often made that qualitative reasoning is akin to human reasoning, but with little or no empirical justification". In this paper we will continue our effort in studying the cognitive validity, also referred to as *cognitive adequacy* (we will define this term later), of existing approaches to QSR based on empirical investigations. Similar to our previous work [7], we will perform a grouping task where all presented items show two different colored spatial regions related in a different way. In [7], all objects were circles. Although this does not prevent subjects from grouping items according to size, distance, direction, or topology, other aspects of space cannot be distinguished. In the investigation presented in this paper we weaken this restriction on the shape of objects and use regular polygons instead. Unlike circles, these polygons can touch each other not only at one point but also at more than one point or along a line. This simple extension of the shape of regions led to surprising results that encourage the development of a new and extended topological calculus for qualitative spatial reasoning.

The remainder of the paper is structured as follows. In Section 2 we define the term cognitive adequacy and give an overview of the psychological methodology used to study cognitive adequacy. In Section 3 we introduce topological spatial relations and calculi based on these relations. Section 4 summarizes our previous investigation on topological spatial relations. In Section 5 we describe an empirical investigation and present the obtained results. Finally, in Section 6 we discuss our results, propose developing a new system of topological relations, and give links to possible future work.

2   Psychological Background

In determining the cognitive adequacy of a QSR formalism, we previously argued that this has to be broken down into at least two sub-aspects, namely the representational and the inferential aspect of the calculus. We termed these two sub-concepts of cognitive adequacy *conceptual adequacy* and *inferential adequacy* [8,7]. Conceptual adequacy refers to the degree in which a set of entities and relations correspond to the mental set of a person when he/she conceptualizes space and spatial configurations. Inferential adequacy refers to the degree in which the computational inference algorithm (e.g. constraint satisfaction, theorem proving, ...) conforms to the human mental reasoning process. A cognitively adequate qualitative reasoning system would therefore be based on a set of spatial relations that humans also use quite naturally, and would draw inferences in a way people do with respect to certain criteria (type of inferences; order of inferences; preferred, easy and hard inferences; in the extreme maybe also in accordance with processing latencies). The determination of the cognitive adequacy would be of interest to basic research in the field of the psychology of knowledge and the psychology of thinking on the one hand, and for computer science and artificial intelligence on the other. But mostly, applied research like cognitive ergonomics and human-computer-interaction (HCI) research would benefit from these results, since these could then justify a decision for one candidate out of a set of concurrent spatial calculi in spatial information systems. In the following we concentrate on the problem of conceptual adequacy of spatial relations and its empirical determination.

Looking for empirical methods to determine the conceptual adequacy, many results can be found in psychological research on conceptual structure. But most theories are about the conceptual structure of entities like natural kind terms or artifacts (see for example the overviews given in Komatsu [9], and Rips [16]). Relational concepts, however, were not investigated very heavily, and therefore psychological methods are tuned to the assessment of the conceptual structure of entities, but not for relations. Other research on spatial relations comes from (psycho-)linguistics where mainly acceptability ratings of natural language expressions like "above", "below", "left", "right" and so on, were obtained (see for example [6,17]). These results are heavily dependent on the investigated language and may, therefore, not reveal everything about people's conceptualization of space; this is due to the fact that different languages may provide different means to express the conceptualizations, and linguistic expressions are "cut" differently to cover spatial configurations, as one can easily verify when comparing English prepositions with those of other languages, even with those of close relatives like German. Therefore, language independent methods are the most promising approach to determine the human conceptualization of space. Mark et al. [10] mentioned some of them, like the graphic examples task, the graphic comparison task, and the grouping task. In particular, the *grouping task* seems to be one prominent empirical method that is easy to establish and to communicate to the participant. The subject is given a sample of spatial configurations that she/he is prompted to group together. The basic idea is that the conceptualization of space guides the grouping of items into categories. These observable categories give important hints on the subjects' conceptualization and may serve to exclude certain sets of relations as cognitively adequate. Subsequent intensional descriptions of the groupings may provide additional hints (1) what informational content was used by the subject to group the items, and (2) to lower the risk of choosing the wrong intension of extension-equivalent categories.

3 Topological Relations for Spatial Reasoning

Relationships that are invariant under certain transformations of the underlying space are called topological relationships. This property makes them very interesting for both GIS and QSR where representation of and reasoning about topological relationships is often used. When representing direction, the question of the underlying frame of reference is ubiquitous. Whether using an intrinsic, extrinsic, or deictic frame of reference is a matter of the represented situation. Looking at a spatial situation from different angles often changes the direction of objects with respect to a certain frame of reference. When representing distance, a typical problem is that of how many times "close" becomes "far", e.g., given a sequence of n objects that are all subsequently close to each other, for which number n is an object far from the first object of the sequence. Difficulties like these do not occur for topological relations, although reasoning with topological relations can also become extremely hard (see, e.g., Grigni et al. [5]).

Systems of topological relations were developed in the area of GIS and QSR independently of each other. In GIS, Egenhofer [2] classified the relationship between two spatial entities according to the nine possible intersections of their interior, exterior, and boundaries, hence called the *9-intersection-model*. When looking only at whether these intersections are empty or non-empty and imposing certain constraints on the nature of the considered spatial entities (two-dimensional, one-piece, non-holed), this results in eight different binary topological relationships (see Figure 1). In QSR, Randell et al. [14] studied the different possibilities for connecting spatial regions, in their approach regular subsets of a topological space (i.e., arbitrary but equal dimension, multiple pieces and holes possible). Among the many possibilities, they selected a set of eight different topological relationships with respect to connectedness which they called *Region Connection Calculus* RCC-8 (see Figure 1). The names of the relations are abbreviations of their meanings: DisConnected (DC), Externally Connected (EC), Partially Overlapping (PO), EQual (EQ), Tangential Proper Part (TPP), Non-Tangential Proper Part (NTPP), and their converses TPP^{-1} and NTPP^{-1}.

Surprisingly, these two completely different approaches to topological relationships lead to exactly (apart from the different constraints on regions) the same set of topological relations. Thus, there seems to be a natural agreement about what is a reasonable level of granularity of topological relations. Within both approaches systems of five topological relations on a coarser level of granularity were developed that are slightly different from each other. They are referred to as *medium resolution* in Egenhofer's case (as opposed to the set of eight relations referred to as *high resolution*), and RCC-5 in Randell et al.'s case. The differences between the two systems were compared in [7] and are shown in Figure 1. All systems of relations consist of jointly exhaustive and pairwise disjoint relations, i.e., between any two regions (that behave according to the given constraints) exactly one of these relations holds. Note that in the following we only refer to the Region Connection Calculus when referring to topological relations. This is only for increasing readability. All statements we make about the RCC-8 relations are of course equally true for the eight relations defined by Egenhofer.

Apart from these systems of topological relations, there is also work on topological relations on a finer level of granularity where more relations can be distinguished. Gotts [3,4], for instance, identified a large number of different topological relations that can be distinguished by their connectedness (the "C" relation). Another example of more expressive topological relations is given by Papadimitriou et al. [13].

4   Previous Empirical Studies on Topological Relations

In this section we summarize the results of our previous empirical investigation [7] on the cognitive adequacy of topological relations. The goal of that

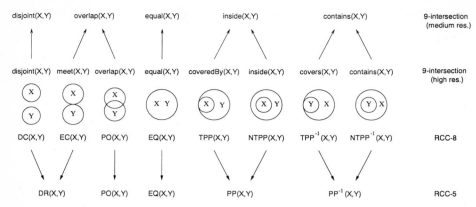

Fig. 1. Comparison of different systems of topological relations: the Region Connection Calculi RCC-8 and RCC-5 and the high and medium resolution relations of Egenhofer's 9-intersection model [7]

investigation was to find first evidence of whether the systems of topological relations developed by Egenhofer [2] and by Randell et al. [14] (see previous section) were cognitively adequate and which level of granularity is more adequate. For this investigation, 20 subjects had to group a set of 96 different items, each showing two different colored circles with differing centers and radii, such that items of the same group are most similar to each other. Examples of these items are shown in Figure 2. For each of the eight RCC-8 relations, 12 ± 3 items were chosen randomly as instances of these relations. After the grouping phase,

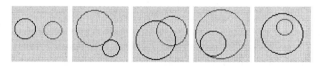

Fig. 2. Items with two different colored circles as used in [7]

subjects had to describe their groupings in a few words. This was done in order to obtain further information on the criteria that guided the grouping behavior.

The cluster analysis of the groupings showed that all instances of the RCC-8 relations were clustered immediately. After a few clustering steps, instances of the relations TPP and TPP^{-1} and of NTPP and NTPP^{-1} were clustered which is due to disregarding the colors of the circles. At no stage of the cluster analysis the RCC-5 relations or Egenhofer's medium resolution relations were present which can be regarded as an empirical evidence that these systems of relations are conceptually inadequate.

The evaluation of the verbal descriptions of the groupings which was done by two independent raters (who agreed in more than 97% of their ratings) showed that for more than 62% of the descriptions only topological terms were used, more than 33% used topological and orientational or topological and metrical terms. Descriptions based on purely orientation or metric terms did not occur.

These results support the claim that the systems of topological relations developed by Egenhofer and Randell et al. might be regarded as cognitively adequate. We use this term very carefully and consider our previous investigation only as a first step towards proving cognitive adequacy. In order to finally prove cognitive adequacy our investigation was not general enough, since all used regions were circles. This does not prevent subjects from grouping items according to distance, size, orientation, or topology. However, for a general assessment of cognitive adequacy of the topological relations, we have to lift the restriction of the shape of regions we were using in our previous investigation. In the following section we present our new empirical investigation where we partially lift this restriction of the shape of regions.

5   The Empirical Investigation

In our new investigation we partially lift the constraints on the shape of the regions. In the former investigation all regions were circles (see Figure 2), whereas in our new investigation the regions are regular polygons such as squares, pentagons or hexagons. The main difference to using circles is that when circles externally connect each other or are tangential proper part of each other, their boundaries always touch each other at a single point, whereas regular polygons with these properties can touch each other at one or more points, along a line, or even at both. Some of the different possibilities of how regular polygons can touch each other are shown in Figure 3.

Fig. 3. Different possibilities of external and internal connection of regular polygons

5.1   Subjects, Method and Procedure

19 students (10 female, 9 male) of the University of Freiburg participated for course credit. Subjects had to accomplish a grouping task (on a Sun Workstation) that consisted of 200 items showing a configuration of a red and a blue regular

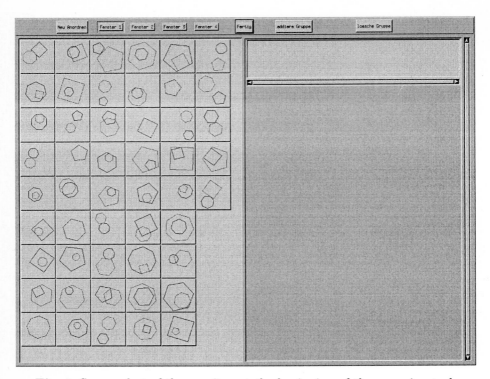

Fig. 4. Screen-shot of the monitor at the beginning of the grouping task

polygon as shown in Figure 3. For the sample of 200 items we randomly selected 25 ± 5 items for each of the eight RCC-8 relations with the following restrictions. We generated regular polygons with four to eight corners, and made sure that there were at least some items for each possibility of how two regions can connect to each other.

After solving some practice trials to get acquainted with the procedure subjects saw the color screen as depicted in Figure 4. The screen was split into two different sections. The left section contained the items and the right section the groupings subjects had to form. The items had to be moved with a mouse from the left to the right. In order not to bias the number of groupings, only one group was displayed on the right in the beginning, and subjects had the possibility to add as many groupings as they wanted. The 200 items in the left section were arbitrarily split into four different windows such that subjects could switch between the four windows at any time.

All subjects had to judge the same 200 items because the aggregation of data had to be done across subjects per item. After the grouping phase, subjects had to describe their groupings by natural language descriptions (in German). We did this to obtain further information on the criteria that guided subjects' grouping behavior. We applied this (unexpected) phase at the end of the empirical session in order to avoid influences of verbalizations on the pure grouping behavior itself. For each subject the complete procedure was stored in a log-file and we developed a tool for tracing the complete procedure afterwards.

5.2   Results of the Investigation

Subjects needed about 45 minutes to group items and to describe their groupings. By aggregating grouping answers over all 19 subjects, we obtained a 200×200 matrix of Hamming distances between items that was the basis of a cluster analysis using the method of average linkage between groups. Unlike the results of [7] where only some subjects did not distinguish between TPP and TPP^{-1} and between NTPP and NTPP^{-1}, this time the result is more unique. As it can be seen in the dendrogram in Figure 5, there is no distinction between NTPP and NTPP^{-1} and between TPP and TPP^{-1} at all. As in our previous study, the clustering of TPP and NTPP with their converses can be attributed to disregarding the relationship of a reference object (RO) with a to-be-localized object (LO). However, there are sub-clusters of TPP or TPP^{-1} items and of EC items. By looking at the items of the different sub-clusters, we found that these sub-clusters belong exactly to the different ways two regular polygons can be externally or internally connected to each other, either at a single point, at two points, or along a line. The clustering of groups on a higher level of the cluster analysis is due to idiosyncratic categorizations or mis-groupings of single items by a few subjects (one subject wrote that he used one group as a clipboard and forgot to assign these items in the end).

Furthermore, the verbalizations of subjects' groupings were analyzed by categorizing them according to whether they included purely topological (T), orientational (O), and metrical information (M). Additionally, we introduced two new categories: for the degree of contact of the boundaries (C), and for mentioning of pure shape (S). We also considered all possible combinations of these five factors. In Table 1, we present the results of the categorization of subjects' verbalizations, as completed by one independent rater.

Table 1 clearly shows that the informational content of nearly all verbalizations incorporated topological information alone or in combination with other information ($42.1\% + 14.6\% + 36.6\% + 1.2\% = 94.5\%$). Metrical and orientational information was only used in combination with topological information. The category "other" in Table 1 consists of verbalizations that do not belong to any combination of T, O, M, C, or S, but were mostly mis-interpretations of the EQ items. The most surprising result, however, is that a substantial number of verbalizations mentioned the degree of contact as an additional modifier of topological relations. This also confirms the findings of the cluster analysis, where EC, TPP and TPP^{-1} configurations had sub-clusters that differed with respect to the degree of contact.

This observation was confirmed by evaluating the log-files we collected during the experiments. The "trace" tool enabled us to recover the final screen the subjects produced when finishing the grouping and the verbalization tasks

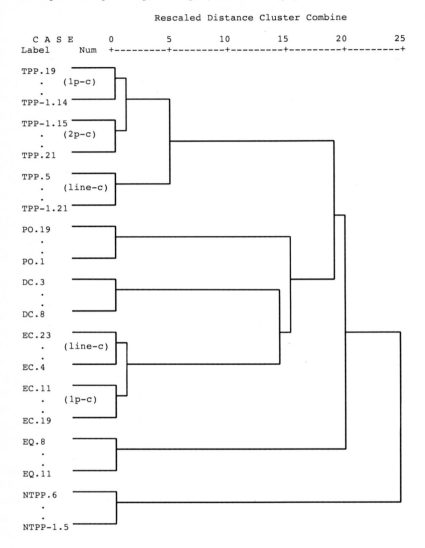

Fig. 5. Dendrogram of the cluster analysis of 200 items based on Hamming distances between items. Note that the labels given on the left are only for identification purposes. The three sub-clusters of the TPP and TPP^{-1} items have the property of one-point contact (1p-c), two-point contact (2p-c), or line-contact (line-c), respectively. For the EC items there are two sub-clusters of items with line contact and one-point contact

Table 1. Categorization of subjects' verbalizations of their groupings (n=164 verbalizations)

Informational content	Percentage
Topology (T)	42.1%
Orientation (O)	–
Metric (M)	–
Contact (C)	1.8%
Shape (S)	–
T + O	14.6%
T + C	36.6%
T + M	1.2%
Other	3.4%

Table 2. Evaluation of the subjects' final screen

#of subjects	description of their groupings
1	TPP and NTPP grouped together, DC plus direction
5	RCC-8*
1	RCC-8* plus directions
1	RCC-8* plus size of overlapping area
1	RCC-8* plus different connections for TPP
6	RCC-8* plus different connections for EC and TPP
3	RCC-8* plus different connections for EC and TPP and PO
1	RCC-8* plus different connections for EC and TPP, and different directions for PO and DC

(see Figure 6). We found that except for one subject, who grouped the items belonging to TPP and NTPP and to TPP^{-1} and NTPP^{-1} together and who distinguished items showing disconnected regions according to their direction, all other subjects did not distinguish the color of the regions presented on the objects, i.e., they did not distinguish between NTPP and NTPP^{-1} and between TPP and TPP^{-1}. Table 2 shows the detailed evaluation of the final screens of the subjects. By RCC-8* we denote the set of RCC-8 relations without distinction of the reference object and the to-be-localized object. Figure 6 gives an example of the most common result, where the subject distinguished different kinds of external and internal connection.

6 Discussion and Future Work

For the investigation presented in this paper, we increased the different possibilities of how subjects can group items by partially lifting the constraints on the shape of the regions we used. Except for one subject, all others grouped the items

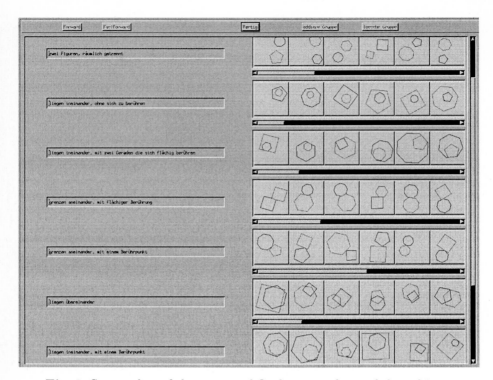

Fig. 6. Screen-shot of the recovered final screen of one of the subjects

according to the distinctions made by the RCC-8 relations. Even those subjects who also took direction or size into account only did this to further refine the RCC-8 relations but never used orientation or shape independently of the topological relations. This result gives further evidence that topological relations in general are an important part of human spatial cognition, and that the level of granularity given by the RCC-8 relations is particularly important. Topological distinctions on a coarser level of granularity do not seem to be cognitively adequate. This confirms the observations made in our previous investigation [7].

Many subjects also distinguished items according to a finer level of granularity than given by the RCC-8 relations. 11 out of 19 subjects distinguished between the different possibilities of how two regions can touch each other from the outside or from the inside. These distinctions can all be made on a purely topological level without taking into account other spatial aspects such as direction, distance, or shape. Thus, it appears that topological relationships are much more important in human spatial cognition than any other relationships between spatial regions. If one wants to develop a new system of topological relations based on the results of our investigation, we propose to refine the relations EC, TPP, and TPP^{-1} into different kinds of connection. Some possibilities are to distinguish connections at a single point, at more than one point, or along a line. It follows from Gotts's work [3,4] that all these relations can be defined in terms of the connected relation which is also used to define the RCC-8 relations.

Although almost all subjects did not distinguish between TPP and TPP^{-1} and between NTPP and NTPP^{-1}, we do not consider the set of six relations we called RCC-8* as cognitively adequate. As already explained, this is due to disregarding the relationship of a reference object (RO) with a to-be-localized object (LO). We believe that when this distinction is explicitly emphasized in the instructions, subjects will group items accordingly. Furthermore, it follows from [15] that reasoning with RCC-8* relations is NP-hard whereas reasoning with the eight RCC-8 relations is tractable [12].

Future work includes further lifting the restrictions on the shape of the presented regions. One question is whether subjects continue to distinguish the number of connections and up to which number. In our investigation some subjects distinguished between connections at a single point, at two points, and along a line. It will be interesting to see whether subjects distinguish connections at $1, 2, 3, \ldots, n$ points, and more than n points and which number n turns out to be the most commonly chosen (maybe it is the magical number 7 ± 2, the capacity of the human short term memory [11]). For arbitrary shapes of regions it is also possible that two regions connect along two or more lines. Another interesting question is how subjects group items when regions are allowed to consist of multiple pieces or to have holes. Answering this question will also allow judging the adequacy of the different restrictions on regions as given by Egenhofer [2] and by Randell et al. [14] (cf. Section 3). So far we have only investigated the conceptual cognitive adequacy of systems of topological relations. It will be interesting to see how humans reason about these relations, and to investigate the inferential cognitive adequacy of the reasoning methods used in qualitative spatial reasoning (cf. [8]).

Acknowledgments

We would like to thank the following people: Niclas Hartmann for implementing the following three tools: The create tool for generating the items according to certain criteria, the computer-aided grouping experiment, and the trace tool for visualization of subjects' grouping behavior, Katrin Balke for running the experiment, and Daniel van den Eijkel, who categorized subjects' verbalizations. This research was supported by grants Str 301/5-2 (MEMOSPACE) and Ne 623/1-2 (FAST-QUAL-SPACE) in the priority program "Spatial Cognition" of the Deutsche Forschungsgemeinschaft.

References

1. Anthony G. Cohn. Qualitative spatial representation and reasoning techniques. In G. Brewka, C. Habel, and B. Nebel, editors, *KI-97: Advances in Artificial Intelligence*, volume 1303 of *Lecture Notes in Computer Science*, pages 1–30, Freiburg, Germany, 1997. Springer-Verlag.

2. Max J. Egenhofer. Reasoning about binary topological relations. In O. Günther and H.-J. Schek, editors, *Proceedings of the Second Symposium on Large Spatial Databases, SSD'91*, volume 525 of *Lecture Notes in Computer Science*, pages 143–160. Springer-Verlag, Berlin, Heidelberg, New York, 1991.

3. Nicholas M. Gotts. How far can we C? defining a 'doughnut' using connection alone. In E. Sandewall J. Doyle and P. Torasso, editors, *Principles of Knowledge Representation and Reasoning: Proceedings of the 4th International Conference (KR94)*, pages 246–257, San Francisco, 1994. Morgan Kaufmann.

4. Nicholas M. Gotts. Topology from a single primitive relation: Defining topological properties and relations in terms of connection. Technical Report 96-23, University of Leeds, School of Computer Studies, 1996.

5. Michelangelo Grigni, Dimitris Papadias, and Christos Papadimitriou. Topological inference. In *Proceedings of the 14th International Joint Conference on Artificial Intelligence*, pages 901–906, Montreal, Canada, August 1995.

6. William G. Hayward and Michael J. Tarr. Spatial language and spatial representation. *Cognition*, 55:39–84, 1995.

7. Markus Knauff, Reinhold Rauh, and Jochen Renz. A cognitive assessment of topological spatial relations: Results from an empirical investigation. In *Proceedings of the 3rd International Conference on Spatial Information Theory (COSIT'97)*, volume 1329 of *Lecture Notes in Computer Science*, pages 193–206, 1997.

8. Markus Knauff, Reinhold Rauh, and Christoph Schlieder. Preferred mental models in qualitative spatial reasoning: A cognitive assessment of Allen's calculus. In *Proceedings of the Seventeenth Annual Conference of the Cognitive Science Society*, pages 200–205, Mahwah, NJ, 1995. Lawrence Erlbaum Associates.

9. Lloyd K. Komatsu. Recent views of conceptual structure. *Psychological Bulletin*, 112:500–526, 1992.

10. David M. Mark, David Comas, Max J. Egenhofer, Scott M. Freundschuh, Michael D. Gould, and Joan Nunes. Evaluation and refining computational models of spatial relations through cross-linguistic human-subjects testing. In A. U. Frank and W. Kuhn, editors, *Spatial Information Theory*, volume 988 of *Lecture Notes in Computer Science*, pages 553–568, Berlin, Heidelberg, New York, 1995. Springer-Verlag.

11. George A. Miller. The magical number seven, plus or minus two: Some limits on our capacity for processing information. *Psychological Review*, 63:81–97, 1956.

12. Bernhard Nebel. Computational properties of qualitative spatial reasoning: First results. In I. Wachsmuth, C.-R. Rollinger, and W. Brauer, editors, *KI-95: Advances in Artificial Intelligence*, volume 981 of *Lecture Notes in Artificial Intelligence*, pages 233–244, Bielefeld, Germany, 1995. Springer-Verlag.

13. Christos H. Papadimitriou, Dan Suciu, and Victor Vianu. Topological queries in spatial databases. In *Proceedings of the 15th ACM Symposium on Principles of Database Systems*, pages 81–92, 1996.

14. David A. Randell, Zhan Cui, and Anthony G. Cohn. A spatial logic based on regions and connection. In B. Nebel, W. Swartout, and C. Rich, editors, *Principles of Knowledge Representation and Reasoning: Proceedings of the 3rd International Conference*, pages 165–176, Cambridge, MA, October 1992. Morgan Kaufmann.

15. Jochen Renz and Bernhard Nebel. On the complexity of qualitative spatial reasoning: A maximal tractable fragment of the Region Connection Calculus. *Artificial Intelligence*, 108(1-2):69–123, 1999.

16. Lance J. Rips. The current status of research on concept. *Mind and Language*, 10:72–104, 1995.

17. Constanze Vorwerg and Gert Rickheit. Typicality effects in the categorization of spatial relations. In C. Freksa, C. Habel, and K. F. Wender, editors, *Spatial cognition. An interdisciplinary approach to representing and processing spatial knowledge*, volume 1404 of *Lecture Notes in Artificial Intelligence*, pages 203–222. Springer-Verlag, Berlin, Heidelberg, New York, 1998.

Interactive Layout Generation with a Diagrammatic Constraint Language[1]

Christoph Schlieder[*] and Cornelius Hagen[**]

[*]University of Osnabrück, Institute for Semantic Information Processing
cschlied@cl-ki.uni-osnabrueck.de
[**]University of Freiburg, Center for Cognitive Science
hagen@cognition.iig.uni-freiburg.de

Abstract. The paper analyzes a diagrammatic reasoning problem that consists in finding a graphical layout which simultaneously satisfies a set of constraints expressed in a formal language and a set of unformalized mental constraints, e.g. esthetic preferences. For this type of problem, the performance of a layout assistance system does not only depend on its use of computational resources (algorithmic complexity) but also on the mental effort required to understand the system's output and to plan the next interaction (cognitive complexity). We give a formal analysis of the instantiation space of a weakly constrained rectangle layout task and propose a measure for the cognitive complexity. It is discussed how the user's control of the presentation order of the different constraint instantiations affects the cognitive complexity.

1 Partially Unformalized Constraint Systems

One of the fields in which constraint-based approaches to spatial configuration have been particularly successful is the generation of document layouts. Assistance systems as the ones described by [17] and [7] allow the user to specify layout properties with a relational constraint language. The system then generates the layout by means of a constraint solver. However, not all layout problems permit this neat form of declarativity which completely hides the search aspect of constraint satisfaction from the user. Often, the user has to interact with the system in order to guide the search process. One rather trivial reason for combining machine reasoning with mental reasoning is given when the search heuristics produce suboptimal results. A human observer who visually inspects a machine-generated layout can sometimes provide precious hints about how to improve the quality of the solution. Of course, this type of *mental postprocessing* constitutes only a last resort. Therefore, in applications such as yellow page layout where design rules are easily formalized, much research effort is spent on eliminating the need for mental postprocessing.

A completely different situation arises in creative layout tasks. These tasks are gov-

---

1. This work was supported by the Deutsche Forschungsgemeinschaft (DFG) in the priority program on spatial cognition (grant Str 301/5-2).

Ch. Freksa et al. (Eds.): Spatial Cognition II, LNAI 1849, pp. 198-211, 2000.

erned by constraints that, at least until now, have resisted complete formalization. Typically, the layouter is able to specify some of the constraints that a solution must satisfy, but these constraints do only partially capture his notion of a "good" solution. Design principles which blend functional considerations with esthetic preferences are known to be very difficult to verbalize, let alone to formalize in a constraint language [3]. However, the mental processing of the unformalized design principles is generally very efficient: a human layouter recognizes without difficulty a solution satisfying such constraints; usually he is also able to rate the quality of the solution.

This type of layout task can be characterized as *partially unformalized constraint problem (PUCP)*: Find a layout that satisfies simultaneously (1) a set F of spatial constraints specified in a constraint language L, and (2) a set U of mental constraints which cannot be expressed in L. The assistance system then helps the user to explore the solution space of the PUCP. Its main function consists in computing instantiations for the constraints in F and in providing suitable navigation operations for moving between instantiations. The user navigates through the instantiation space checking the constraints in U until a satisfying layout is found. In other words, the system acts as a medium which guarantees that during the navigation the user will not take layouts into consideration which violate any of the formalized constraints (Fig. 1).

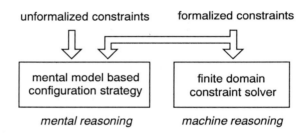

Fig. 1 Partially unformalized constraint problem

Because of the essential role that mental reasoning plays in solving a PUCP, a performance analysis of the assistance system cannot ignore the time that the user spends on evaluating a layout and planning the next interaction as it could be done in a mental postprocessing scenario. To mark this difference, we speak of interaction based on *mental coprocessing*.

Mental coprocessing is a special type of interactive constraint processing. It differs from other interaction schemes studied in the area of spatial constraint solvers in two ways:

(1) The user cannot provide the system with an instantiation. Mental and formal constraints combine to form a problem that is too complex to be solved mentally. This is in clear difference to the scenario described by Kurlander and Feiner [11] where constraints are inferred from example solutions specified by the user.

(2) The formal constraint system is weakly constrained. In other words, many instantiations exist and the problem consists in finding a way to present the large space

of alternative solutions to the user. Interactive constraint solvers typically rely on the assumption that the task is strongly constrained and a unique or no instantiation exits. A classical example is Borning's constraint solver ThingLab [2].

As a consequence, the problem for mental coprocessing consists in finding a good way to present a large instantiation space to the user. This problem has been addressed first not in spatial, but in temporal reasoning. Hoebel, Lorenzen, and Martin [8] visualizes the solutions found by their temporal planner, i.e. alternative plans satisfying a set of (formal) temporal constraints. Their approach amounts to simultaneously present all instantiations. Clearly, this is feasible only for very small spaces (< 100 instantiations).

The purpose of this paper is to study the case where a simultaneous presentation of all solutions is not feasible because of the size of the instantiation space. Our analysis refers to a concrete layout problem (section 2) which is based on a simple diagrammatic constraint language that is described in section 3. We argue that the cognitive complexity of a PUCP critically depends on the degree to which the user is allowed to control the order in which constraint instantiations are produced (section 4). A criterion for an optimal degree of cognitive control is formulated (section 5) and applied to the layout problem (section 6).

2 Constrained-Based Document Layout

Constraint languages for document layout allow to express relations between rectangular graphical objects such as textboxes and images. Fig. 2 shows some of the relational constraints used in the seminal work of Weitzman and Wittenburg [17]. These spatial relations describe the ordering of the boxes (e.g. left-of) or their alignment (e.g. top-aligned) without specifying metrical information. We consider a constraint language that encompasses all possible relations of this kind. The basic idea consists in projecting the configuration of rectangles onto the horizontal and the vertical axis thereby obtaining two configurations of intervals.

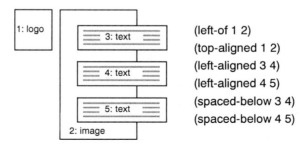

Fig. 2 Constraints for a document layout

Constraints between intervals are then expressed by means of a system of relations that is widely used in AI research on temporal and spatial reasoning: the *interval relations* introduced by Allen [1]. The system $I = \{<, m, o, s, d, f, =, fi, di, si, oi, mi, >\}$ consists of the 13 relations between two intervals $X = [x_1, x_2]$ and $Y = [y_1, y_2]$ that can be distinguished in terms of the ordering of x_1, x_2, y_1, and y_2. Disjunctive information is represented by sets of interval relations. $X \{o,s,d\}\, Y$, for instance, is a shorter notation for $X o Y \vee X s Y \vee X d Y$. Every element of $INT = 2^I$ can thus be interpreted as a *generalized interval relation*, i.e. a disjunction of interval relations. Inversion and composition are defined componentwise for relations from INT. Some relations deserve special attention: the universal relation **1** corresponding to I and the contradictory relation **0** corresponding to the empty set. The 13 interval relations are represented by singleton subsets that are called the *basic relations* of INT.

We use generalized interval relations for stating constraints between intervals. An *interval constraint system* for a set of intervals $\{X_1,...,X_n\}$ is an $n \times n$ matrix $C = (r_{ij})$ with entries $r_{ij} \in INT$. The entry r_{ij} specifies the generalized interval relation that must hold between interval X_i and X_j. If the entry is **1** it does not constrain the position of the intervals — any relation could hold. In standard terminology, the r_{ij} constitute the constraint variables that are instantiated with values from the domain I. An assignment of values to the constraint variables of $C = (r_{ij})$ is given by a matrix $B = (b_{ij})$, whose entries are interval relations $b_{ij} \in I$ satisfying $b_{ij} \in r_{ij}$. We call B a consistent assignment or *instantiation* of C iff $\forall i \forall j \forall k\ b_{ik} \in (b_{ij} \bullet b_{jk})$. In other words, consistency means that the assignment is compatible with the composition table. Since the constraint variables range over a finite domain, namely the 13 interval relations, the satisfiability problem is decidable. Although for general interval constraint systems deciding satisfiability is known to be NP-complete, the problem becomes polynomial as soon as the relations of the constraint system are restricted to certain subsets of INT [13]. In these cases a simple $O(n^3)$ constraint propagation algorithm (path consistency) can be used to decide satisfiability, or — more interesting — to compute a constraint instantiation. In the general case the propagation algorithm is used as forward checking heuristic during backtracking [12].

3 A Simple Diagrammatic Constraint Language

As was already mentioned, constraints between two rectangles A and B are represented by constraints between the two pairs of intervals that result from projecting A and B onto the horizontal and vertical axis: A_x, B_x and A_y, B_y respectively. Formally, the *rectangle relation* $A\ r_x{:}r_y\ B$ with $r_x, r_y \in I$ holds iff $A_x r_x B_x$ and $A_y r_y B_y$ (cf. Fig. 3). The system of rectangle relations consists of $13 \times 13 = 169$ relations $R = \{<{:}<, <{:}m,$ $<{:}o, ..., oi{:}>, mi{:}>, >{:}>\}$. In order to express disjunctive information, rectangle relations are generalized analogously to the interval relations. However, to guarantee tractability, we do not consider arbitrary relations from 2^R but only those that can be written as a product of generalized interval relations[2]. The product of $r_1, r_2 \in INT$ is $r_1 \times r_2 :=$ $\{b_1{:}b_2 | b_1 \in r_1 \wedge b_2 \in r_2\}$. In this sense $\{f{:}d, f{:}o, m{:}d, m{:}o\} = \{f,m\} \times \{d,o\}$ is a *generalized rectangle relation* whereas $\{f{:}d, f{:}o, m{:}o\}$ is not. We denote the set of these relations by REC. Because generalization is restricted, relations from REC can be written in product form, e.g. A {f,m}:{d,o} B for A {f:d, f:o, m:d, m:o} B (cf. Fig. 3).

The composition of generalized rectangle relations reduces to the composition of generalized interval relations. For relations $r_{11}{:}r_{12}$ and $r_{21}{:}r_{22}$ from REC $(r_{11}{:}r_{12}) \bullet (r_{21}{:}r_{22}) = (r_{11} \bullet r_{12}){:}(r_{21} \bullet r_{22})$. Now that composition is defined, constraint systems can be introduced in complete analogy to interval constraint systems: A *rectangle constraint system* for a set of rectangles $\{X_1,...,X_n\}$ is an $n \times n$ matrix $C = (r_{ij})$ with entries $r_{ij} \in$ REC. An assignment of values to the constraint variables of $C = (r_{ij})$ is given by a matrix $B = (b_{ij})$ whose entries are rectangle relations $b_{ij} \in R$ satisfying $b_{ij} \in r_{ij}$. The assignment B is an *instantiation* of C iff $\forall i \forall j \forall k\ b_{ik} \in (b_{ij} \bullet b_{jk})$. Note that because all relations are from REC, a rectangle constraint system can be solved by solving two independent interval relation systems.

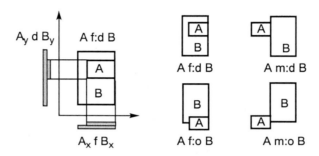

Fig. 3 Projection of layout boxes

Rectangle constraint systems differ from the finite domain constraint systems traditionally used for visual interaction. Constraint variables represent qualitative relations, not coordinates from a finite range of integers such as in [2] or symbolic constraints as in [7]. The type of qualitative abstraction underlying rectangle constraint systems can be described formally in the framework of relational algebras[3] [12]. Constraint-based reasoning in relational algebras has been studied thoroughly by AI research on *qualitative spatial reasoning* [4]. Most of the following analysis generalizes to reasoning with arbitrary spatial relational algebras.

4 Navigation in Instantiation Space

We consider a PUCP for which the set F of formal constraints is given by a rectangle constraint system. Only the user can check whether an instantiation of F also satisfies

---

2. Actually, more is needed for tractability. In each projection, the generalized interval relations must be members of a tractable subalgebra of 2^R (see [13]).

3. The technical difference between symbolic constraints such as those used in [7] and qualitative constraints is that the latter express information about relational composition and use path-consistency (not the weaker arc-consistency) as propagation method.

the additional set U of unformalized constraints. If $\Im$ denotes the set of instantiations of F, then the mental operation of checking the constraints in U realizes a mapping $m: \Im \to \{\text{true, false}\}$ where $m(I)$ = true if the instantiation $I \in \Im$ satisfies U, and $m(I)$ = false otherwise. In the following we regard this mental process as computationally intransparent. Although the details of the interaction between mental and machine reasoning can be designed very differently, all assistance systems for PUCP share the same *basic interaction cycle*. It consists of a loop which combines mental operations (step 1) and computations performed by the system (step 2):

> repeat
>> (1) Select a navigation operation $\omega \in \Omega$.
>> (2) Find an alternative instantiation I_ω of F.
> until $m(I_\omega)$ is true

Searching through the instantiations of F is realized by iterative execution of step 2. The user can influence search order through the choice of a navigation operation. In the document layout scenario, he could for example point at a layout box that the system should reposition, or, more specifically, drag the layout box to the place he prefers. In both cases, one option ω out of a set Ω of possible options is selected. However, the user has more control over the search process in the latter than in the former case. Quantitatively, this difference is reflected by a bigger cardinality of Ω: the number of layout boxes is multiplied by the number of positions in the layout grid. How much search control should the user be given? Obviously, more control is an advantage as long as it does not increase the cognitive complexity of the task. If the user executes c basic interaction cycles spending an average time T_ω on selecting the navigation operation then the *cognitive complexity*, i.e. the total time spent on mental reasoning, is given by

$$T_{\text{mental}} = c \cdot T_\omega$$

This means that there is a trade-off between spending time on evaluating different layouts generated by the system (c interaction cycles) and spending time on a single layout deciding how to modify it (T_ω navigation time). As a tool for describing the trade-off we introduce the *navigation graph*. The graph encodes what instantiation $I_\omega \in \Im$ is generated when the navigation operation $\omega \in \Omega$ is selected. Generally, this depends on the instantiation $J \in \Im$ to which the navigation operation is applied. In other words, each $\omega \in \Omega$ realizes a mapping $\omega: \Im \to \Im$. The navigation graph $N = (V, E)$ is specified by $V = \Im$ and $E = \{(I, J) \in \Im \times \Im | I = \omega(J), \omega \in \Omega\}$. Note that navigation graphs can become very large: $\Im$ can grow exponentially with n, the number of layout objects. Just think of the interval constraint system $C = (r_{ij})$ with entries $r_{ij} = \{<, >\}$ for $i \neq j$ and $r_{ii} = \{=\}$. Its navigation graph has $n!$ vertices. The number of edges of a navigation graph can vary considerably. Fig. 4 illustrates this with three regular graphs for $|\Im| = 2^n$: the cycle C_{2^n}, the hypercube H_{2^n}, and the complete graph K_{2^n}.

C_{2^n} and K_{2^n} describe the two extremes where the user exerts no or complete search control respectively. In the first case, the assistance system generates instantiations in a fixed order. As the user is forced to follow the tour C_{2^n}, he spends constant time for the navigation decisions but it may take him as many as $O(2^n)$ interaction cycles to reach his goal (diameter of C_{2^n}).

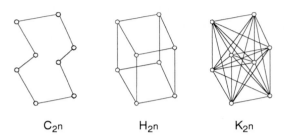

$$C_{2}n \qquad H_{2}n \qquad K_{2}n$$

Fig. 4 Different types of navigation graphs

With K_{2^n} the user has complete search control. Any instantiation can be reached in a single interaction cycle. However, he must spend considerable time to decide which of the $O(2^n)$ navigation alternatives (degree of K_{2^n}) to choose. A compromise between the two extremes is realized by H_{2^n} which implements partial search control. There are $O(n)$ different navigation operations between which the user has to decide and with $O(n)$ interactions he reaches any instantiation.

navigation graph	diameter	degree	search control
$C_{2}n$	$O(2^n)$	$O(1)$	none
$H_{2}n$	$O(n)$	$O(n)$	partial
$K_{2}n$	$O(1)$	$O(2^n)$	complete

Fig. 5 Navigation graph and search control

5 Lessons from Mental Model Theory

The optimal balance between diameter and degree of the navigation graph should minimize the cognitive complexity of the whole task $T_{mental} = c \cdot T_\omega$. Clearly, the optimum depends on how T_ω relates to the degree, or equivalently, to the cardinality of Ω . Our argument is based on (1) the formal analysis of the problem space given in the previous section, (2) empirical evidence about the use of mental models in spatial reasoning. We briefly summarize the relevant findings.

Reasoning with spatial constraints has been studied for more than 20 years in the *spatial-relational inference* paradigm (see [5]). Tasks in this paradigm consist of the non-universal entries of a constraint matrix which are presented in verbal form to subjects of the experiment. In our case: "The red interval touches the green interval from the left" for X m Y. Subjects are then asked to specify instantiations for certain or all constraint variables ("Which relation must/could hold between the blue and the yellow interval?"). It was found that most human reasoners adopt a model-theoretic rather than proof-theoretic strategy for solving spatial relational inference tasks. They try to build alternative spatial layouts in visuo-spatial working memory and check which relations hold in the layouts rather then proving with inference rules that certain relations must/could hold. This finding led Johnson-Laird to formulate *mental model theory*, an explanatory framework unifying results about model-based reasoning strategies [9]. Results from the experiments conducted by Rauh and Schlieder [14] clearly indicate that interval constraint systems are mentally solved by reasoning with mental models. Human reasoners seem not to be able to effectively use constraint propagation — a mental proof strategy. Instead, they immediately try to find an instantiation by means of simple spatial layout strategies. The layout strategies for constructing interval configurations have been analyzed and described in [15]. It turns out that they are very efficient but incomplete, i.e. the strategies do not always produce an interval configuration. If the mental layout strategies succeed, they produce a first mental model, the *preferred mental model,* which constitutes the starting point for the further reasoning process. Interestingly, subjects agree considerably on which instantiation of an interval constraint system they prefer [10]. Fig. 6 shows all instantiations of the interval constraint system that consists of three non-universal constraints A di B, $B > C$, and C m D. According to the data of an experiment from [14], interval configuration 37 constitutes the mental model preferred by most subjects. Only few subjects preferred the somewhat similar configuration 36. None of the other configurations acted as preferred mental model.

Human reasoners seem not to possess different sets of layout strategies for constructing different instantiations of an interval constraint system. They obtain alternative instantiations by transforming the preferred mental model. As a consequence, cognitive complexity increases *linearly* with the number of mental models constructed, or equivalently, with the number of transformations applied. If the task is solved by mental reasoning without the possibility to externalize working memory contents then only few of the possible instantiations are found. This is where assistance systems enter. They provide operations for navigating through instantiation space that support the process of *mental model transformation*. The system checks which transformations produce an instantiation and gives the user a complete overview of all applicable transformations. In order to decide which navigation operation to take next, the user must anticipate the result of this operation and evaluate whether it brings him closer to a layout that satisfies the unformalized constraints. Generally, he can only partially anticipate the result. Based on a mental model of part of the layout, it is often impossible to see that a local transformation of the layout has global consequences.

Moving a single layout box to a new position can lead to the repositioning of all other layout boxes. Because the transformation of mental models is the basic reasoning strategy at the user's disposition for anticipating the effects of his navigation operation and because the cognitive complexity of mental model transformation increases linearly with the number of models that are constructed, we expect to find $T_\omega = O(|\Omega|)$.

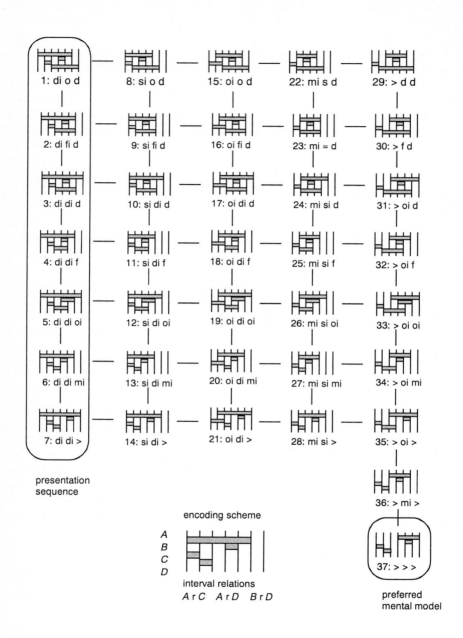

Fig. 6 Instantiation space for the interval constraint system
A di B, $B > C$, C m D

6 Neighborhood Navigation Operations

As we have argued in the last section, $T_\omega = O(|\Omega|)$. In order to obtain a navigation graph that minimizes the cognitive complexity, we will have to choose Ω in such a way that $|\Omega| = O(n)$ where n denotes the number of layout objects (rectangles, intervals). Additionally, the navigation operations should correspond to transformations of mental models that can be computed with little cognitive effort. On the level of single interval relations, cognitively simple transformations are known to exist. They implement the *conceptual neighborhood* of interval relations [6]. Interval relations r_1 and r_2 are said to be conceptual neighbors if a configuration of two intervals X and Y satisfying $X\ r_1\ Y$ can be continuously transformed into a configuration of intervals X' and Y' satisfying $X'\ r_2\ Y'$ such that during the transformation no configuration arises in which a relations different from r_1 and r_2 holds. For example, there is no continuous transformation of $X < Y$ into X o Y which avoids a stage where X m Y: the relations $<$ and m as well as m and o are conceptual neighbors whereas $<$ and o are not. The edges of the graph in Fig. 7 specify conceptual neighborhood for the interval relations

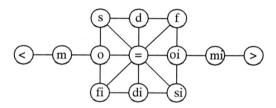

Fig. 7 Conceptual neighborhood graph

We generalize this idea to configurations of more than two intervals. Distance in the neighborhood graph of Fig. 7 is denoted by $d(r_1, r_2)$, e.g. $d(\mathrm{m, si}) = 4$. For two instantiations $A = (a_{ij})$ and $B = (b_{ij})$ of an interval constraint system the *conceptual distance matrix* is defined as

$$\Delta(A, B) = (d(a_{ij}, b_{ij})).$$

The most strict notion of *instantiation neighborhood* considers instantiations A and B neighbors iff in the conceptual distance matrix $\Delta(A, B)$ only two entries, r_{ij} and r_{ji}, corresponding to the relation between interval X_i and X_j are 1 whereas all other entries vanish. In the instantiation space of Fig. 5 configuration 4 and 11 are conceptual neighbors; 4 and 18 are not. Transformations that map an instantiation into a neighboring one seem good candidates for navigation operations that are cognitively simple to compute. It is not difficult to show that the maximum number of transformations into conceptually neighbored configurations is bounded by $8 \cdot n = O(n)$. This means that the number of neighbors of an instantiation is small enough to allow a mental evaluation of all navigation alternatives. *Neighborhood navigation* relies on a set of navigation operations that move only from an instantiation to its conceptual neighbors.

Actually, a slightly more general notion of instantiation neighborhood was used to compute the navigation graph in Fig. 6. With our definition, the edge between instantiation 15 and 22 does not correspond to a neighborhood transformation since relations between more than two intervals change at once. However, the notion of neighborhood transformations can be generalized to capture such cases. Neighborhood navigation implements a local navigation scheme. It is not only local with respect to the instantiation space but also local with respect to the geometry of each single instantiation. Neighborhood transformations only change relations between objects that are close to each other.

In order to change relations between layout objects that are far apart, a sequence of transformations is needed. For this purpose, *presentation sequences* where all neighborhood transformation refer to the same interval are especially useful. Fig. 5. shows an example consisting of a sequence involving the instantiations 1, 2, ..., 7. The algorithmic problem of generating presentation sequences consists in modifying a path-consistency constraint solver for relational algebras such that it generates instantiations in a particular ordering. To achieve this goal the constraint variables must be assigned values in a particular order. A best first strategy can be used for this purpose: (1) assign variables that denote relations which involve the selected interval first, (2) assign values that denote relations which are conceptual neighbors of the last assignment first.

Because of their local nature, neighborhood navigation operations are easily visualized. Fig. 8 illustrates the structure of the visual interface of a system assisting the user to solve a PUCP in the document layout scenario. The formalized constraints are symbolized by flag icons which stand for rectangle relations: b1 {<}:{f} b2, b3 {s}:{>} b4, b4 {s}:{>} b5. To navigate through the instantiation space, the user selects a layout box. The system then computes the applicable neighborhood transformations and visualizes them using arrow icons. By selecting one of the arrow icons, the user communicates his navigation decision to the system which alters the layout respecting the formalized constraints. A prototype assistance system based on neighborhood navigation has been implemented. It serves as a testbed for evaluating the different sets of navigation operations that result from using definitions for instantiation neighborhood of different strength. Which notion of neighborhood minimizes cognitive complexity is an empirical question. However, the analysis of the cognitive complexity of partially unformalized constraint systems presented in this paper provides the conceptual framework for studying the issue.

7 Conclusion

The analysis of the cognitive complexity of navigation in instantiation space implies that the user should be given a number of navigation choices that does not increase more than linearly with the number of layout boxes. One way to implement this principle is to allow the user to select a box and to let the system generate an appropriate presentation sequence. This raises the issue of efficiently controlling the order in which a path-consistency constraint solver computes instantiations. In our prototype system, this is done by using a best-first strategy for ordering variable-value assignments. Future work will move beyond single presentation sequences and try to formulate heuristics which allow to decompose a connected component of the instantiation space into sequences of presentation sequences.

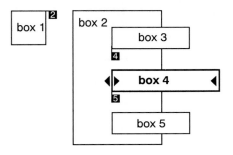

Fig. 8 Visualization of navigation operations

We conclude with a discussion of the limitations of local navigation and a brief sketch of how to overcome them. The most obvious limitation is due to the fact that, in general, the instantiation space consists of several connected components, not just one as in our example. Local navigation allows to navigate within a component, but since it moves only to conceptually neighboring instantiations, it will never reach any other component. A second limitation of local navigation consists in its relative slowness. With a single interaction step, the user can move only at unit distance from the present position in the navigation graph. It seems natural to provide the layout assistance system in addition to local navigation operations with operations for *global navigation*. Such operations should permit the user to move with a single interaction step distances $d > 1$ including the case $d = \infty$ which corresponds to a movement between connected components.

As we have seen, local navigation can be realized on the interface level by simply selecting a layout box. Many interactive constraint systems provide for a more complex type of graphical interaction such as the dragging and dropping of a layout box (e.g. [2], [7], [11]). Although these systems are designed for iteratively finding a solution for a strongly constrained layout task, global interactions (drag and drop) are useful also in the context of weakly constrained tasks. Selection of a layout box implements navigation within instantiation space while dragging can temporally cause navigation to leave instantiation space and to enter configuration space, i.e. the unconstrained configurations of layout boxes. *Configuration space navigation* raises a computational problem that was not present in *instantiation space navigation*. When the user produces a configuration that violates some layout constraints, then the layout assistance system must generate the instantiation that lies closest to the configuration under a suitable similarity metric on the configuration space.

Since the system's behavior becomes more complex, we expect that it is more difficult for the user to mentally anticipate the effects of his actions in configuration space navigation than in instantiation space navigation. More precisely, configuration space navigation requires the user to make a kind of analogical inference. The user knows the effect of a navigation operation n (dragging a particular box to a particular position) on a specific instantiation i_1: it causes the system to produce instantiation i_2. Currently, the system is displaying instantiation i_3 and the user has to infer what instantiation i_4 the system produces when n is applied. In other words, the user has to solve an analogy of the type $i_1 : i_2 = i_3 : i_4$. We plan to study this type of analogical inference in its connection of global navigation strategies in the near future.

References

1. J. Allen, Maintaining knowledge about temporal intervals, *Communications of the ACM*, 26, pp. 832-843, 1983.
2. A. Borning, Graphically defining new building blocks in ThingLab, In: E. Glinert (ed.) *Visual programming environments: paradigms and systems*, IEEE Computer Societey Press: Los Alamitos, CA, 450-469, 1990.
3. R. Coyne, M. Rosenman, A. Radford, M. Balachandran and J. Gero, *Knowledge-based design systems*, Addison-Wesley, Reading MA, 1990.
4. A. G. Cohn, Qualitative spatial representation and reasoning techniques. In: Proceedings KI-97: Advances in Artificial Intellligence, Springer, Berlin, pp. 1-30, 1997.
5. J. Evans, S. Newstead and R. Byrne, *Human reasoning: The psychology of deduction*, Lawrence Erlbaum, Hillsdale, NJ, 1993.
6. C. Freksa, Temporal reasoning based on semi-intervals, Artificial Intelligence, 54, pp. 199-227, 1992.
7. W. Graf, A. Kroender, S. Neurohr and R. Goebel, Experience in integrating AI and constraint programming methods for automated yellow pages layout, *Künstliche Intelligenz*, 2, 79-85, 1998.
8. L. Hoebel, W. Lorenzen and K. Martin, Integrating graphics and abstract data to visualize temporal constraints, *SIGART Bulletin*, 9, 18-23, 1998.
9. P. Johnson-Laird and R. Byrne, *Deduction*. Lawrence Erlbaum, Hillsdale, NJ, 1991.
10. M. Knauff, R. Rauh and C. Schlieder, Preferred mental models in qualitative spatial reasoning: A cognitive assessment of Allen´s calculus. In *Proceedings Conference of the Cogntive Science Society*, Lawrence Erlbaum, Mahwah, NJ, pp. 200-205, 1995.
11. D. Kurlander and S. Feiner, Inferring constraints from multiple snapshots, *ACM Transactions on Graphics*, 12, 277-304, 1993.
12. P. Ladkin and A. Reinefeld, Fast algebraic methods for interval constraint problems. *Annals of Mathematics and Artificial Intelligence*, 19, 1997.
13. B. Nebel and H. Bürckert, Reasoning about temporal relations: A maximal tractable subclass of Allen´s interval algebra. *Journal of the ACM*, 42, pp. 43-66, 1995.
14. R. Rauh and C. Schlieder, Symmetries of model construction in spatial relational inference, In *Proceedings Conference of the Cogntive Science Society*, Lawrence Erlbaum, Mahwah, NJ, pp. 638-643, 1997.

15. C. Schlieder and B. Berendt, Mental model construction in spatial reasoning: A comparison of two computational theories. In U. Schmid, J. Krems and F. Wysotzki (Eds.). *Mind modelling: A cognitive science approach to reasoning, learning and discovery*. Pabst Science Publishers, Berlin, 1998.

16. C. Schlieder, Diagrammatic transformation processes on two-dimensional relational maps, *Journal of Visual Languages and Computing*, 9, pp. 45-59, 1998.

17. L. Weitzman and K. Wittenburg, Grammar-based articulation for multimedia document design, *Multimedia Systems Journal*, 4, pp. 99-111, 1996.

Inference and Visualization of Spatial Relations*

Sylvia Wiebrock, Lars Wittenburg, Ute Schmid, and Fritz Wysotzki

Methods of Artificial Intelligence, Technical University of Berlin
FR 5–8, Franklinstraße 28/29, D 10587 Berlin
sppraum@cs.tu-berlin.de
http://ki.cs.tu-berlin.de/~sppraum

Abstract. We present an approach to spatial inference which is based
on the procedural semantics of spatial relations. In contrast to qualitative reasoning, we do not use discrete symbolic models. Instead, relations between pairs of objects are represented by parameterized homogeneous transformation matrices with numerical constraints. A textual
description of a spatial scene is transformed into a graph with objects
and annotated local reference systems as nodes and relations as arcs.
Inference is realized by multiplication of transformation matrices, constraint propagation and verification. Constraints consisting of equations
and inequations containing trigonometric functions can be solved using
machine learning techniques. By assigning values to the parameters and
using heuristics for the placement of objects, a visualization of the described spatial layout can be generated from the graph.

1 Introduction

Understanding descriptions of spatial layouts implies that questions about spatial relations which were not included in the description can be answered, that
the described scene can be visualized, or that it can be judged whether a depiction corresponds to the described layout. We propose an AI approach which
realizes the first two claims: inference of spatial relations and construction of
visualizations. From an application perspective, understanding of linguistic descriptions is a step towards a more natural human-machine communication, for
example in the domains of robot instructions (*take the box from the* **right** *table*)
and graphical user interfaces (*put the mailbox* **next to** *the file-manager window*). Furthermore, it might lead to a tool for visualization of textual descriptions (Jörding and Wachsmuth 1996). From a cognitive science perspective, our
approach might provide a formal backbone to the theory of mental models in
text understanding (Johnson-Laird 1983; Claus et al. 1998). Mental models are
proposed to be the representation of a described situation that is constructed in
addition to a propositional representation.

* This research was supported by the Deutsche Forschungsgemeinschaft (DFG) in the
project "Modelling Inferences in Mental Models" (Wy 20/2–2) within the priority
program on spatial cognition ("Raumkognition").

Ch. Freksa et al. (Eds.): Spatial Cognition II, LNAI 1849, pp. 212–224, 2000.

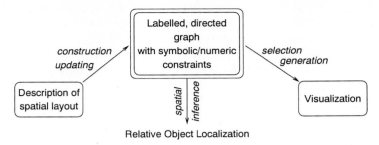

Fig. 1. Overview of the approach

In contrast to qualitative approaches to spatial reasoning, our approach is not based on discrete, symbolic models. Instead, we propose a metric approach with an underlying procedural semantics. While Clementini et al. (1997) demonstrate the advantages of qualitative reasoning, there are some critical issues. Finding the granularity appropriate for the problem and representing object orientations necessary to handle *intrinsic* relations are some of them. Besides, results presented in Musto et al. (2000) indicate that information loss during relation composition can be unacceptably high for long chains of qualitative inferences. In their paper, they use a mixed approach: measured metric values are abstracted to qualitative ones. In the example given, they delay the discretization to overcome this problem, and first gather the metric values for some steps before switching to qualitative values.

An overview of our quantitative approach is given in Fig. 1 (see also Claus et al. (1998); Schmid et al. (1999) for a more detailed description). Input into the system is the linguistic description of a spatial layout as a sequence of propositions. From this description, a graph is constructed incrementally: each object introduced in the text is represented as a node which is associated with a canonical coordinate system. The origin of the coordinate system represents the geometric center of the object. For objects with intrinsic axes, the positive x-axis represents the intrinsic right, the positive y-axis the intrinsic front and the positive z-axis the intrinsic top. Extensions of objects are represented by variables for their height, and their width and depth, or radius, respectively. Currently, objects are restricted to cuboids and cylinders.

Relations between pairs of objects A and B are represented by constraints for their transformation matrices (Ambler and Popplestone 1975). A homogeneous 4×4 transformation matrix contains a vector necessary to "translate" the origin of an object A to the origin of an object B, and a rotational part that aligns A's coordinate system with B's. The constraints restrict parameters of the transformation. For example, to represent *right(A, B)*, "B is intrinsically right of A", the x-value of the translation vector has to be at least half the width of A plus half the width of B to ensure nonoverlapping. Further constraints might restrict the admissible deviation on the front-back axis, for example with respect to a defined sector for *right* (Hernández 1994, see Fig. 4(a)). Note that constraints are equations and inequations defined on *parameterized* relational representations. Each given relation is introduced as a labelled directed arc into the graph. Formally, the resulting graph represents the class of all models satisfying the given constraints. That is, our formalization of a mental model, i.e. the graph as a relational description together with the constraints on its parameters, describes *sets* of real spatial situations. This corresponds to the fact that language is under-determined with respect to depictions and to the claim that mental models represent classes of *possible* situations (Johnson-Laird 1983). Note that as a consequence the inference mechanism manipulates implicitly sets of possible real situations satisfying the constraints (i.e. the given relations)!

To infer relations between objects which were not given in the description, a path between these objects has to be found and the matrices along this path have to be multiplied. The resulting matrix with the constraints on its elements can then be compared with the predefined spatial relations. If it satisfies the constraints for such a relation, the relative position of the objects can be verbalized, otherwise it can at least be visualized. As is well known in robotics, transformations involve trigonometric functions in general so that known methods for constraint solving cannot be applied for inference. For this general case we propose to use a machine learning technique to induce the resulting relations from training examples.

Generating a visualization of the described scene involves new problems, because we have to find a globally consistent instantiation of all variables in the graph that also satisfies the general constraint (not included in the graph) that objects must be pairwise nonoverlapping. In the following, we will present our approach to spatial inference, machine learning, and visualization in more detail.

2   Inference

Inferring the spatial relation between two objects consists in propagating the constraints with respect to the positions of those objects and then checking which (if any) of the defined relations hold. The most important questions are *which* relations have to be inferred and *when* to infer them. This depends largely on the assumptions we make of the mental model. Some possibilities are discussed in Claus et al. (1998), Hörnig et al. (2000), and Hörnig et al. (1999).

While these questions are open, we organize the graph according to efficiency considerations. The graph contains all the relations explicitly given plus the arcs for the background knowledge that every object is in the room (thus ensuring that the graph is always connected), and those arcs that have been inferred after a query from the user. Inference is done when the relation between A and B is explicitly queried by the user and there is no arc between them in the graph, or when a relation between A and B is explicitly given and both objects are already represented in the graph. In the latter case, we start the inference process to ensure *that the constraints on the (new) arc contain all the information available about the relative positions of A and B*. As the graph is connected, we can find a path between A and B, compute the transformation matrix for it, and equate with the matrix for the given relation.

The problems concerning constraint handling are due to the fact that the constraints consist of equations and inequations with parameters and trigonometric functions. Furthermore, we cannot always compare two constraints for the same variable. Therefore, the inference process often consists in gathering constraints for the variables rather than computing the intersection of those constraints and simplifying the problem. The occurrence of trigonometric functions means that for *unknown* rotations, algebraic methods cannot in general be applied to decide the satisfaction of the constraint inequations.

In our first approach described in Claus et al. (1998), sizes of objects were only constrained by the relations, e.g. in Fig. 3, we know that $2C.d + S.r \leq \Delta_x^{(S,W1)}$. Most constraints were purely symbolic. Considering that we need to find "realistic" values for the visualization (cf. section 3), we have augmented the object definitions to include default values and admissible intervals for the extensions. If we also use default values for object orientations, checking constraint satisfaction can often be done by case analysis using instantiations of the variables. This is shown in the example below. For the general case we have experimented with machine learning programs. Some results are presented in Sec. 2.2.

2.1   Example for the Inference of Spatial Relations

To keep things simple, consider the objects and relation definitions for the 2D case shown in Fig. 2: persons are represented by circles with a radius $r \in [0.2, 0.4]$, fridges are rectangles with a width $w \in [0.3, 0.4]$ and a depth $d \in [0.3, 0.4]$, cupboards are rectangles with a width $w \in [0.4, 1.0]$ and a depth $d \in [0.2, 0.5]$, lamps are circles with a radius $r \in [0.1, 0.5]$, and the room (not shown in the figure) is a rectangle with a width $w \in [1, 5]$ and a depth $d \in [1.5, 5]$. Persons, fridges, and cupboards have intrinsic front and right sides[1], while the orientation of the lamp's coordinate system is arbitrary.

For this example, the only defined relations are those for the four basic directions, where the center points of the located object must be on the corresponding axis in the coordinate system of the relatum, and the additional relation *at_wall*, where the default is to place an object directly against the wall, but inference of the relation is also possible when the distance is smaller than 0.5. The default orientation for objects standing at the wall is to orient their front sides (if they have an intrinsic front) away from it. The coordinate system for the walls are chosen such that the y-axes are oriented counterclockwise along the walls, while the positive x-axes always point outside the room (see Fig. 2).

Suppose we are given the propositions

(1) *right(S,C)* ("The cupboard C stands right of Stefanie S.")

(2) *at_wall(W1,C)* ("The cupboard stands at the wall $W1$.", (1) and (2) are the linearization of "The cupboard stands at the wall to the right of Stefanie.")

---

[1] The question of handedness is left out in this example.

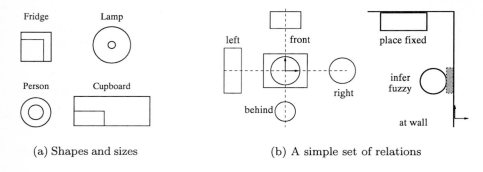

(a) Shapes and sizes   (b) A simple set of relations

Fig. 2. Example for objects and relations

(3) *front(S,F)* ("The fridge F stands in front of Stefanie.")

(4) *at_wall(W2,F)* ("The fridge stands at the wall $W2$.", (3) and (4) are the linearization of "The fridge stands at the wall in front of Stefanie." Thus we know that $W1 \neq W2$.)

(5) *left(F,L)* ("The lamp L stands left of the fridge.")

(6) *right(C,L)* ("The lamp stands right of the cupboard.")

Each of the first five propositions introduces a new object. Thus there cannot already be a path in the graph between the two objects, and no inference is necessary. The constraints for the first five propositions are

$$\Delta_x^{(S,C)} \geq S.r + C.d \qquad \Delta_y^{(S,C)} = 0 \qquad \theta^{(S,C)} = 90°$$

for (1) with the default orientation: front side away from the wall.

$$\Delta_x^{(W1,C)} = -C.d \qquad \text{default placement directly at the wall for (2)}$$
$$\Delta_y^{(S,F)} \geq S.r + F.d \qquad \Delta_x^{(S,F)} = 0 \qquad \theta^{(S,F)} = 180°$$

for (3) with the default orientation: front side away from the wall.

$$\Delta_x^{(W2,F)} = -F.d \qquad \text{default placement directly at the wall for (4),}$$
$$\text{and} \qquad \Delta_x^{(F,L)} \leq -F.w - L.r \qquad \Delta_y^{(F,L)} = 0 \qquad \theta^{(F,L)} = 0°$$

default orientation aligned with the fridge as the relatum, because the lamp has no intrinsic sides, for (5). When adding proposition (6), however, there is already a path in the graph (see Fig. 3).[2] We must now compute the transformation matrix for the path $L \rightarrow F \rightarrow S \rightarrow C$.[3] Equating the transformation matrix for the relation *right(C,L)* with the computed matrix we get

---

[2] The node for the room and the arcs to the room node are not shown.

[3] In the graph there is only an arc C → S, but arcs can easily be inverted.

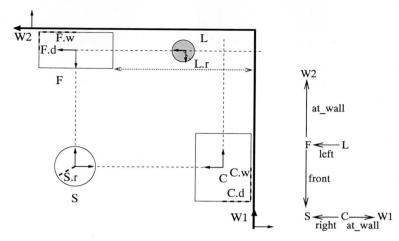

Fig. 3. Situation after propositions (1) to (5)

$$\begin{pmatrix} \cos\theta & -\sin\theta & \Delta_x^{(C,L)} \\ \sin\theta & \cos\theta & 0 \\ 0 & 0 & 1 \end{pmatrix} = \begin{pmatrix} 0 & -1 & \Delta_y^{(S,F)} \\ 1 & 0 & \Delta_x^{(F,L)} + \Delta_x^{(S,C)} \\ 0 & 0 & 1 \end{pmatrix}$$

Therefore we know that $\theta = 90°$, $\Delta_x^{(C,L)} = \Delta_y^{(S,F)}$, and $\Delta_x^{(F,L)} = -\Delta_x^{(S,C)}$ and can substitute for the variables on the left sides. These constraints express the fact that the lamp must be positioned on the point where the left/right axis of C's coordinate system crosses the left/right axis of F's coordinate system.

Another situation arises, when a query is given to the system. For instance, we might ask the question $at\_wall(W?,L)$. This means we have to check for every wall, whether the lamp is standing at it. This is done by finding a path between the lamp and the corresponding wall, computing the transformation matrix and then checking whether the constraints for the relation at_wall are satisfied.

For the wall $W1$ we can compute $\qquad T^{(W1,L)} = T^{(W1,C)} \times T^{(C,L)}$.
This matrix consists of a rotation matrix for 180°,
$$\Delta_x^{(W1,L)} = -C.d \qquad\qquad \text{and} \qquad\qquad \Delta_y^{(W1,L)} = \Delta_y^{(S,F)} + \Delta_y^{(W1,C)}$$
As shown in Fig. 2(b), the only constraint is $\quad \Delta_x^{(W1,L)} \geq -L.r - 0.5$,
i.e. the distance of the lamp w.r.t. the wall must be at most 0.5.
Substituting for $\Delta_x^{(W1,L)}$ gives
$$-C.d \geq -L.r - 0.5 \qquad\qquad \Leftrightarrow \qquad\qquad L.r \geq C.d - 0.5.$$
If we assign $C.d$ its maximal value, we get $\qquad L.r \geq 0.5 - 0.5 = 0$
and from $L.r > 0$, we can conclude that the relation holds for all possible values of $C.d$ and $L.r$. Thus we know that $at\_wall(W1,L)$ is true.
Analogously, we can prove that $at\_wall(W2,L)$ holds.

Now consider the case $at\_wall(W3,L)$. Using the background knowledge (a room has 4 walls, $W1$ is opposite $W3$), we can compute

$$T^{(W3,L)} = T^{(W3,W1)} \times T^{(W1,L)} \qquad \text{(one possible path)}.$$

We get a rotation matrix for $0°$,

$$\Delta_x^{(W3,L)} = C.d - 2R.w, \qquad \text{and} \qquad \Delta_y^{(W3,L)} = -\Delta_y^{(S,F)} - \Delta_y^{(W1,C)},$$

where $R.w$ is the width of the room. Again, we have to check whether

$$\Delta_x^{(W3,L)} \geq -L.r - 0.5. \qquad \text{Substituting for } \Delta_x^{(W3,L)} \text{ gives}$$

$$C.d - 2R.w \geq -L.r - 0.5 \qquad \Leftrightarrow \qquad L.r \geq 2R.w - C.d - 0.5.$$

Checking for the upper bound of the right expression, we get

$$L.r \geq 10 - 0.2 - 0.5 = 9.3.$$

This cannot be true, therefore we know that the lamp *need not* stand at $W3$. To complete, we check for the lower bound of the right expression and get

$$L.r \geq 2 - 0.5 - 0.5 = 1.$$

This is false, too. Thus we know that the lamp *cannot* stand at wall $W3$. The computation for $W4$ is analogous.

2.2   Machine Learning

In the general case, the constraints for an intrinsic relation between two objects contain the extensions of both objects, the elements of the translation vector, and the elements of the rotation matrix. Even in the restricted case, where all objects stand upright, and rotation is only allowed around the z-axis, the constraints for a relation involve seven variables. For example, the binary relation $right(x,y)$ is transformed into a 7-dimensional constraint, represented by a region in the 7-dimensional parameter space. What we are really interested in is the border in this space between the region(s) where the relation holds and the inadmissible regions where it does not hold. An approximation of this border can be found by machine learning programs.

In our case, such programs are fed a set of training data each consisting of a 7-dimensional parameter vector, which uniquely describes a spatial constellation of the two objects, and a class value (e.g. R for right and N for not right, see Fig. 4(a)). From these training samples, a classifier is constructed that can decide for arbitrary constellations whether an object falls into the *right* region of another object or not. Using this classifier is in general faster than checking the (in)equations for the parameter values for high dimensional parameter spaces (in the simple example below, there are 12 inequations). For the learning task, in principle any standard algorithm for classification learning, which separates class regions in a continuous parameter space, can be applied. As high accuracy in approximating the class boundaries is wanted in our case, an artificial neuronal net of the Perceptron type was preferred. We have used *Dipol92* (Schulmeister and Wysotzki 1997), a hybrid (statistical/neural) algorithm that computes for a given training set and given numbers of clusters for every class, e.g. R and N (the numbers of clusters are the only parameters to be chosen by the user), the discriminating hyperplanes for each pair of clusters.

By decomposing each class into a set of clusters, a piecewise linear approximation of the class boundaries is computed, which can be converted into a decision function for the constraint satisfaction problem, which could then be used in the inference process.

A critical issue is the distribution and number of training data to be generated to obtain satisfactorily low error rates. The task was to try whether we can get smaller errors while using fewer data for the training set when we *selectively* generate data near the boundary to be learned. For the sample problem shown below, we had admissible intervals for all seven parameters, and also preferred intervals where more data had to be generated. This was motivated by the assumption that medium sized objects occur more often (i.e. have higher probability) than very small or very big ones, and the distance values $Dx(A, B)$ can only reach their maximum when both objects are positioned at opposite walls. The seven parameter values were then generated independently (i.e. a small $A.w$ value can be paired with a large $A.d$ value), and the class value (R or N) was computed using the (in)equations defining the constraint (relation) *right*. As *right* is defined such that B must be completely included in the *right* region of A, only $\approx 20\%$ of the data were in class R when data were generated randomly, but respecting the given admissible and preferred intervals for each parameter. This method was called REF. But since our task is learning class boundaries by exploration ("active learning"), i.e. by *generating* training sets, it is desirable to have equal a priori probabilities for both classes to get optimal decision functions (classifiers).

This problem is discussed in Stolp et al. (1999). Among other methods we tried out, EPS2 was best. Here, data are generated according to the REF method and then filtered. This means, only those data vectors $(A.w, A.d, B.w, B.d, Dx(A, B), Dy(A, B), \theta)$, where changing any one parameter value by $+\epsilon$ or $-\epsilon$

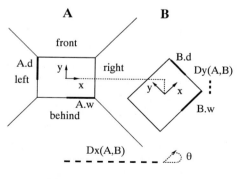

(a) *Right* for two rectangles

Data		train	test
gen.	ϵ	error	error
1000 points for training			
REF	/	0.50%	6.64%
EPS2	1.5	1.20%	3.24%
2500 points for training			
REF	/	1.36%	5.20%
EPS2	1.5	5.28%	2.66%
25000 points for training			
REF	/	3.04%	3.75%
EPS2	1.5	8.10%	1.93%

(b) Results with Dipol

Fig. 4. Learning *right* for two rectangles with Dipol

changed the class (from N to R or vice versa) were kept. This has the advantage that all values are near the critical border, both classes get the same number of training samples, and still the preferred intervals for every variable are respected. An appropriate value for ϵ was found in some training runs. The number of clusters for each class was set to 50 for both R and N. The table in Fig. 4(b) shows the train errors (errors on the data set used for training) and test errors (errors on a different test set of data for the learned classifier). For testing, a set with 100 000 points generated with REF was used throughout. As can be seen in the table, the generalization is quite good for both methods. For reasonably large sets of training data, the errors on the training set and on the test set are nearly equal for REF. Due to the fact that all data points are near the border, EPS2 has higher errors on the training set, but surprisingly small errors on the test set. Our main goal, to achieve good error rates with a small number of training data, was fulfilled. Using EPS2 for generating data, we got a lower test error with 1000 training vectors than by using REF with 25.000 training vectors. A more detailed account is given in Stolp et al. (1999) and Wiebrock and Wysotzki (1999). Currently, we are working on the generation of depictions using the learned regions where the relations hold. Depending on the efficiency of this method, we may use it in the future as an alternative to the algorithm described below.

3   Generating Depictions

When generating depictions, we are faced with several problems: all variables in the graph must be instantiated, in addition to the explicitly represented constraints we must ensure that two objects do not overlap, and we must choose a perspective for the visual representation. The last point leads back to the organization of the mental model (see Sec. 2). As we are using VRML to visualize the scenes, we have to compute positions in a global coordinate system for every object. The perspective is set afterwards and can be changed interactively. We generate the depiction parallel to the graph, i.e. immediately update the depiction for every new proposition. For every object, default sizes are given, and for every relation, default positions are provided.

Classical constraint propagation algorithms like Waltz (1975) work for a finite number of possible instantiations for each variable. The algorithm for Parcon, a program for handling objects in GUIs described in Uhr et al. (1997), is a variant for intervals over real numbers. The algorithm handles disjunctive constraints by copying the constraint network and working in parallel on all copies. For our domain, where we have lots of disjunctive constraints (i.e. nonoverlapping), this would not be efficient, because the number of models to check grows exponentially in the number of objects. Therefore we propose a simpler variant involving backtracking. We can divide the variables occuring in the constraints into three classes: 1) object sizes, 2) relative object rotations, and 3) distance parameters. The order is meant to indicate the "variability" of the values: object sizes are least likely to be changed, and distance values are most variable. In the im-

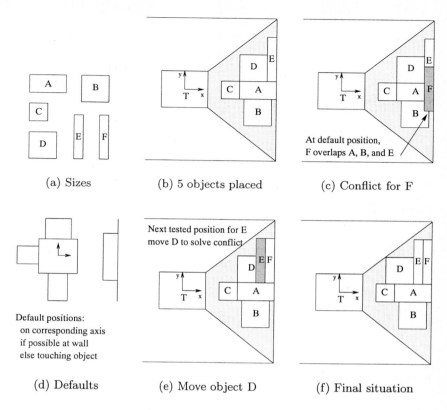

Fig. 5. Example problem with six objects

plemented algorithm outlined in Tab. 1, only distance values are manipulated. Rotations are restricted to multiples of 90° and are fixed.

For every object, we know its size and coordinates, the constraints for its position, and the objects the position depends on. We also keep a list *posl* of possible positions. If an object is to be placed, we first try the default (= preferred) position. If there is a collision, we move the object front/left/back/right by the amount necessary to avoid the collision and add those positions to *posl*. For the actual position, we store the name(s) of the colliding object(s) and mark the position as visited. For the colliding objects, too, we try wether by moving them we find a new position to be tested. To avoid cycles, object positions are stored. We consider positions equal if they differ by less than a given ϵ value. This limits the number of possible positions and ensures termination while for extreme examples we don't find all possible solutions.

As an example for the algorithm, consider the following problem: there are six objects A to F. Each object is right of T, i.e. must be placed in the grey area.

Table 1. Overview of the placing algorithm

type Obj = {name, extensions, **LINE list** *bounds, **Obj list** *dep_obs}
         /* bounds: list of lines bordering the admissible area
             dep_obs: objects with parameters mentioned in bounds */
type Pos = {x, y, **Obj list** *coll} /* coll: objects colliding at position (x,y) */
type Scene = {**Obj** *obj, **Pos** *act_pos, **Pos** *pref_pos,
             Obj list *coll, **Pos list** *posl } /* posl is initialized to pref_pos */
             /* colliding objects at act_pos, and alternative positions posl */
type Config = {**Scene list** *placed, **Scene list** *add}
bool place_obj (**Scene list** *placed, **Scene** *curr)
/* try to add new object without repositioning objects already placed */
{ **repeat** collision := false;
     pos := first unvisited position in curr→posl;
     forall obj in placed overlapping curr at pos **do**
     { move curr front/left/back/right to avoid conflict and add positions to curr→posl;
        move obj front/left/back/right to avoid conflict and add positions to obj→posl;
        /* positions are only added if they are not already on the list*/
        add obj to curr→coll; collision := true; }
     mark pos as visited;
   until not collision **or** no more unvisited positions in curr→posl;
   return collision; }
bool place_list (**Scene list** *placed, **Scene list** *add)
/* try to place each object in add without repositioning objects already placed */
{ curr := first object in add;
   repeat success := true;
     if place_obj (placed, curr)
     { move curr from add to placed;
        if not place_list (placed, add)
        { mark act_pos in curr as visited; success := false
           move curr from placed to add; } }
     else success := false;
   until success **or** no more unvisited positions in curr→posl;
   return success; }
bool place_all (**Scene list** *placed, **Scene list** *add, **Config list** *all)
/* try to place each object in add; if necessary objects are removed from placed list;
   all tested configurations are stored to avoid loops */
{ conf = {placed, add};
   repeat success := true;
     if not place_list(conf.placed, conf.add)
     { unplaced := first element of conf.add; success := false;
        forall coll in unplaced→posl **do**
        { new_conf := conf with all scenes for objects in coll moved from placed to add;
           insert new_conf into all such that all is sorted by expected effort; }
        /* least number of remaining objects, shortest collision list */
        conf := next element of all; }
   until success **or** not conf; /* no more untested configurations */
   if success { placed := conf.placed; add := conf.add; }
   return success; }

Additional constraints are: *behind(A,B)* (B is behind A) and *left(A,C)* (C is left of A). Suppose that all objects have the same orientation as T. Object sizes are shown in Fig. 5(a) and default positions in Fig. 5(d). When objects A to E are placed, we have the situation shown in Fig. 5(b). Now trying to place F at the default position gives the conflict shown in Fig. 5(c). To resolve the conflict, object E is removed, because A and B are interdependent. F is positioned at E's place. Now we have to re-place E. The next position tried is to the left of its former position, see Fig. 5(e). This gives a conflict with D. D is moved by the width of E, and E is included, see Fig. 5(f). To compute the scene depicted here, there were between 29 and 73 positions stored for the objects (314 for the six objects together), and 174 constellations were tested.

4   Conclusions

In the text above, we have considered three different representations of a situation: the propositional description of the spatial constellation, the graph as the structural representation constructed from it, and a visual representation of the situation. Models of spatial representation based on the work of Kosslyn, e.g. Schlieder and Behrendt (1998), postulate that beside the mental *model* a mental *image* is constructed that can be used for inspection. Unlike the mental model, this image is necessarily bound to a fixed perspective. This can make updating the model harder when the texts induce a change of perspective. Some results discussed in Hörnig, Eyferth, and Gärtner (2000) indicate that the visual representation is only available as long as the perspective does not change. Possibly the cognitive effort for updating both representations simultaneously would be too high, otherwise.

For our AI model, the question of a global reference frame is relevant for the updating and inference process where we have to decide which relations are represented explicitly in the graph and available without inference. Several possibilities are discussed in Hörnig et al. (1999), and Hörnig et al. (2000). Choosing a perspective for the generated depiction is a related problem. Another one concerns finding good heuristics for the instantiation of variables, especially with respect to the plausibility of the resulting depiction. While there is some evidence for the admissible regions and default positions for simple spatial relations, it is still unclear how to generalize those results for complex scenes.

References

Ambler, A. P. and R. J. Popplestone (1975). Inferring the Positions of Bodies from Specified Spatial Relationships. *Artificial Intelligence 6*, 157–174.

Claus, B., K. Eyferth, C. Gips, R. Hörnig, U. Schmid, S. Wiebrock, and F. Wysotzki (1998). Reference Frames for Spatial Inferences in Text Comprehension. In C. Freksa, C. Habel, and K. F. Wender (Eds.), *Spatial Cognition - An Interdisciplinary Approach to Representing and Processing Spatial Knowledge*, Springer.

Clementini, E., P. D. Felice, and D. Hernández (1997). Qualitative representation of positional information. *Artificial Intelligence 95*, 317–356.

Hernández, D. (1994). *Qualitative Representation of Spatial Knowledge*. Springer.

Hörnig, R., B. Claus, and K. Eyferth (1999). In Search for an Overall Organizing Principle in Spatial Mental Models: A Question of Inference. In S. O'Nuallain and M. Hagerty (Eds.), *Spatial Cognition; Foundations and Applications*. John Benjamins. Forthcoming.

Hörnig, R., K. Eyferth, and H. Gärtner (2000). Orienting and Reorienting in Egocentric Mental Models. This volume.

Johnson-Laird, P. N. (1983). *Mental Models: Towards a Cognitive Science of Language, Inference and Consciousness*. Cambridge: Cambridge University Press.

Jörding, T. and I. Wachsmuth (1996). An Antropomorphic Agent for the Use of Spatial Language. In *Proceedings of ECAI'96-Workshop on Representation and Processing of Spatial Expressions*, S. 41–53.

Musto, A., K. Stein, A. Eisenkolb, T. Röfer, W. Brauer, and K. Schill (2000). From Motion Observation to Qualitative Motion Representation. This volume.

Schlieder, C. and B. Behrendt (1998). Mental model construction in spatial reasoning: a comparison of two computational theories. In U. Schmid, J. Krems, and F. Wysotzki (Eds.), *Mind modelling: a cognitive science approach to reasoning, learning, and discovery*, Lengerich, S. 133–162. Pabst Science Publishers.

Schmid, U., S. Wiebrock, and F. Wysotzki (1999). Modelling Spatial Inferences in Text Understanding. In S. O'Nuallain and M. Hagerty (Eds.), *Spatial Cognition; Foundations and Applications*. John Benjamins. Forthcoming.

Schulmeister, B. and F. Wysotzki (1997). DIPOL – A Hybrid Piecewise Linear Classifier. In G. Nakhaeizadeh and C. C. Taylor (Eds.), *Machine Learning and Statistics - The Interface*, S. 133–152. Wiley.

Stolp, R., B. Weber, M. Müller, S. Wiebrock, and F. Wysotzki (1999). Zielgerichtete Trainingsmethoden des Maschinellen Lernens am Beispiel von DIPOL. In P. Perner (Ed.), *Maschinelles Lernen, FGML'99*. IBaI Report.

Uhr, H., P. Griebel, M. Pöpping, and G. Szwillus (1997). Parcon: ein schneller Solver für grafische Constraints. *KI 97*(1), 40–45.

Waltz, D. (1975). Understanding Line Drawings of Scenes with Shadows. In P. H. Winston (Ed.), *Psychology of Computer Vision*. New York: McGraw-Hill.

Wiebrock, S. and F. Wysotzki (1999). Lernen von räumlichen Relationen mit CAL5 und DIPOL. Fachberichte des Fachbereichs Informatik No. 99-17, TU Berlin.

A Topological Calculus for Cartographic Entities [1]

Amar Isli[1], Lledó Museros Cabedo[2], Thomas Barkowsky[1], and Reinhard Moratz[1]

[1] University of Hamburg, Department for Informatics,
Vogt-Kölln-Str. 30, 22527 Hamburg, Germany
{isli,barkowsky,moratz}@informatik.uni-hamburg.de

[2] Universitat Jaume I, Departamento de Informática,
Campus del Riu SEc, TI-1126-DD, Castellon, Spain
museros@inf.uji.es

Abstract. Qualitative spatial reasoning (QSR) has many and varied applications among which reasoning about cartographic entities. We focus on reasoning about topological relations for which two approaches can be found in the literature: region-based approaches, for which the basic spatial entity is the spatial region; and point-set approaches, for which spatial regions are viewed as sets of points. We will follow the latter approach and provide a calculus for reasoning about point-like, linear and areal entities in geographic maps. The calculus consists of a constraint-based approach to the calculus-based method (CBM) in (Clementini et al., 1993). It is presented as an algebra alike to Allen's (1983) temporal interval algebra. One advantage of presenting the CBM calculus in this way is that Allen's incremental constraint propagation algorithm can then be used to reason about knowledge expressed in the calculus. The algorithm is guided by composition tables and a converse table provided in this contribution.

1   Introduction

Geographic maps are generally represented as a quantitative database consisting of facts describing the different regions of the map. For many applications, however, the use of such maps is restricted to the topological relations between the different regions; in other words, such applications abstract from the metric considerations of the map.

Topological relations are related to the connection between spatial objects. Their important characteristic is that they remain invariant under topological transformations, such as rotation, translation, and scaling. A variety of topological approaches to spatial reasoning can be found in the literature, among which the following two trends:

1. Approaches for which the basic spatial entity is the *spatial region*. These approaches consider spatial regions as non-empty, regular regions; therefore, points,

---

[1] This work is supported by the Deutsche Forschungsgemeinschaft (DFG) in the framework of the Spatial Cognition Priority Program (grants Fr 806-7 and Fr 806-8).

Ch. Freksa et al. (Eds.): Spatial Cognition II, LNAI 1849, pp. 225-238, 2000.

lines, and boundaries cannot be considered as spatial regions (Randell et al., 1992; Gotts, 1996; Bennett, 1994; Renz & Nebel, 1998). The basic relation for these is $C(x,y)$, i.e. x connects with y. $C(x,y)$ holds when the topological closures of x and y share a point. Using the C relation, the approaches define a set of atomic topological relations, which are generally mutually exclusive and complete: for any two regions, one and only one of the atomic topological relations holds between them. One important theory in this group is the *region connection calculus* (RCC) developed in (Randell et al., 1992).

2. Approaches which consider a region as a set of points. The basic entities for these approaches consist of points, lines and areas. The topological relations between regions are defined in terms of the intersections of the interiors and boundaries of the corresponding sets of points (Egenhofer & Franzosa, 1991; Pullar & Egenhofer, 1988; Egenhofer, 1991; Clementini & Di Felice, 1995). Each approach has its own set of atomic topological relations.

The aim of this work is to develop a calculus suitable for reasoning about topological relations between point-like, linear, and areal cartographic entities. We cannot follow the first trend since it does not consider point-like and linear entities as regions. We will indeed follow the second trend.

An important work in the literature on reasoning about topological relations of cartographic entities is the CBM calculus (*calculus-based method*) described in (Clementini & Di Felice, 1995), originating from Egenhofer's work on intersections of the interiors and boundaries of sets of points (Egenhofer & Franzosa, 1991; Pullar & Egenhofer, 1988; Egenhofer, 1991). The objects (features) manipulated by the calculus consist of points, lines and areas commonly used in *geographic information systems* (GIS); that means that all kinds of features consist of closed (contain all their accumulation points) and connected (do not consist of the union of two or more separated features) sets. The knowledge about such entities is represented in the calculus as facts describing topological relations on pairs of the entities. The relation of two entities could be *disjoint, touch, overlap, cross,* or *in*.

In geographic maps, it is often the case that one has to distinguish between the situation that a region is completely inside another region, the situation in which it touches it from inside, and the situation in which the two regions are equal. For instance, a GIS user might want to know whether Hamburg is strictly inside Germany or at (i.e., touches) the boundary of the country with the rest of the earth. The CBM calculus makes use of a boundary operator which, when combined with the *in* relation, allows for the distinction to be made. A close look at the calculus shows that, indeed, it is suited for conjunctive-fact queries of geographic databases (i.e. queries consisting of facts describing either of the five atomic relations or a boundary operator on a pair of features). In this contribution, we provide an Allen style approach (Allen, 1983) to the calculus. Specifically, we present an algebra which will have nine atomic relations resulting from the combination of the five atomic relations and the boundary operator of the CBM calculus. Our main motivation is that we can then benefit from Allen style reasoning:

1. We can make use of Allen's constraint propagation algorithm to reason about knowledge expressed in the algebra. This means that composition tables recording

the composition of every pair of the atomic relations have to be provided for the algebra, as well as a converse table.

2. The algebra will benefit from the incrementality of the propagation algorithm: knowledge may be added without having to revise all the processing steps achieved so far.

3. Disjunctive knowledge will be expressed in an elegant way, using subsets of the set of all nine atomic relations: such a disjunctive relation hold on a pair of features if and only if either of the atomic relations holds on the features. This is particularly important for expressing uncertain knowledge, which is closely related to the notion of conceptual neighborhoods (Freksa, 1992).

2   The Calculus

As mentioned in the introduction, our aim is to propose a constraint-based approach to the CBM calculus developed by Clementini, Di Felice, and Oosterom (Clementini & Di Felice, 1995; Clementini et al., 1993). For that purpose, we will develop an algebra as the one Allen (1983) presented for temporal intervals, of which the atomic relations will be the three relations resulting from the refinement of the *in* relation, together with the other four atomic relations of the CBM calculus. We will provide the result of applying the converse and the composition operations to the atomic relations: this will be given as a converse table and composition tables. These tables in turn will play the central role in propagating knowledge expressed in the algebra using Allen's constraint propagation algorithm (Allen, 1983).

We will use the topological concepts of boundary, interior, and dimension of a (point-like, linear or areal) feature. We thus provide some background on these concepts, taken from (Clementini et al., 1993). The boundary of a feature h is denoted by δh; it is defined for each of the feature types as follows:

1. δP: we consider the boundary of a point-like feature to be always empty.
2. δL: the boundary of a linear feature is the empty set in the case of a circular line, the two distinct endpoints otherwise.
3. δA: the boundary of an area is the circular line consisting of all the accumulation points of the area.

The interior of a feature h is denoted by h°. It is defined as h° = h - δh. Note that the interior of a point-like entity is equal to the feature itself.

The function *dim*, which returns the dimension of a feature of either of the types we consider, or of the intersection of two or more such features, is defined as follows (the symbol $\varnothing$ represents the empty set):

If $S \cdot \varnothing$ then

$$dim(S) = \begin{cases} 0 & \text{if S contains at least a point and no lines and no areas} \\ 1 & \text{if S contains at least a line and no areas} \\ 2 & \text{if S contains at least an area} \end{cases}$$

else dim(S) is undefined.

2.1   The Relations

We can now define the topological relations of our algebra. We use the original relations *touch, overlap, cross* and *disjoint* of the CBM calculus. As we cannot use the boundary operator which allows the CBM calculus to distinguish between the three subrelations *equal, completely-inside,* and *touching-from-inside* of the *in* relation (which are not explicitly present in the calculus) the three subrelations will replace the superrelation in the list of atomic relations. In addition, the new relations *completely-inside* and *touching-from-inside* being asymmetric, we need two other atomic relations corresponding to their respective converses, namely *completely-inside$_i$* and *touching-from-inside$_i$*. The definitions of the relations are given below. The topological relation r between two features h1 and h2, denoted by (h1, r, h2), is defined on the right hand side of the equivalence sign in the form of a point-set expression.

Definition 1. The *touch* relation:
$(h_1, touch, h_2) \leftrightarrow h°_1 \cap h°_2 = \varnothing \wedge h_1 \cap h_2 \bullet \varnothing$

Definition 2. The *cross* relation:
$(h_1, cross, h_2) \leftrightarrow$
$\quad dim(h°_1 \cap h°_2) = max(dim(h°_1), dim(h°_2)) - 1 \wedge h_1 \cap h_2 \bullet h_1 \wedge h_1 \cap h_2 \bullet h_2$

Definition 3. The *overlap* relation:
$(h_1, overlap, h_2) \leftrightarrow$
$\quad dim(h°_1) = dim(h°_2) = dim(h°_1 \cap h°_2) \wedge h_1 \cap h_2 \bullet h_1 \wedge h_1 \cap h_2 \bullet h_2$

Definition 4. The *disjoint* relation:
$(h_1, disjoint, h_2) \leftrightarrow h_1 \cap h_2 = \varnothing$

Definition 5. We define the *equal, completely-inside,* and *touching-from-inside* relations using the formal definition of the *in* relation:
$(h_1, in, h_2) \leftrightarrow h_1 \cap h_2 = h_1 \wedge h°_1 \cap h°_2 \bullet \varnothing$
Given that (h_1, in, h_2) holds, the following algorithm distinguishes between the *completely-inside*, the *touching-from-inside,* and the *equal* relations:
if (h_2, in, h_1) then $(h_1, equal, h_2)$
else if $h_1 \cap \delta h_2 \bullet \varnothing$ then $(h_1, touching-from-inside, h_2)$
else $(h_1, completely-inside, h_2)$

Definition 6. The *completely-inside$_i$* relation:
$(h_1, completely-inside_i, h_2) \leftrightarrow (h_2, completely-inside, h_1)$

Definition 7. The *touching-from-inside$_i$* relation:
$(h_1, touching-from-inside_i, h_2) \leftrightarrow (h_2, touching-from-inside, h_1)$

Figure 1 shows examples of the *completely-inside* and *touching-from-inside* situations.

At this point we have defined the atomic relations of the new calculus which are *touch, cross, overlap, disjoint, equal, completely-inside, touching-from-inside, completely-inside$_i$, and touching-from-inside$_i$.* Now, we will prove that these relations are mutually exclusive, that is, it cannot be the case that two different relations hold between two features. Furthermore, we will prove that they form a full covering of all possible topological situations, that is, given two features, the relation between them must be one of the nine defined here. To prove these two characteristics we construct the *topological relation decision tree* depicted in Fig. 2.

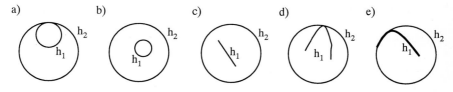

Fig. 1. Examples of the *completely-inside* and *touching-from-inside* situations: examples with two areal entities, a) representing (h_1, *touching-from-inside*, h_2) and b) (h_1, *completely-inside*, h_2); examples of situations with a linear entity and an areal entity, c) representing (h_1, *completely-inside*, h_2), and d) and e) representing (h_1, *touching-from-inside*, h_2) situations

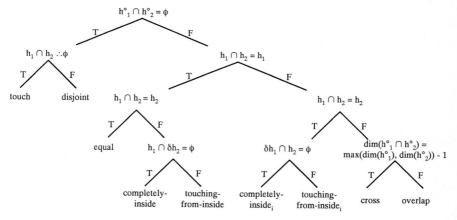

Fig. 2. Topological relation decision tree

Proof. Every internal node in this topological relation decision tree represents a Boolean predicate of a certain topological situation. If the predicate evaluates to true then the left branch is followed, otherwise the right branch is followed. This process is repeated until a leaf node is reached which will indicate which of the atomic topological relations this situation corresponds to. Two different relations cannot hold between two given features, because there is only one path to be taken in the topological relation decision tree to reach a particular topological relation. And there can be no cases outside the new calculus, because every internal node has two branches, so for every Boolean value of the predicate there is an appropriate path and every leaf node has a label that correspond to one of the atomic topological relations. □

Definition 8. A general relation of the calculus is any subset of the set of all atomic relations. Such a relation, say R, is defined as follows:

$$(\forall h_1, h_2) \, ((h_1, R, h_2) \Leftrightarrow \bigvee_{r \in R} (h_1, r, h_2))$$

The next step is to define the operations that we can apply to these relations, namely *converse* and *composition*.

2.2   The Operations

Definition 9. The *converse* of a general relation R is denoted as $R^{\cup}$. It is defined as:

$$(\forall h_1, h_2) \, ((h_1, R, h_2) \Leftrightarrow (h_2, R^{\cup}, h_1))$$

Definition 10. The *composition* $R1 \otimes R2$ of two general relations R1 and R2 is the most specific relation R such that:

$$(\forall h_1, h_2, h_3) \, ((h_1, R1, h_2) \wedge (h_2, R2, h_3) \Rightarrow (h_1, R, h_3))$$

Table 1 and Tables 5 - 22 provide the converse and the composition for the atomic relations of the algebra. For general relations R1 and R2 we have:

$$R1^{\cup} \quad = \bigcup_{r \in R1} \{r^{\cup}\}$$

$$R1 \otimes R2 = \bigcup_{\substack{r_1 \in R1 \\ r_2 \in R2}} r_1 \otimes r_2$$

Table 1. The converse table

r	$r^{\cup}$
Overlap	Overlap
Touch	Touch
Cross	Cross
Disjoint	Disjoint
Completely-inside	Completely-inside$_i$
Touching-from-inside	Touching-from-inside$_i$
Completely-inside$_i$	Completely-inside
Touching-from-inside$_i$	Touching-from-inside
Equal	Equal

We denote by XY-U, with X and Y belonging to {P, L, A}, the universal relation, i.e. the set of all possible atomic relations, between a feature h_1 of type X and a feature h_2 of type Y (we use P for a point-like feature, L for a linear feature, and A for an areal feature). For instance, PP-U is the set of all possible topological relations between two point-like entities. These universal relations are as follows:

PP-U = $\{$ *equal, disjoint* $\}$
PL-U = $\{$ *touch, disjoint, completely-inside* $\}$
PA-U = $\{$ *touch, disjoint, completely-inside* $\}$
LP-U = $\{$ *touch, disjoint, completely-inside$_i$* $\}$
LL-U = $\{$ *touch, disjoint, overlap, cross, equal, touching-from-inside,*
      completely-inside, touching-from-inside$_p$ completely-inside$_i$ $\}$
LA-U = $\{$ *touch, cross, disjoint, touching-from-inside, completely-inside* $\}$
AP-U = $\{$ *touch, disjoint, completely-inside$_i$* $\}$
AL-U = $\{$ *touch, cross, disjoint, touching-from-inside$_p$ completely-inside$_i$* $\}$
AA-U = $\{$ *touch, overlap, disjoint, equal, touching-from-inside,*
      completely- inside, touching-from-inside$_p$ completely-inside$_i$ $\}$

Note that the sets LP-U, AP-U, and AL-U are the converse sets of PL-U, PA-U, and LA-U, respectively.

Given any three features h_1, h_2, and h_3 such that (h_1, r_1, h_2) and (h_2, r_2, h_3), the composition tables should be able to provide the most specific implied relation R between the extreme features, i.e. between h_1 and h_3. If we consider all possibilities with h_1, h_2, and h_3 being a point-like feature, a linear feature, or an areal feature, we would need 27 (3^3) tables. However, we construct only 18 of these tables from which the other 9 can be obtained. The 18 tables to be constructed split into 6 for h_2=point-like feature, 6 for h_2=linear feature and 6 for h_2=areal feature: when feature h_1 is of type X, feature h_2 of type Y, and feature h_3 of type Z, with X, Y, and Z belonging to {P, L, A}, the corresponding composition table will be referred to as the XYZ table. In tables 2, 3, and 4 we show the tables constructed and their numbers of entries.

Table 2. Number of entries of the constructed tables for h_2=point-like entity

TABLE	NUMBER OF ENTRIES
PPP table	$|$PP-U$|$ x $|$PP-U$|$= 4
PPL table	$|$PP-U$|$ x $|$PL-U$|$= 6
PPA table	$|$PP-U$|$ x $|$PA-U$|$= 6
LPL table	$|$LP-U$|$ x $|$PL-U$|$= 9
LPA table	$|$LP-U$|$ x $|$PA-U$|$= 9
APA table	$|$AP-U$|$ x $|$PA-U$|$=9
TOTAL NUMBER OF ENTRIES:	43

Table 3. Number of entries of the constructed tables for h_2=linear entity

TABLES	NUMBER OF ENTRIES
PLP table	$|$PL-U$|$ x $|$LP-U$|$= 9
PLL table	$|$PL-U$|$ x $|$LL-U$|$= 27
PLA table	$|$PL-U$|$ x $|$LA-U$|$= 15
LLL table	$|$LL-U$|$ x $|$LL-U$|$= 81
LLA table	$|$LL-U$|$ x $|$LA-U$|$= 45
ALA table	$|$AL-U$|$ x $|$LA-U$|$= 25
TOTAL NUMBER OF ENTRIES:	202

Table 4. Number of entries of the constructed tables for h_2=areal entity

TABLES	NUMBER OF ENTRIES
PAP table	|PA-U| x |AP-U|= 9
PAL table	|PA-U| x |AL-U|= 15
PAA table	|PA-U| x |AA-U|= 24
LAL table	|LA-U| x |AL-U|= 25
LAA table	|LA-U| x |AA-U|= 40
AAA table	|AA-U| x |AA-U|= 64
TOTAL NUMBER OF ENTRIES:	177

Let us consider the case h_2=linear entity. The six tables to be constructed for this case are the PLP, PLL, PLA, LLL, LLA, and ALA tables. From these six tables, we can get the other three, namely the LLP, ALP, and ALL tables. We illustrate this by showing how to get the $r_1 \otimes r_2$ entry of the LLP table from the PLL table. This means that we have to find the most specific relation R such that for any two linear features L_1 and L_2, and any point-like feature P, if (L_1, r_1, L_2) and (L_2, r_2, P) then (L_1, R, P). We can represent this as:

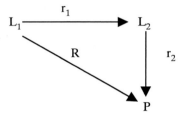

From the converse table we can get the converses $r1^{\cup}$ and $r2^{\cup}$ of r_1 and r_2, respectively. The converse $R^{\cup}$ of R is clearly the composition $r2^{\cup} \otimes r1^{\cup}$ of $r2^{\cup}$ and $r1^{\cup}$, which can be obtained from the PLL table:

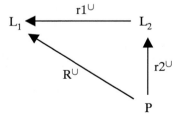

Now R is the converse of $R^{\cup}$: $R = (R^{\cup})^{\cup}$.

Below we present the composition tables, in which the relation *touch* is denoted by T, *cross* by C, *overlap* by O, *disjoint* by D, *completely-inside* by CI, *touching-from-inside* by TFI, *equal* by E, *completely-inside$_i$* by CI_i, and *touching-from-inside$_i$* by TFI_i.

Table 5. The PPP composition table

r_2 / r_1	E	D
E	E	D
D	D	$\{E, D\}$

Table 6. The PPL composition table

r_2 / r_1	T	D	CI
E	T	D	CI
D	PL-U	PL-U	PL-U

Table 7. The PPA composition table

r_2 / r_1	T	D	CI
E	T	D	CI
D	PA-U	PA-U	PA-U

Table 8. The LPA composition table

r_2 / r_1	T	D	CI
T	$\{T, C, TFI\}$	$\{T, C, D\}$	$\{C, TFI, CI\}$
D	LA-U	LA-U	LA-U
CI_i	$\{T, C\}$	$\{T, C, D\}$	$\{C, CI, TFI\}$

Table 9. The LPL composition table

r_2 / r_1	T	D	CI
T	$\{T, O, C, E, TFI, TFI_i\}$	$\{T, D, O, C, TFI_i, CI_i\}$	$\{T, O, C, TFI, CI\}$
D	$\{T, D, O, C, CI, TFI\}$	LL-U	$\{T, D, O, C, TFI_i, CI\}$
CI_i	$\{T, O, C, TFI_i, CI_i \}$	$\{T, D, O, C, TFI_i, CI_i\}$	$\{O, C, E, TFI, CI, TFI_i, CI_i\}$

Table 10. The APA composition table

r_2 / r_1	T	D	CI
T	$\{T, O, E, TFI, TFI_i\}$	$\{T, O, D, CI_i, TFI_i\}$	$\{O, TFI, CI\}$
D	$\{T, O, D, TFI, CI\}$	AA-U	$\{T, O, D, CI, TFI\}$
CI_i	$\{O, TFI_i, CI_i\}$	$\{T, O, D, CI_i, TFI_i\}$	$\{O, E, TFI, CI, TFI_i, CI_i\}$

Table 11. The PLP composition table

r_2 / r_1	T	D	CI_i
T	PP-U	D	D
D	D	PP-U	D
CI	D	D	PP-U

Table 12. The PLA composition table

r_1 \ r_2	T	C	D	CI	TFI
T	{T, D}	PA-U	D	CI	{T, CI}
D	PA-U	PA-U	PA-U	PA-U	PA-U
CI	{T, D}	PA-U	D	CI	{T, CI}

Table 13. The PLL composition table

r_1 \ r_2	T	C	O	D	E	CI	TFI	CI_i	TFI_i
T	PL-U	PL-U	PL-U	D	T	CI	{T, D}	D	{T, D}
D	PL-U	PL-U	PL-U	PL-U	D	PL-U	PL-U	D	D
CI	{T, D}	PL-U	PL-U	D	CI	CI	CI	PL-U	PL-U

Table 14. The LLA composition table

r_1 \ r_2	T	C	D	TFI	CI
T	{T, C, D, TFI}	LA-U	{T, C, D}	{T, C, TFI, CI}	{C, TFI, CI}
D	LA-U	LA-U	LA-U	LA-U	LA-U
O	{T, C, D}	LA-U	{T, C, D}	{C, TFI, CI}	{C, TFI, CI}
C	{T, C, D}	LA-U	{T, C, D}	{C, TFI, CI}	{C, TFI, CI}
E	T	C	D	TFI	CI
TFI	{T, D}	LA-U	D	{TFI, CI}	CI
CI	{T, D}	LA-U	D	{TFI, CI}	CI
TFI_i	{T, C}	C	{T, C, D}	{TFI, CI}	{C, TFI, CI}
CI_i	{T, C}	C	{T, C, D}	{TFI, CI}	{C, TFI, CI}

Table 15. The ALA composition table

r_1 \ r_2	T	C	D	CI	TFI
T	AA-U\{CI, CI_i}	AA-U\{E, CI, CI_i}	AA-U\{E, CI, TFI}	AA-U\{T, D, E, CI_i, TFI_i}	AA-U\{D, E, CI_i, TFI_i}
C	AA-U\{E, CI, CI_I}	AA-U	AA-U\{E, CI, TFI}	{O, TFI, CI}	{O, TFI, CI}
D	AA-U\{E, CI_i, TFI_i}	AA-U\{E, CI_i, TFI_i}	AA-U	AA-U\{E, CI_i, TFI_i}	AA-U\{E, CI_i, TFI_i}
CI_i	AA-U\{T, D, E, CI, TFI}	{O, TFI_i, CI_i}	AA-U\{E, CI, TFI}	AA-U\{T, D}	{O, TFI_i, CI_i}
TFI_i	AA-U\{D, E, CI, TFI}	{O, TFI_i, CI_i}	AA-U\{E, CI, TFI}	{O, TFI, CI}	{O, E, TFI, TFI_i}

Table 16. The LLL composition table

r_1 \ r_2	T	C	O	D	E	CI	TFI	CI$_i$	TFI$_i$
T	LL-U\ {CI$_i$}	{T, C, O, D, TFI, CI}	{T, C, O, D, TFI, CI}	{T, C, O, D, TFI$_i$, CI$_i$}	T	{T, C, O, TFI, CI}	{T, C, O, TFI, CI}	D	{T, D}
C	{T, C, O, D, TFI$_i$, CI$_i$}	LL-U	LL-U\ {E, CI$_i$, TFI$_i$}	LL-U\ {E, CI, TFI}	C	{C, O, TFI, CI}	{C, O, TFI, CI}	{T, C, D}	{T, C, D}
O	{T, C, O, D, TFI$_i$, CI$_i$}	LL-U\ {E, CI, TFI}	LL-U	LL-U\ {E, CI, TFI}	O	{O, TFI, CI}	{O, TFI, CI}	{T, O, D, TFI$_i$, CI$_i$}	{T, O, D, TFI$_i$, CI$_i$}
D	{T, C, O, D, TFI, CI}	LL-U\ {E, CI$_i$, TFI$_i$}	LL-U\ {E, CI$_i$, TFI$_i$}	LL-U	D	LL-U\ {E, CI$_i$, TFI$_i$}	LL-U\ {E, CI$_i$, TFI$_i$}	D	D
E	T	C	O	D	E	CI	TFI	CI$_i$	TFI$_i$
CI	D	{T, C, D}	{T, O, D, TFI, CI}	D	CI	CI	CI	LL-U\ {C}	{T, O, D, TFI, CI}
TFI	{T, D}	{T, C, D}	{T, O, D, TFI, CI}	D	TFI	LL-U\ {C, E}	{TFI, CI}	{T, O, D, TFI$_i$, CI$_i$}	LL-U\ {C, CI, CI$_i$}
CI$_i$	{T, C, O, TFI$_i$, CI$_i$}	{C, O, TFI$_i$, CI$_i$}	{O, TFI$_i$, CI$_i$}	LL-U\ {E, CI, TFI}	CI$_i$	LL-U\ {T, C, D}	{O, TFI$_i$, CI$_i$}	CI$_i$	CI$_i$
TFI$_i$	{T, C, O, TFI$_i$, CI$_i$}	{C, O, TFI$_i$, CI$_i$}	{O, TFI$_i$, CI$_i$}	LL-U\ {E, CI, TFI}	TFI$_i$	{O, TFI, CI}	{O, E, TFI, TFI$_i$}	CI$_i$	{CI, CI$_i$}

Table 17. The PAP composition table

r_1 \ r_2	T	D	CI$_i$
T	PP-U	D	D
D	D	PP-U	D
CI	D	D	PP-U

Table 18. The PAL composition table

r₁ \ r₂	T	C	D	CIᵢ	TFIᵢ
T	PL-U	PL-U	D	D	{T, D}
D	PL-U	PL-U	PL-U	D	D
CI	D	PL-U	D	PL-U	PL-U

Table 19. The PAA composition table

r₁ \ r₂	T	O	D	E	CI	TFI	CIᵢ	TFIᵢ
T	{T, D}	PA-U	D	T	CI	{T, CI}	D	{T, D}
D	PA-U	PA-U	PA-U	D	PA-U	PA-U	D	D
CI	D	PA-U	D	CI	CI	CI	PA-U	PA-U

Table 20. The LAL composition table

r₁ \ r₂	T	C	D	CIᵢ	TFIᵢ
T	LL-U	{T, D, C, O, TFI, CI}	{T, D, C, O, TFIᵢ, CIᵢ}	D	{T, D, O}
C	{T, D, C, O, TFIᵢ, CIᵢ}	LL-U	{T, D, C, O, TFIᵢ, CIᵢ}	{T, D, C, O, TFIᵢ, CIᵢ}	{T, D, C, O, TFIᵢ, CIᵢ}
D	{T, D, C, O, TFI, CI}	{T, D, C, O, TFI, CI}	LL-U	D	D
CI	D	{T, D, C, O, TFI, CI}	D	LL-U	{T, D, C, O, TFI, CI}
TFI	{T, D, O}	{T, D, C, O, TFI, CI}	D	{T, D, C, O, TFIᵢ, CIᵢ}	{T, D, C, O, E, TFI, TFIᵢ}

Table 21. The LAA composition table

r₁ \ r₂	T	O	D	E	CI	TFI	CIᵢ	TFIᵢ
T	{T, D, C, TFI}	LA-U	{T, D, C}	T	{C, CI, TFI}	{T, C, CI, TFI}	D	{T, D}
C	{T, D, C}	LA-U	{T, D, C}	C	{C, CI, TFI}	{C, CI, TFI}	{T, C, D}	{T, C, D}
D	LA-U	LA-U	LA-U	D	LA-U	LA-U	D	D
CI	D	LA-U	D	CI	CI	CI	LA-U	LA-U
TFI	{T, D}	LA-U	D	TFI	CI	{CI, TFI}	{T, C, D}	{C, T, D, TFI}

Table 22. The AAA composition table

r_1 \ r_2	T	O	D	E	CI	TFI	CI_i	TFI_i
T	{T, D, O, E, TFI, TFI_i}	{T, D, O, TFI, CI}	{T, D, O, TFI_i, CI_i}	T	{O, CI, TFI}	{O, T, CI, TFI}	D	{D, T}
O	{T, D, O, TFI_i, CI_i}	AA-U	{T, D, O, TFI_i, CI_i}	O	{O, CI, TFI}	{O, TFI, CI}	{T, D, O, TFI_i, CI_i}	{T, D, O, TFI_i, CI_i}
D	{T, D, O, TFI, CI}	{T, D, O, TFI, CI}	AA-U	D	{T, D, O, TFI, CI}	{T, D, O, TFI, CI}	D	D
E	T	O	D	E	CI	TFI	CI_i	TFI_i
CI	D	{T, D, O, TFI, CI}	D	CI	CI	CI	AA-U	{T, D, O, CI, TFI}
TFI	{T, D}	{T, D, O, CI, TFI}	D	TFI	CI	{CI, TFI}	{T, D, O, TFI_i, CI_I}	{T, D, O, E, TFI_i, TFI}
CI_I	{O, CI_i, TFI_i}	{O, CI_i, TFI_i}	{T, D, O, CI_i, TFI_i}	CI_i	{E, O, CI, TFI, CI_i, TFI_i}	{O, TFI_i, CI_i}	CI_i	CI_i
TFI_i	{T, O, CI_i, TFI_i}	{O, CI_i, TFI_I}	{T, D, O, CI_i, TFI_i}	TFI_i	{O, CI, TFI}	{O, E, TFI, TFI_i}	CI_i	{TFI_i, CI_i}

3   Conclusions

We have proposed a constraint-based approach to the CBM calculus in (Clementini et al., 1993). What we have obtained is a calculus consisting of atomic relations and of the algebraic operations of converse and composition. As such, the calculus is an algebra in the same style as the one provided by Allen (1983) for temporal intervals.

The objects manipulated by the calculus are point-like, linear and areal features, contrary to most constraint-based frameworks in the qualitative spatial and temporal reasoning literature, which deal with only one type of feature (for instance, intervals in (Allen, 1983)). One problem raised by this was that the calculus had 27 composition tables, fortunately of moderate sizes. We have shown in this work that the use of 18 of these tables is sufficient, as from these 18 we can derive the other nine.

Reasoning about knowledge expressed in the presented calculus can be done using a constraint propagation algorithm alike to the one in (Allen, 1983), guided by the 18 composition tables and the converse table. Such an algorithm has the advantage of being incremental: knowledge may be added without having to revise the processing steps achieved so far.

Acknowledgments

We would like to thank Christian Freksa, Teresa Escrig, and Julio Pacheco for helpful discussions and suggestions during the preparation of this work. We thank the anonymous reviewer for the critical comments. This work has been partially supported by CICYT under grant number TAP99-0590-C02-02.

References

Allen, J. F. (1983). Maintaining knowledge about temporal intervals. *Communications of the ACM, 26*(11), 832-843.

Bennett, B. (1994). Spatial reasoning with propositional logics. In J. Doyle, E. Sandewall, & P. Torasso (Eds.), *Principles of knowledge representation and reasoning: Proceedings of the Fourth International Conference KR94* (pp. 51-62). Morgan Kaufmann.

Clementini, E., & Di Felice, P. (1995). A comparison of methods for representing topological relationships. *Information Sciences, 3*, 149-178.

Clementini, E., Di Felice, P., & van Oosterom, P. (1993). A small set of formal topological relationships suitable for end-user interaction. In D. Abel, & B. C. Ooi (Eds.), *Advances in spatial databases - Third International Symposium, SSD'93, Singapore* (pp. 277-295). Berlin: Springer.

Egenhofer, M. (1991). Reasoning about binary topological relations. In O. Gunther & H.-J. Schek (Eds.), *Second Symposium on Large Spatial Databases, Zurich, Switzerland* (pp. 143-160). Berlin: Springer.

Egenhofer, M., & Franzosa, R. (1991). Point-set topological spatial relations. *International Journal of Geographical Information Systems, 5*(2), 161-174.

Freksa, C. (1992). Temporal reasoning based on semi-intervals. *Artificial Intelligence, 54*(1-2), 199-227.

Gotts, N. M. (1996). *Topology from a single primitive relation: defining topological properties and relations in terms of connection* (Technical Report No. 96_23). University of Leeds: School of Computer Studies.

Pullar, D., & Egenhofer, M. (1988). Toward formal definitions of topological relations among spatial objects. *Third International Symposium on Spatial Data Handling, Sydney, Australia, August 1988.*

Randell, D. A., Cui, Z., & Cohn, A. G. (1992). A spatial logic based on regions and connection. *Proc 3rd Int. Conf on Knowledge Representation and Reasoning, Boston, October, 1992.*

Renz, J., & Nebel, B. (1998). Spatial reasoning with topological information. In C. Freksa, C. Habel, & K. F. Wender (Eds.), *Spatial cognition - An interdisciplinary approach to representation and processing of spatial knowledge.* Berlin: Springer.

The Influence of Linear Shapes
on Solving Interval-Based Configuration Problems

Reinhold Rauh[1] and Lars Kulik[2]

[1]Albert Ludwig University of Freiburg
Institute of Computer Science and Social Research
Center for Cognitive Science
reinhold@cognition.iig.uni-freiburg.de
[2]University of Hamburg
Department for Informatics
kulik@informatik.uni-hamburg.de

Abstract. Spatial configuration problems can be considered as a special kind of inference tasks, and can therefore be investigated within the framework of the well-established mental model theory of human reasoning. Since it is a well-known fact that content and context affects human inference, we are interested to know to what extent abstract properties of linear shape curves conform to previous findings of interval-based reasoning. This investigation is done on a formally grounded basis. The main issue of this paper concerns the question whether the shape of linear curves in general and salient points on the curves in particular have an influence on solving interval-based configuration problems. It has been shown in previous experiments that there are preferred mental models if the linear structure consists of a straight line segment. The reported experiment demonstrates under which conditions arbitrary shaped curves reveal similar and different effects. To distinguish different types of points on a curve a classification of points based on ordering geometry is introduced. It turns out that only those shape features are employed in solving configuration-based problems that can be characterized on the basis of ordering geometry. Curves supplied with salient points also lead to strongly preferred configurations corroborating the notion of preferred mental models. Differences to the obtained types of preferred solutions in comparison to former investigations are discussed and possible explanations are given.

1 Solving Configuration Problems with Mental Models

Dealing with configuration problems is part of various everyday activities including loading the trunk with luggage as well as developing the layout of a personal home page in the World Wide Web. Many professionals, e.g. architects, designers, and engineers, also have to deal heavily with spatial configuration problems. Thus, configuration problems seem to be a frequent and relevant class of problems that often require a lot of thinking. How do people solve such problems?

Ch. Freksa et al. (Eds.): Spatial Cognition II, LNAI 1849, pp. 239–252, 2000.
© Springer-Verlag Berlin Heidelberg 2000

This question has been addressed by the psychology of thinking where several theories of human inference were developed. A prominent one is the theory of mental models originally developed as a theory of text comprehension on the one hand and as a theory of human reasoning on the other (Johnson-Laird, 1983). The core assumptions of mental model theory were formulated within the realm of deductive inference, but can easily be generalized to a broader class of inference tasks, such as solving configuration problems. This is due to the fact that thinking with mental models can be described as diagrammatic thinking in possibilities.

The core idea of *mental model theory* is that people transform an external situation of the real world into an integrated mental representation—the *mental model*. This special kind of mental representation is the basis and starting point of a sequential inference process employed in solving given inference tasks like configuration problems. A mental model can be seen as a relational structure in human working memory that constitutes a model (in the usual logical sense) of the premises. Taken this way, human inference can be broken down into three distinct phases, which are often called the phase of comprehension, the phase of description, and the phase of validation (Johnson-Laird & Byrne, 1991). In the following, we will refer to these phases as *model construction, model inspection,* and *model variation* in order to clarify the character and function of these phases (for details see Knauff, Rauh, Schlieder, & Strube, 1998).

In the first phase (*model construction*) reasoners use knowledge about the semantics of spatial expressions (together with long-term background knowledge) to construct an internal model of the *"state of affairs"* described by the premises. In this stage of the reasoning process the given premises are integrated into a unified mental model. According to the theory, merely the mental model has to be kept in memory, whereas the premises need not be taken into account. It is important to point out that spatial descriptions are often underdetermined, i.e., there is more than one model that is consistent with the given premises. For this reason, we have to distinguish *determinate* tasks in which only a single model can be constructed from *indeterminate* tasks that have multiple models. The influence of this difference on the difficulty of reasoning tasks is one of the most prominent and most important findings of the mental model theory. During the *inspection phase*, the mental model is scanned for relations which are not explicitly given. In the *variation phase*, people try to find alternative models of the premises if necessary. Deductive inference is such an example where model variation can occur: From a relation that is contingently true in the initial model, a putative conclusion can be formed that states that this relation is necessarily true given the premises are true. However, it may not be true in all models of the premises. If such a contradictory example has been found, the putative conclusion has to be rejected. If not, the statement has to be accepted as valid conclusion (Johnson-Laird & Byrne, 1991).

Many findings, partially obtained in the near past, show that (i) there is evidence for integrated representations, (ii) that these integrated representations have natural initial starting points, and (iii) that reasoning is best described as a sequential process of generation and inspection of mental models. For example, Knauff, Rauh, and Schlieder (1995) showed that there are "preferred mental models". Therefore, in multiple model cases the sequence of generated models is not random. On the contrary, the construction of a first mental model seems to be a general cognitive process that works the same way for most people. This finding was confirmed in a replication study by Kuß, Rauh, and Strube (1996). Schlieder and Berendt (1998) explained the obtained preferred mental models and their inherent properties within the framework of Kosslyn's (1994) theory of the imagery system, and of Logie's (1995) conceptualization of the visuo-spatial working memory, respectively.

Another important point is the well-established fact that content and context of a reasoning task influences the type of answers reasoners give. This was shown in many reasoning experiments, the most prominent example being the empirical divergences in conditional reasoning in the abstract Wason selection task (Wason, 1966) compared to content-embedded versions of this task (e.g., Gigerenzer & Hug, 1992; Johnson-Laird, Legrenzi, & Legrenzi, 1972). In order to extend our previous findings to non-straight objects like natural or geographic entities (e.g., rivers, roads, or railway tracks), we use linear shape curves in our configurations problems.

Before we proceed to the testing of the influence of linear shape curves on the solving of configuration problems, we describe the formally sound material that was used in our experiment.

2   Interval Relations in Spatial Descriptions

Ordering relations (like "right-of", "left-of", "in-front-of" and "behind") are typical for configuration problems in two-dimensional space. Investigations concerning configuration problems rely on a unique and well-defined semantics of these predicates. The work of Eschenbach and Kulik (1997) shows that depending on the considered reference frame there is a multitude of different geometric constellations underlying these relations. Knauff, Rauh, Schlieder, and Strube (1998) also emphasize the ambiguity of these relations and the undesirable consequences when used in psychological experiments on reasoning. This result is further supported by the study of Eschenbach, Habel, and Leßmöllmann (in press) where combinations of these relations are considered. The analysis shows that there are basically two readings for combinations resulting in different spatial models. Therefore, we confine ourselves to investigate ordering information on linear structures in two-dimensional space. We use the set of thirteen qualitative relations introduced by Allen (1983) in our experiment. These relations are jointly exhaustive and pairwise disjoint. They have a clear geometric semantics, are easy to acquire by humans, and are well investigated from a formal and computational point of view (e.g., Ligozat, 1990; Nebel & Bürckert, 1995; Schlieder, 1999). Furthermore, they are also of interest in application areas like Geographic Information Systems (e.g., Molenaar & De Hoop, 1994), automatic document analysis (Walischewski, 1997), and diagrammatic layout generation (Schlieder & Hagen, this volume).

In Table 1, we list these 13 relations together with the verbalizations we use in the verbal descriptions of our configuration problems. The ordering of startpoints and endpoints form the basis of the model-theoretical foundation. From the graphical examples, it is easy to verify that the interval relations can be used to express any qualitative relationship that may exist between two (one-dimensional) objects.

Table 1. The 13 qualitative interval relations, associated natural language expressions, one graphical realization, and ordering of startpoints and endpoints (adapted and augmented according to Allen, 1983)

Symbol	Natural language description	Graphical realization	Point ordering (s=startpoint, e=endpoint)
X < Y	X lies to the left of Y		$s_X < e_X < s_Y < e_Y$
X m Y	X touches Y at the left		$s_X < e_X = s_Y < e_Y$
X o Y	X overlaps Y from the left		$s_X < s_Y < e_X < e_Y$
X s Y	X lies left-justified in Y		$s_Y = s_X < e_X < e_Y$
X d Y	X is completely in Y		$s_Y < s_X < e_X < e_Y$
X f Y	X lies right-justified in Y		$s_Y < s_X < e_X = e_Y$
X = Y	X equals Y		$s_X = s_Y < e_Y = e_X$
X fi Y	X contains Y right-justified		$s_X < s_Y < e_Y = e_X$
X di Y	X surrounds Y		$s_X < s_Y < e_Y < e_X$
X si Y	X contains Y left-justified		$s_X = s_Y < e_Y < e_X$
X oi Y	X overlaps Y from the right		$s_Y < s_X < e_Y < e_X$
X mi Y	X touches Y at the right		$s_Y < e_Y = s_X < e_X$
X > Y	X lies to the right of Y		$s_Y < e_Y < s_X < e_X$

The set consists of the following 13 relations: *before* (<) and its inverse *after* (>), *meets* (m) and *met by* (mi), *overlaps* (o) and *overlapped by* (oi), *finishes* (f) and *finished by* (fi), *during* (d) and *contains* (di), *starts* (s) and *started by* (si), and *equal* (=), that is inverse to itself.

Combining two relations r_i and r_j gives the composition $c(r_i, r_j)$ that specifies the possible relationships between an interval X and Z given the qualitative relationship between X and Y, and Y and Z. For instance, given that X *meets* Y and Y *is during* Z then the following relations between X and Z are possible: X *overlaps* Z or X *is during* Z or X *starts with* Z. Since Allen's theory contains thirteen relations, there are 144 compositions $c(r_i r_j)$, if the trivial "=" relation is omitted. As can be seen in the example mentioned above, there are compositions (exactly 72 of the 144) that have multiple solutions. From these compositions it is easy to construct spatial descriptions that are the basis of inference tasks known in the psychology of reasoning as three-term series problems (e.g., Johnson-Laird, 1972).

3   A Geometry of the Shape of Linear Curves

In order to investigate configuration problems not only in one-dimensional but also in two-dimensional space we consider curves instead of straight lines as underlying structure for solving interval-based configurations. As expounded in section 2, we obtain that the specification of two relations of three intervals X, Y and Z, for instance between X and Y as well as between Y and Z, usually does not determine the relation between X and Z. Since we consider curves, the question arises to what extent shape features of these curves influence the preferred mental models for determining the third relation between X and Z. We present a classification of local shape features to investigate whether particular shape features induce different strategies when a reasoner has to place the intervals on the curve.

There are several possibilities as to which features could be considered relevant. Hence, one purpose of the experiment is to provide some hints as to which shape features of a curve play an important role and which features can be neglected. Eschenbach, Habel, Kulik, and Leßmöllmann (1998) developed a formal framework primarily concerned with a description of shape features of planar objects. Since a boundary of a planar object is a linear object, more precisely a closed curve, this approach provides a characterization of shape features of curves in general. The approach suggests a taxonomy of shape features and distinguishes three types of points: inflection points, vertices, and smooth points. In this experiment we test whether these points play any role when placing intervals on curves.

The geometric framework characterizes the spatial features of curves employing the axiomatic method. An axiomatic system does not define basic terms like 'point', 'curve' or 'between'. Instead, it constitutes a system of constraints which determines the properties of these basic terms and specifies their relations. Therefore, an axiomatic specification of shape features of a curve reveals their underlying spatial structure. The formal framework assumes three types of entities: points, curves, and half-planes. Points denote the location of local shape features of curves like kinks or inflection points. Half-planes are required to describe the type of shape feature of a curve at a certain point. They can be characterized by straight lines because every straight line divides the (two-dimensional) plane into two halves, called half-planes. The entities are related using two basic relations: incidence and betweenness. The incidence relation describes points lying on a curve or on a half-plane and betweenness specifies the conditions when a point is between two others. For details, we refer to Eschenbach et al. (1998).

Since our goal is to identify the underlying structure of shape properties of curves, ordering geometry is used as basis in this framework. Ordering geometry makes very weak assumptions about the underlying spatial structure and is sufficient for characterizing basic shape features. Therefore, this approach does not rely on concepts of differential geometry like differentiability, tangents or real numbers which demonstrates that these concepts are not essential to characterize fundamental shape features.

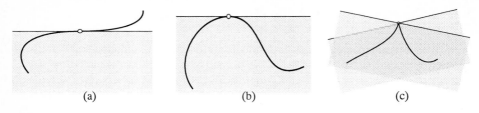

Fig. 1. (a) Inflection Point: The part of the curve which is bent to the right is included in the half-plane. Therefore, every part of the curve which is bent to the left and starts at the inflection point leads out of the half-plane. (b) Smooth Point: One uniquely determined half-plane touches a smooth point with its boundary and contains at least a part of the curve. (c) Vertex: Two half-planes whose boundaries touch the given point and contain at least a part of the curve including the vertex as inner point

3.1 A Classification of Points on a Curve

The main idea is to build up complex curves out of simple primitives, namely segments and arcs. The essential difference between arcs and segments can be summarized as follows: For every triple of points on a segment holds that one of them is between the others, whereas for any three points on an arc holds that none of them is between the others.

Segments and arcs can be combined in various ways leading to different types of connection points. Ordering geometry supplied with half-planes is able to distinguish three different types of points: inflection points, smooth points, and vertices.

- *Inflection points.* An inflection point occurs for instance at a location where a curve which is bent to the right is followed by a curve which is bent to the left. More formally, in this case it is not possible to find any half-plane such that the inflection point of a curve lies on the boundary of the half-plane and at least a part of the curve including the inflection point as inner point is totally included in the half-plane (see Figure 1a). If the part of the curve which is bent to the right is included in the half-plane, then every part of the curve which is bent to the left leads out of the half-plane.

- *Smooth points.* Inner points of segments and arcs are smooth points. In differential geometry uniquely determined tangents ensure that a curve described by real coordinates is differentiable at that point. This condition can be expressed using half-planes only. A point is a smooth point of a curve if there is a uniquely determined half-plane whose boundary touches only the considered point and at least part of the curve surrounding the point is completely contained in this half-plane (see Figure 1b). Since half-planes are sufficient to capture smoothness the notion of differentiability and the use of real coordinates is not essential.

- *Vertices.* If a curve has a kink then two parts of the curve are connected at this point representing a non-continuous change of direction. In this case there are at least two half-planes whose boundaries touch the given point and contain at least part of the curve including the point as inner point. This point is called a vertex of the curve (see Figure 1c).

Fig. 2. Special Smooth Points: The left figure shows a segment that is continuously extended by an arc. The right figure shows two parts of a circle with different radii which are connected smoothly

The use of half-planes allows a canonical classification of points on a curve: Depending on whether there is no half-plane, exactly one half-plane or more than one half-plane containing the curve and touching the considered point with its boundary, the point is either an inflection point, a smooth point, or a vertex.

3.2 Discontinuities of Higher Order

To check whether the underlying spatial structure of configuration problems on curves is based on ordering geometry or if further assumptions have to be considered we threw in some points which are not captured by ordering geometry. This technique follows the principle of adequate specification proposed by Habel and Eschenbach (1997). According to this principle a description employs only those assumptions that are necessary. This results in a sparse characterization of a spatial structure.

The curves were supplied with two types of smooth points having second order discontinuities: Points representing a big change in curvature or points indicating a discontinuity in curvature. The latter case occurs for instance at points connecting two arcs of different curvature or at points where an arc is continued by a segment in a smooth way (see Figure 2).

The main reason why these points are not differentiated in the presented formal framework is rooted in the use of ordering geometry. The ordering information coded in half-planes is able to indicate whether a curve is differentiable at a point or not since the boundaries of half-planes correspond to tangents in differential geometry. The differentiation of discontinuities of second order that is to say if the second derivative of a curve at given points exists or does not exist requires a spatially richer structure. In general, a finer level of granularity indicates that a larger inventory of assumptions about the spatial structure is necessary.

Summarizing this discussion we obtain that if the smooth points with second order discontinuities also play an important role in solving configuration-based problems then a finer classification scheme is required.

3.3 Strategies Based on Salient Points for Solving Configuration Tasks

There are several possible strategies to use points in configurations tasks. We have focused the analysis mainly on three strategies, namely the enclosing strategy, the excluding strategy, and the boundary strategy:

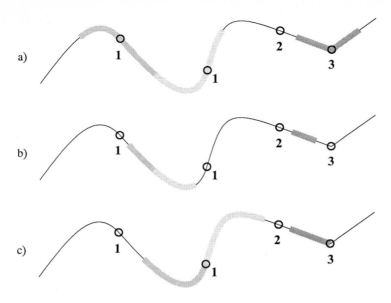

Fig. 3. Instances of configurations where all three intervals conform to (a) the enclosing strategy, (b) the excluding strategy, and (c) the boundary strategy. The shape curve has salient points, that are annotated as follows: 1 = inflection point, 2 = smooth point with second order discontinuity, 3 = vertex

- *Enclosing strategy.* A reasoner uses the enclosing strategy if she/he tries to include a single salient point in an interval such that the enclosed point lies more or less in the middle of the interval (see Figure 3a).
- *Excluding strategy.* The reasoner uses the excluding strategy if she/he explicitly avoids salient points and produces intervals excluding them (see Figure 3b).
- *Boundary strategy.* In case of the boundary strategy the reasoner employs at least one salient point as boundary for the intervals in solving the configuration task (see Figure 3c).

The vertices, the inflection points, and the discontinuities of second order play no significant role in the configuration task if conditions like the following arise: More than one point is included in an interval, or the reasoner uses more general shape aspects of the curve. The latter case can occur for example, if the curve has some "hills" and "valleys" which are used to place the intervals on the curve.

To investigate the three strategies the curves are essentially composed in three different ways: Curves consisting only of segments, only of arcs or of a mixture of arcs and segments. Regarding the three types of distinguished points we adhere to the following three possibilities: A curve can contain in addition to smooth points either vertices, inflection points, or vertices and inflection points. To impede strategies like the 'hill-valley' strategy discussed in the previous section we introduced curves with a falling and rising tendency.

4   Experiment: Influence of Shape Curves on Model Construction

To test for the influence of linear shape curves on configuration tasks, we designed the following experiment. As in former experiments of the MeMoSpace project (Knauff, Rauh, & Schlieder, 1995; Knauff, Rauh, Schlieder, & Strube, 1998; Rauh & Schlieder, 1997), this computer-aided experiment was divided into three blocks: a *definition phase*, a *learning phase*, and an *inference phase*. The reasons for using the procedure are discussed extensively in Knauff, Rauh, and Schlieder (1995). It allows for exerting control over conceptual and inferential aspects and of referring the obtained results exclusively to the inference process, keeping the conceptual aspects constant.

4.1 Participants

Thirty students (15 female, 15 male) of the University of Freiburg, ranging in age from 20 to 45 years ($\bar{x}$ = 26,3 years), were paid for participation in the experiment.

4.2 Materials

On the basis of the formalization mentioned above, 30 linear shape curves were constructed. All shape curves incorporated at least 10 vertices, in order to leave the possibility to use vertices for all six points of the three intervals. 21 shape curves additionally incorporated between 1 and 13 inflection points, and 3 additional curves also had one to two smooth points with second order discontinuities (see Figure 4 for a sample shape curve). An additional shape curve was generated that served as a practice trial in the beginning of the inference phase

From the 72 indeterminate three-term series problems, we chose 30 problems (i) displaying the same stable preferences as in the studies of Knauff et al. (1995) and Kuß et al. (1996), (ii) that did not incorporate the inverse relation of the first premise in order to avoid chunking strategies (see Rauh, 2000), and (iii) ensured that an equal number of occurrences of interval relations in spatial descriptions was obtained across inference tasks. A sample three-term series problem is the following (abbreviated as *d-oi* according to the symbols of Table 1, where the first symbol denotes the relation in the first premise, and the second symbol the relation in the second premise): *"The red interval is completely in the green interval. The green interval overlaps the blue interval from the right."*

The complete list of used three-term series problems is given in Table 2. To each problem, a linear shape curve was randomly assigned. These pairs were the same for all participants.

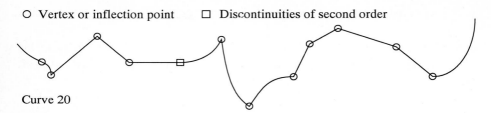

Curve 20

Fig. 4. One sample linear shape curve (No. 20) that was given to the participant after she/he had read the three-term series problem. Participants marked startpoints and endpoints of the three intervals on the curve with colored pens. Annotations are only given for the reader's convenience

4.3 Procedure

In the *definition phase*, participants read descriptions of locations of a red and a blue interval using the 13 qualitative relations (in German). Each verbal description was presented with a short commentary about the location of the beginnings and endings of the intervals and a diagram with a red and blue interval that matched the description.

The *learning phase* consisted of trial blocks, where participants initially read one sentence describing the spatial relationship of the red and blue interval. They then had to determine the startpoints and endpoints of a red and a blue interval using mouse clicks. After confirming the final choices, the participant was told whether the choices were correct or false. If they were false, additional information about the correct answer was given. Trials were presented in blocks of all 13 relations in randomized order. The learning criterion for one relation was accomplished if the participant gave correct answers in 3 consecutive blocks of the corresponding relation. The learning phase stopped as soon as the last remaining relation reached the learning criterion. Participants needed 15 to 30 minutes to accomplish the learning phase. No shape curves were used in the definition phase and the learning phase.

In the *inference phase*, participants had to solve 30 spatial *three-term series problems* (plus one practice trial), that were compiled in a booklet. According to the separated-stages paradigm (Potts & Scholz, 1975), premises were presented on an extra sheet of paper in a self-paced manner. After the participant read the premises, they turned the page over, and then had to specify the startpoints and endpoints of all three intervals on the linear shape curve by using colored pens.

The sequence of the thirty tasks was counterbalanced across participants according to a sequentially counterbalanced Latin square. The whole experiment took approximately 1 hour and was conducted on an individual basis.

4.4 Results

For the hypothesis testing analyses, a general significance level of 5% is adopted. All 30 participants successfully passed the learning phase, and all data collected in the inference phase could be further analyzed. Individual performance showed considerable variation, ranging from 40.0% to 100.0% correct answers.

Existence of Preferred Solutions. The test for the existence of preferred solutions showed clear results: For all 30 three-term series problems we got significant χ^2 values, indicating that participants preferred certain solutions over other ones. The preference rate was even higher than in the study of Knauff et al. (1995): 540 of 900 solutions (= 60.0%) were taken by the preferred solutions; with respect to correct solutions (= 783), the preference rate was 68.97%.

The concordance to the preferred solutions in the Knauff et al. study is only moderate: 19 of 30 three-term series problems were solved the same way (see Table 2 for differences). This indicates an influence of shape curves on constructing spatial configurations.

Table 2. Differences in the preferred solutions to the study of Knauff, Rauh, and Schlieder (1995)

3ts problem[*]	<-oi	<-mi	di-s	s-o	s-fi	d-fi	d-oi	f-si	>-m	>-o	>-s
Preferred relation											
in Knauff et al.	o	o	o	o	o	o	oi	oi	oi	oi	oi
in this experiment	<	<	di	<	<	<	>	>	>	>	>

[*] The other three-term series problems were m-d, m-f, m-oi, o-o, o-si, fi-oi, fi-mi, fi->, d-si, di-<, di-mi, f-o, f-di, si-<, si-m, oi-fi, oi-d, mi-s, mi-d.

Relevance of Salient Points. With respect to the relevance of the salient points, the absolute occurrence frequencies using salient points as startpoints or endpoints of intervals are listed in Table 3. Since the salient points could be also hit by chance, it is necessary to determine the probability under the null hypothesis that the reasoner did not consider them. Since it was not possible to determine the sampling distribution of occurrences under the null hypothesis analytically, we ran Monte Carlo simulations to get an estimation of it. The Monte Carlo simulations were implemented as follows: Depending on the used interval relations in the premises, up to six points of the three intervals were positioned randomly on the respective shape curve. If a point was around 1 mm to the left or right of a salient point it counted as hit (The analyses of the participants' drawings were done the same way). Monte Carlo simulations (100,000 runs) revealed that only the vertices were used reliably more often than expected by chance ($p < .0434$, i.e. the probability of observing 587 hits if the participants had not considered them). Inflection points and smooth points with second order discontinuities, however, were used so seldom that the number of hits did not reach significance.

Table 3. Usage of salient points as startpoints or endpoints of intervals

Points	Absolute Frequency[*]	PMC-estimation
Vertices	587	.0434
Inflection points	162	.3915
Smooth points with second order discont.	4	.9278

[*] n = 5400 (6 points x 30 tasks x 30 participants)

Regarding strategies, each interval was classified whether it conformed either to the boundary strategy, to the enclosing strategy, or to the excluding strategy. The number of occurrences are displayed in Table 4. The use of strategies can be observed rather often: Conforming to the results of the salient points, the boundary strategy, the most frequently applied strategy, reached significance. Also, the observed number of the enclosing strategy instances is well beyond chance level, whereas the excluding strategy seems not to be used systematically by our participants. Summarizing, the salient points also seem to exhibit an indirect influence via the enclosing strategy in positioning intervals in the configuration tasks.

Table 4. Usage of strategies

Strategy	Absolute Frequency[*]	PMC-estimation
Boundary	625	.0001
Enclosing	520	.0001
Excluding	264	.5821
No strategy / unknown other	1291	

[*] n = 2700 (3 intervals x 30 tasks x 30 participants)

5   General Discussion and Conclusions

As the results clearly show, we found ample evidence that preferred solutions exist in solving configuration tasks on linear shape curves. The degree of concordance was even higher than in experiments with straight linear segments. One possible explanation could be that linear shape curves put further demands on the processing capacities that could result in a more conforming solving behavior.

The concordance with the preferred mental models of Knauff et al. (1995) and Kuß et al. (1996) is rather moderate. From this one could conclude that there is an influence of shape curves during the model construction and inspection process. A closer look at the experimental procedure, however, reveals that participants could have remembered the premises and positioned the first two intervals on the configuration sheet and used this external partial representation for processing the second premise to determine the position of the third interval.

The relevant properties of linear shape curves seem to be vertices, and, to some degree, inflection points—properties that are captured very well by the proposed geometric formalization. Since smooth points with second order discontinuities played no significant role in the produced configurations, the formalization has not to be augmented to describe this class of points. This leads to the important consequence that only those shape features are employed in solving configuration-based problems which can be characterized on the basis of ordering geometry. The points with second order discontinuities are employed in the same way as other smooth points. Therefore, there is no need to assume a spatially richer formalization to enable a finer classification. This implies in particular that any metrical information, any data about coordinates or any knowledge of higher order smoothness is not required since the shape geometry is based on ordering geometry which does not rely on these concepts. This is consistent with the linguistic analysis of shape nouns given in Eschenbach et al. (1998) where the geometrical system also turns out to be sufficient to characterize the essential shape concepts underlying the shape nouns. Both results support the conjecture that this planar shape geometry supplies a general inventory for the description of relevant shape features of curves.

For the future, it seems desirable to disentangle the contribution of properties of shape curves and the use of external partial representations within configuration tasks in further empirical investigations.

6   Acknowledgments

The reported research was supported by grants *Str 301/5* (Project *"MeMoSpace"*) and *Ha 1237/7* (Project *"Axiomatik räumlicher Konzepte"*) in the priority program *"Spatial cognition"* of the German National Research Foundation (Deutsche Forschungsgemeinschaft, DFG). We are in particular indebted to Carola Eschenbach, Christopher Habel, Cornelius Hagen, Markus Knauff, Paul Lee, Florian Röhrbein, Christoph Schlieder, Gerhard Strube, and an anonymous reviewer for their valuable comments. This paper also benefits from fruitful discussions in the Hamburg Working Group on Spatial Cognition. We greatly appreciate the help of Daniel van den Eijkel for programming the Monte Carlo simulations. We also thank Katrin Balke, Thomas Kuß, Goran Sunjka, and Dirk Zugenmaier for their assistance in several parts of the empirical investigation, and Patrick Mueller for proof-reading an earlier draft.

7   References

Allen, J. F. (1983). Maintaining knowledge about temporal intervals. *Communications of the ACM, 26,* 832–843.

Eschenbach, C., Habel, C., Kulik, L., & Leßmöllmann, A. (1998). Shape nouns and shape concepts: A geometry for 'corner'. In C. Freksa, C. Habel, & K. F. Wender (Eds.), *Spatial cognition. An interdisciplinary approach to representing and processing spatial knowledge* (pp. 177–201). Berlin, Heidelberg: Springer.

Eschenbach, C., Habel, C., & Leßmöllmann, A. (in press). Multiple frames of reference in interpreting complex projective terms. In P. Olivier (Ed.), *Spatial language: Cognitive and computational aspects.* Dordrecht: Kluwer.

Eschenbach, C., & Kulik, L. (1997). An axiomatic approach to the spatial relations underlying 'left'–'right' and 'in front of'–'behind'. In G. Brewka, C. Habel, & B. Nebel (Eds.), *KI-97: Advances in Artificial Intelligence* (pp. 207–218). Berlin: Springer.

Gigerenzer, G., & Hug, K. (1992). Domain-specific reasoning: Social contracts, cheating, and perspective change. *Cognition, 43,* 127–171.

Habel, C., & Eschenbach, C. (1997). Abstract structures in spatial cognition. In C. Freksa, M. Jantzen, & R. Valk (Eds.), *Foundations of Computer Science. Potential – Theory – Cognition* (pp. 369–378). Berlin: Springer.

Johnson-Laird, P. N. (1972). The three-term series problem. *Cognition, 1,* 58–82.

Johnson-Laird, P. N. (1983). *Mental models. Towards a cognitive science of language, inference, and consciousness*. Cambridge, MA: Harvard University Press.

Johnson-Laird, P. N., & Byrne, R.M.J. (1991). *Deduction*. Hove: Lawrence Erlbaum Associates.

Johnson-Laird, P. N., Legrenzi, P., & Legrenzi, M. S. (1972). Reasoning and a sense of reality. *British Journal of Psychology, 63*, 395–400.

Knauff, M., Rauh, R., & Schlieder, C. (1995). Preferred mental models in qualitative spatial reasoning: A cognitive assessment of Allen's calculus. In *Proceedings of the Seventeenth Annual Conference of the Cognitive Science Society* (pp. 200–205). Mahwah, NJ: Lawrence Erlbaum Associates.

Knauff, M., Rauh, R., Schlieder, C., & Strube, G. (1998). Mental models in spatial reasoning. In C. Freksa, C. Habel, & K. F. Wender (Eds.), *Spatial cognition. An interdisciplinary approach to representing and processing spatial knowledge* (pp. 267-291). Berlin: Springer.

Kosslyn, S. M. (1994). *Image and brain*. Cambridge, MA: MIT Press.

Kuß, T., Rauh, R., & Strube, G. (1996). Präferierte mentale Modelle beim räumlich-relationalen Schließen: Eine Replikations- und Validierungsstudie. In R. H. Kluwe & M. May (Eds.), *Proceedings der 2. Fachtagung der Gesellschaft für Kognitionswissenschaft* (pp. 81–83). Hamburg: Universität Hamburg.

Ligozat, G. (1990). Weak representations of interval algebras. *Proceedings of the Eighth National Conference on Artificial Intelligence* (Vol. 2, pp. 715–720). Menlo Park, CA: AAAI Press / MIT Press.

Logie, R. H. (1995). *Visuo-spatial working memory*. Hillsdale, NJ: Lawrence Erlbaum.

Molenaar, M., & De Hoop, S. (Eds.). (1994). *Advanced geographic data modelling: Spatial data modelling and query languages for 2D and 3D applications*. Delft: Nederlands Geodetic Commission.

Nebel, B., & Bürckert, H.J. (1995). Reasoning about temporal relations: A maximal tractable subclass of Allen's interval algebra. *Communication of the ACM, 42*, 43–66.

Potts, G. R., & Scholz, K.W. (1975). The internal representation of a three-term series problem. *Journal of Verbal Learning and Verbal Behavior, 14*, 439–452.

Rauh, R. (2000). Strategies of constructing preferred mental models in spatial relational inference. In W. Schaeken, G. De Vooght, A. Vandierendonck, & G. d'Ydewalle (Eds.), *Deductive reasoning and strategies* (pp. 177–190). Mahwah, NJ: Lawrence Erlbaum Associates.

Rauh, R., & Schlieder, C. (1997). Symmetries of model construction in spatial relational inference. In *Proceedings of the Nineteenth Annual Conference of the Cognitive Science Society* (pp. 638–643). Mahwah, NJ: Lawrence Erlbaum Associates.

Schlieder, C. (1999). The construction of preferred mental models in reasoning with interval relations. In G. Rickheit & C. Habel (Eds.), *Mental models in discourse processing and reasoning* (pp. 333–357). Amsterdam: Elsevier Science Publishers.

Schlieder, C., & Berendt, B. (1998). Mental model construction in spatial reasoning: A comparison of two computational theories. In U. Schmid, J. F. Krems, & F. Wysotzki (Eds.), *Mind modelling: A cognitive science approach to reasoning, learning and discovery* (pp. 133–162). Lengerich, Germany: Pabst Science Publishers.

Schlieder, C., & Hagen, C. (2000). Interactive layout generation with a diagrammatic constraint language. This volume.

Walischewski, H. (1997). Learning and interpretation of the layout of structured documents. In G. Brewka, C. Habel, & B. Nebel (Eds.), *KI-97: Advances in Artificial Intelligence* (pp. 409–412). Berlin, Heidelberg, New York: Springer.

Wason, P. C. (1966). Reasoning. In M. B. Foss (Ed.), *New horizons in psychology*. Harmondsworth: Penguin Books.

Transfer of Spatial Knowledge
from Virtual to Real Environments

Patrick Péruch, Loïc Belingard, and Catherine Thinus-Blanc

Centre de Recherche en Neurosciences Cognitives
CNRS, 31 Chemin Joseph Aiguier, 13402 Marseille cedex 20, France
{peruch,thinus}@lnf.cnrs-mrs.fr
http://lnf.cnrs-mrs.fr/crnc/space/espace.htm

Abstract. The transfer of spatial knowledge from virtual to real environments is one important issue in spatial cognition research. Up to now, studies in this domain have revealed that the properties of spatial representations are globally the same in virtual and real environments, and in most cases transfer of spatial information from one kind of environment to the other occurs. Although these results suggest that virtual environments contain much of the spatial information used in real environments, it seems difficult or even impossible to draw any clear conclusion about the spatial information which is transferred and about the conditions of transfer. Being able to quantitatively and/or qualitatively predict and observe such a transfer would broaden the possibilities of training and our knowledge of the cognitive processes involved in spatial behavior. In a first step, arguments in this sense are developed on the basis of a review of some recent studies concerned with the transfer of spatial knowledge between virtual and real environments. In a second step, empirical data are reported, that illustrate the interest and limits of such studies.

1 Introduction: Spatial Cognition and Spatial Knowledge

In large spaces people are frequently required to move towards unseen goals, and therefore they must plan their movements. To do so, spatial knowledge about the environment is required, which may be in the form of a physical map or a mental (cognitive) map. Many studies have documented the processes involved in the mental representation of realworld environments (see Golledge, 1987, for a review), and have shown that different levels of spatial knowledge, from "route" to "survey" (Siegel & White, 1975) are elaborated and required according to the task to be performed. Reproducing a familiar route is possible with only a route-type mental representation, while taking a shortcut or selecting a new route is supposed to require a survey-type mental representation. Several alternative models have been proposed recently. For instance, Montello (1998) argues for a quantitative rather than qualitative evolution in the acquisition of spatial knowledge, and Thinus-Blanc and Gaunet (1999) suggest that spatial representations are necessary not only for planning spatial behavior, but they also control the organization of the spontaneous acquisition of information. The features of cognitive maps have been extensively investigated in different situations

Ch. Freksa et al. (Eds.): Spatial Cognition II, LNAI 1849, pp. 253-264, 2000.
© Springer-Verlag Berlin Heidelberg 2000

and in various populations. Among others, a strong aspect of cognitive maps is their inaccuracy: it is well known that mental representations are deformed, distorted (e.g., Tversky, 1981). Such features are for instance expressed in errors made on direction and/or distance estimates between places or landmarks. Also, studies on experts are particularly interesting since they may give us a general idea of what could be an optimal situation. One well-known example is reported in Pailhous (1970)[1], who distinguished primary (or basic) and secondary networks in mental representations of Parisian taxi-drivers. The primary network, mainly composed of large streets (about 10%), roughly corresponds to the skeleton of the mental representation. One of the studies reported in this chapter (Section 4, Study 1) is based on this idea.

Humans acquire spatial knowledge both while traveling through environments and through the use of maps, photographs, videotapes, verbal (oral or written) descriptions, and, most recently, virtual environments. The literature provides abundant evidence that the nature of the cognitive map formed depends to some degree on the information available during knowledge acquisition (e.g., Thorndyke & Hayes-Roth, 1982). For instance, Presson and Hazelrigg (1984) make the distinction between primary (or direct, such as navigation) and secondary (or symbolic) learning. Learning from a map results in a mental spatial representation which has a specific orientation, while learning from navigation results in a more flexible mental representation.

2   Virtual Environment Technology (VET)

In their Preface of a Special Section of the Journal "Humans Factors" devoted to Virtual Environments (1998), Barfield and Williges define Virtual Environment (VE) and Virtual Environment Technology (VET). "The term VE represents a family of computer-generated virtual representations of human visual, proprioceptive, haptic, auditory, and olfactory displays". ..."Under the general rubric of VET, there is all the technology related to virtual reality, augmented reality, visualization, head-mounted displays, desktop computer displays, wall/room projections, perspective displays, computer walk-throughs, stereoscopic displays, wearable computers, computer-based simulations, and so forth".

2.1   VET Systems

Two main categories of systems are currently used: desk-top systems (which display the virtual environment on a fixed computer screen) and immersive-display systems (e.g., the environment is displayed on two small screens of a head-mounted display).

---

[1]   Pailhous, J. (1970). La représentation de l'espace urbain: L'exemple du chauffeur de taxi. Paris: Presses Universitaires de France. For an English summary of Pailhous's work see (1984) The representation of urban space: its development and its role in the organisation of journeys. In R. Farr & S. Moscovici (Eds.), Social Representations. Cambridge: Cambridge University Press. See also Golledge (1987).

In desk-top systems the direction of gaze is altered by translations and rotations from an input device (e.g., joystick or mouse), while in immersive-display systems the direction of gaze is linked to head-movements. Moreover, vestibular information is available from immersive-display systems only. Finally, due to head-movements the scene occupies a wider angle in immersive-display systems than in desk-top systems, though image quality is generally poorer. In summary, both systems have specific advantages and limitations, and their use is tightly related to the nature of questions under study (see for example Arthur, Hancock, & Chrysler, 1997; Pausch, Shackelford, & Proffitt, 1993; Ruddle, Payne, & Jones, 1999). The same line of reasoning can be applied to the output interfaces that produce the movements, that is, keyboard, joystick, treadmill, etc.

2.2   VET and Spatial Cognition

Among others, one increasing application of VET is spatial cognition studies. The potential advantages and drawbacks of this technology have been reported by several authors (see Darken, Allard, & Achille, 1998; Loomis, Blascovitch, & Beall, 1999; Mallot, Gillner, van Veen, & Buelthoff, 1998; Péruch & Gaunet, 1998; Péruch, Gaunet, Thinus-Blanc, & Loomis, 2000; Wilson, 1997). The main advantages of VET is the possibility it gives to create environments of varying complexity, to make on-line measurements during (interactive) navigation, to control many spatial learning parameters, such as the amount of exposure to the environment and the number, position, and nature of landmarks. However, the present state-of-art of this technology has several drawbacks: lack of realistic environmental modelling, slow image generation and rendering, narrow field of view, optical distortions, poor spatial resolution, etc. In spite of these, the relevance of VET for spatial cognition studies becomes more and more evident.

Although VET is relatively recent, cognitive mapping studies using it have already investigated several different issues. In brief, as is often the case when a new technology first comes into existence, investigators have first tried to replicate (or/and extend) some basic experiments dealing with cognitive mapping. An example is the study by Ruddle, Payne, and Jones (1997), in which most of the findings obtained by Thorndyke and Hayes-Roth (1982) in real settings were confirmed within virtual environments: participants were found to be equally good at performing direction and relative distance estimates. Such research confirms that the same processes operate in virtual (in general, purely visual) conditions and, at the same time, provides further validation of using VET for research in spatial cognition.

Since VET permits the isolation of sensory modalities, other studies are concerned with the respective role of external (for instance visual) vs internal (for instance proprioceptive or vestibular) information in spatial learning (see for instance Chance, Gaunet, Beall, & Loomis, 1998; Klatzky, Loomis, Beall, Chance, & Golledge, 1998).

Some other studies have investigated the properties of the cognitive maps that are elaborated in virtual environments, through navigation and way-finding tasks (for a discussion see also Darken et al., 1998). With natural environments, navigation results in a representation that does not depend on a particular orientation while the use of maps is tightly related to the orientation in which they have been presented (see for instance Presson & Hazelrigg, 1984). Results supporting this hypothesis have been found with navigation in virtual environments though other data are more or less divergent (see for instance Tlauka & Wilson, 1996) or more or less contradictory (see for instance May, Péruch, & Savoyant, 1995; Péruch & Lapin, 1993; Rossano & Moak, 1998; Richardson, Montello, & Hegarty, 1999).

Another important issue in spatial cognition is the status and role of landmarks. The following questions have been investigated: "What is a landmark? How is it used? Why is a cue, among so many others, selected and used as a landmark?". Using VET, Tlauka and Wilson (1994) have investigated to what extent landmarks are decisive in the acquisition of route knowledge in a virtual environment: performance was higher in the landmark group than in the non-landmark group. Steck and Mallot (1998) have addressed the question of the role of global and local landmarks in navigation, in situations in which landmark information was manipulated. Moreover, Jacobs, Thomas, Laurance, and Nadel (1998) have conducted experiments in a virtual arena. Human adult participants learned to find an invisible target that remained in a fixed location relative to distant landmarks. Removing these landmarks had no important effect, while merging them (changing their topographical relations) dramatically decreased performance.

Finally, some studies are concerned with the transfer between virtual and real environments (see for example Bliss, Tidwell, & Guest, 1997; Darken & Banker, 1998; Waller, Hunt, & Knapp, 1998; Witmer, Bailey, & Knerr, 1996). They demonstrate that virtual environments contain much of the essential spatial information that is utilised by people in real environments.

3   Transfer between Virtual and Real Environments

Transfer studies make a distinction between transfer of skill (for instance Regian, Shebilske, & Monk, 1992), and transfer of spatial knowledge. This last case can be considered either as the conservation of a type of spatial knowledge from a learning to a test situation (Witmer et al., 1996), or as the transfer of spatial information from one sensory modality to an other.

The need to develop a set of tasks to support research on training applications of VET (i.e., the Virtual Environment Performance Assessment Battery or VEPAB, see Lampton, Knerr, Goldberg, Bliss, Moshell, & Blau, 1994) reveals that the transfer of skill or/and of spatial knowledge is complex. For instance, Witmer et al. (1996) suggest that although studies about the training potential of VEs have shown how task performance improves with practice, they have not indicated how skills acquired in a VE affect real world performance. The authors invoke various reasons, some of them have been already cited in section 2.1 as drawbacks of VET: VEs have deficiencies that diminish the training transfer, the visual simulation may be imprecise, the mode of production of movement is more or less artificial, and people may be affected by simulator sickness.

Waller et al. (1998) report a general assumption found in the literature, according to which knowledge or skills acquired in a VE will transfer to the real world.

Referring to Witmer et al. (1996), the authors acknowledge that one key aspect of transfer is exposure to a VE, which can substitute for actual exploration of the real world. However, the authors stress the need to examine the variables that mediate the training effects of VEs: fidelity of the interface (the mapping between the VE and the mental environment of the trainee), environmental fidelity (the mapping between the real-world environment and the VE), and training time. Waller et al. (1998) define fidelity as the "extent to which the VE and interactions with it are indistinguishable from the participant's observations and of interactions with a real environment" (p. 130). Although information about a real-world environment is never preserved perfectly in either the training environment or the trainee's mental representation (e.g., Tversky, 1981), some structures are preserved in the mapping between the three domains: fidelity is concerned with the quality of these mappings. Finally, the authors underline that a slight increase in fidelity may be very expensive. Developers are usually confronted to the following compromise: which technological variables may be most easily sacrificed without degrading the trainee performance?

Rose, Attree, Brooks, Parslow, Penn, and Ambihaipahan (1998) confirm that transfer needs for further systematic investigation. At the present time, literature reveals various intended outcomes of the training process: simple sensorimotor performance, complex sensorimotor skills, spatial knowledge of an environment, vigilance, memory, and complex problem solving. In such conditions, comparisons are difficult about the extent and type of transfer. Moreover, Rose et al. (1998) indicate that only a few authors analyse transfer in terms of the well established literature on the transfer of training (for instance Cormier & Hagman, 1987), or of the more extensive literature on the psychology of learning. Some cognitive models (e.g., Parenté & Hermann, 1996) stress the importance of the similarity, between the real and virtual situations, of stimulus and response elements but also of the cognitive strategies. In their paper, Rose et al. (1998) attempt to systematically investigate the nature of the transfer in terms of the extent and robustness of what transfers.

In summary, an overview of some recent studies on transfer shows that transfer of skill and/or of spatial knowledge occurs largely (e.g., Bliss et al., 1997; Rose et al., 1998; Waller et al., 1998) or partially (e.g., Darken & Banker, 1998; Wilson, Foreman, & Tlauka, 1997; Witmer et al., 1996). However, due to the large variability of situations and tasks, it is difficult to known exactly what type of information is actually transferred and the nature of the underlying cognitive processes. According to what has been said before, the main observation is that in some conditions we observe that training improves performance. Additionally, although one can conclude to some equivalence of spatial information in real and VEs, performance (for instance in direction estimates) is generally better in real than in virtual environments.

4   Transfer Studies Conducted in Marseille

In line with the above remarks, in the present section we briefly report two studies conducted in Marseille laboratory. The first one was aimed at evaluating the effects of the amount and quality of information on the transfer of a spatial representation and on its transfer from a VE to the corresponding actual situation. The second one compared the accuracy of representations acquired either in a VE or in the actual environment.

Virtual models of the campus of the CNRS-Marseille were used. The campus covers about 300 x 200 m of floor space and comprises roads, parkings, buildings, lawns and trees. The VET device is a desk-top PC-based graphics workstation (Pentium 133 MHz in Study 1, Pentium II 450 MHz in Study 2). The environments (colored in Study 1 and textured in Study 2) have been modeled with 3D Studio (Study 1) and 3D Studio Max (Study 2), and the real-time rendering program uses RenderWare 2.1. The scenes (in an horizontal field of view of 45 degrees) are displayed on a 21" monitor with a resolution of 640 x 480 pixels in Study 1, and of 800 x 600 pixels in Study 2. The displays are controlled with a keyboard (Study 1) and a 3D SpaceMouse (Study 2) and, according to the complexity of the scene, the frame rate varies from 10 to 20 images per second.

4.1   Study 1: Varying the Amount and/or Quality of Available Information

This study aimed at examining if the increase of amount and/or quality of available information facilitates spatial learning, and how. Thirty adult participants (fifteen females and fifteen males), who did not know the campus, learned a virtual version of it before being tested in the real situation. During the learning phase participants freely explored the environment without time limit. They were encouraged to memorize 6 locations marked by objects and appearing on 6 pictures that were shown permanently. The participants were never shown a top-view of the environment. Three versions of the campus were compared, one for each group of ten participants (Fig. 1).

In all these versions, the buildings were represented by grey, not textured, volumes. Since only the geometry of the environment was maintained, the degree of realism was very weak. In the most detailed or Rich version (Fig. 1a) the buildings, the lawns, the hedges, and the main roads (linking the locations) were represented. The Medium version (Fig. 1b) comprised only some buildings and the main roads. Finally, the Poor version (Fig. 1c) comprised only a subset of the buildings. After learning (from 15 to 40 minutes) the participants were blindfolded and transported by car to the different locations. From each location, they had to indicate the direction, the travelled distance (using the shortest path), and the direct distance of the other locations. Each participant performed ninety tests (6 locations occupied x 5 locations tested x 3 measures).

The results (Table 1) show that direction and travelled distance errors were smaller in the Rich and Medium conditions than in the Poor condition; these results were confirmed by statistical analyses. No difference was observed between the conditions on the direct distance estimates. Performance was equivalent in the Medium and Rich conditions, revealing that there was no effect of the amount of available information. In brief, the best performance occured in the presence of roads linking the locations. Such roads may have played the role of primary grid (Pailhous, 1970), that is, may have structured the mental representation of the environment. In summary, these results show that some transfer of spatial knowledge from a virtual to a real environment is possible even in very schematic virtual conditions. The second study deals with the transfer of spatial knowledge using a more realistic model of the campus.

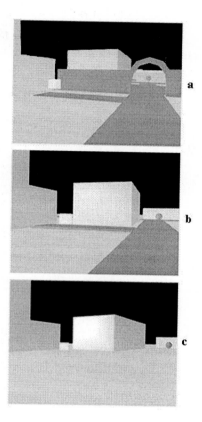

Fig. 1. Perspective views in the different models of the campus of the CNRS-Marseille, taken from the same place. The Rich (a), Medium (b), and Poor (c) views come from the models used in Study 1. Two of the objects marking locations (the white cube on the left and the grey sphere on the right) are visible on each of the views

Table 1. Average (and SE of the mean) performance values by condition in Study 1

	Absolute direction error (degrees)	Travelled distance error (normalized)	Direct distance error (normalized)
Rich	42.5 (10)	.21 (.10)	.20 (.09)
Medium	47.0 (11)	.19 (.09)	.18 (.08)
Poor	62.0 (13)	.26 (.18)	.22 (.12)

4.2 Study 2: Learning and Testing in Virtual vs Real Conditions

A central question investigated here was to evaluate to what extent virtual training (in desk-top mode) can be compared to real training (with all internal and external information). Moreover, to our knowledge no similar study has been conducted in an outside environment: investigations have been carried out only in buildings. Thus, this study aimed at examining the conditions of transfer of spatial information from an exterior virtual (but realistic) environment to a real one.

Fig. 2. Views of the campus of the CNRS-Marseille from Study 2. View (a) shows a part of the virtual environment, while views (b) and (c) correspond approximately to the same views of Study 1 in the virtual and in the real campus, respectively

Two groups of eight adult female and male participants who did not know the campus explored it freely for 20 minutes. The Virtual group explored the virtual campus (Fig. 2a and 2b) while the Real group explored the real campus (Fig. 2c). During the exploration, each participant was requested to search for 5 target locations, corresponding to 5 pictures that were shown permanently. The pictures had been taken in the virtual campus for the Virtual group and in the real campus for the Real group, but they corresponded to the same locations. As in Study 1, the participants were never shown a top-view of the environment. At the end of the exploration, both groups were tested in the virtual environment first and then in the real environment. In the Virtual tests the participants were dropped at each of the locations, while in the Real tests they were first requested to find each of the locations using the shortest path. From each location, the participants had to indicate the direction, direct distance, and travelled distance of all other locations. Each participant performed sixty tests (5 locations occupied x 4 locations tested x 3 measures).

The results (Table 2) show that the Real group performed better than the Virtual group, but the performance was significantly different on Virtual tests only; these results were confirmed by statistical analyses. The performance of the Real group was the same on both Virtual and Real tests, indicating that transfer from the real to the virtual environment globally occured. No improvement of performance was observed from the virtual to the real tests (except on direction error), suggesting that learning was optimal in the real world. By contrast, the performance of the Virtual group improved from the virtual to the real tests on all aspects of performance: the fact that all sensory information was available in the real world, on the one hand, and that participants had more experience of the environment because they walked to the real test conditions, on the other hand, has probably facilitated the transfer of spatial information from the virtual to the real environment.

Table 2. Average (and SE of the mean) performance values by condition in Study 2

Group	Test	Absolute error direction (degrees)	Absolute travelled distance error (meters)	Absolute direct distance error (meters)
Virtual	Virtual	66.8 (4.2)	101.5 (7.0)	72.0 (5.5)
	Real	30.8 (3.1)	72.5 (5.3)	51.0 (4.5)
Real	Virtual	27.8 (2.3)	58.0 (3.5)	40.5 (2.5)
	Real	19.7 (1.5)	56.5 (3.5)	41.5 (2.5)

4.3   Discussion

In summary, data from Study 1 show that the quality of available virtual information is more important than its amount to construct a spatial representation that has to be transferred to the actual environment. In addition, data from Study 2 suggest that although learning in the actual environment results in better quality representations than training in a VE, the transfer of spatial knowledge from a virtual to a real environment is possible to some extent. The experimental conditions in the two above studies were somewhat different, but performance can be compared (at least on direction estimates). In Study 2, direction estimates were better in the Real tests for the Virtual group (30.8 degrees) than those observed in the Rich condition of Study 1 (42.5 degrees). This means that an accurate mental representation is easier to acquire in a realistic virtual environment than in very schematic virtual conditions, although in this last case performance can be optimized by roads linking the locations.

These studies are probably among the first that evaluate the conditions of transfer of spatial knowledge between exterior, virtual and real environments. It appears that training in a pure visual mode (using a desk-top system) may be sufficient to acquire a coherent mental representation. Moreover, this representation is less well elaborated but may be as performant as the representation that may be acquired in the real environment. It is likely that a representation coming from a virtual experience may benefit from real-world experience. In other words, spatial knowledge acquired in a virtual environment (on a pure visual basis) could be used optimally in real conditions (that is, in situations more natural and rich with respect to the variety of available information). In such conditions, a good transfer of spatial knowledge would require several alternated experiences, both in virtual and real environments.

5   Conclusion: Future Work on Transfer

Studies concerned with the transfer of spatial knowledge from virtual to real environments demonstrate that the properties of spatial representations are not radically different in virtual environments than in real ones. These results suggest that virtual environments contain much of the spatial information used in real environments. However, drawing a clear conclusion about the spatial information which is transferred and about the conditions of transfer remains problematic. Being able to predict such a transfer would broaden the possibilities of training and enlarge our knowledge of the cognitive processes involved in spatial cognition.

In summary, VET has been only recently used in spatial cognition research, so transfer studies are just beginning. Gathering more experimental evidence would certainly strengthen the relevance of using VET in human spatial cognition studies. Other steps are of course necessary before strong conclusions can be drawn. Obviously, significant progress will be made only if researchers have a good knowledge of human spatial behavior. Among others, one goal for future research would be to try to define optimal conditions of transfer. It is likely that, among others, two kinds of improvements will play a decisive role here. First, more and more complex and realistic VEs will be available. Second, VET will be more and more combined with other techniques related to movement (whole-body interfaces) and/or to navigation (Geographical Positioning Systems).

Acknowledgements

This work was supported by Grant JC-981121-A000 from the Direction Générale pour l'Armement (DGA), and by Grant 2000364-03 from the Fondation pour la Recherche Médicale (FRM). We thank Jehan-Charles Malotaux who conducted Study 1, Jack Loomis for valuable discussions, Roy Ruddle for providing helpful remarks, Michel Taillard for developing the graphics softwares, and an anonymous reviewer.

References

Arthur, E.J., Hancock, P.A., Chrysler, S.T. (1997) The perception of spatial layouts in real and virtual worlds. Ergonomics, 40, 69-77.

Barfield, W., Williges, R.C. (1998) Special Section Preface. Human Factors, 40, 351-353.

Bliss, J.P., Tidwell, P.D., Guest, M.A. (1997) The effectiveness of virtual reality for administering spatial navigation training for firefighters. Presence, 6, 73-86.

Chance, S.S., Gaunet, F., Beall, A., Loomis, J.M. (1998) Locomotion mode affects the apprehension of spatial layout: The contribution of vestibular and proprioceptive inputs. Presence, 7, 168-178.

Cormier, S.M., Hagman, J.D. (1987) Transfer of learning. Contemporary research and applications. London: Academic Press.

Darken, R.P., Allard, T., Achille, L.B. (1998) Spatial orientation and wayfinding in large-scale virtual spaces: An introduction. Presence, 7, 101-107.

Darken, R.P., Banker, W.P. (1998) Navigating in Natural Environments: A Virtual Environment Training Transfer Study. Proceedings of VRAIS '98, 12-19.

Golledge, R.G. (1987) Environmental cognition. In Stokols, D. and Altman, I. (Eds.), Handbook of Environmental Psychology, Vol 1. New York: Wiley, pp. 131-174.

Jacobs, W.J., Thomas, K.G.F., Laurance, H.E., Nadel, L. (1998) Place learning in virtual space II: Topographical relations as one dimension of stimulus control. Learning and Motivation, 29, 288-308.

Klatzky, R.L., Loomis, J.M., Beall, A.C., Chance, S.S., Golledge, R.G. (1998) Spatial updating of self-position and orientation during real, imagined, and virtual locomotion. Psychological Science, 9, 293-298.

Lampton, D.R., Knerr, B.W., Goldberg, S.L., Bliss, J.P., Moshell, J.M., Blau, B.S (1994) The Virtual Environment Performance Assessment Battery (VEPAB): Development and evaluation. Presence, 3, 145-157.

Loomis, J.M., Blascovich, J.J., Beall, A.C. (1999) Immersive virtual environment technology as a basic research tool in psychology. Behavior Research Methods, Instruments, & Computers, 31, 557-564.

Mallot, H.A., Gillner, S., van Veen, H.A.H.C., Buelthoff, H.H. (1998) Behavioral experiments in spatial cognition using virtual reality. In C. Freksa, C. Habel, and K.F. Wender, (Eds.), Spatial Cognition. An interdisciplinary approach to representing and processing spatial knowledge. Berlin: Springer, pp. 447-467.

May, M., Péruch, P., Savoyant, A. (1995) Navigating in a virtual environment with map-acquired knowledge: Encoding and alignment effects. Ecological Psychology, 7, 21-36.

Montello, D.R. (1998) A new framework for understanding the acquisition of spatial knowledge in large-scale environments. In R.G. Golledge and M.J. Egenhofer (Eds.), Spatial and temporal reasoning in geographic information systems. New York: Oxford University Press, pp. 143-154.

Parenté, R., Hermann, D. (Eds.) (1996) Transfer and generalisation of learning. Retraining cognition. Techniques and applications. Gaithersburg: Aspen Publications.

Pausch, R., Shackelford, M.A., Proffitt, D. (1993) A user study comparing head-mounted and stationary displays. Proceedings of IEEE Symposium on Research Frontiers in Virtual Reality, 41-45.

Péruch, P., Gaunet, F. (1998) Virtual environments as a promising tool for investigating human spatial cognition. Current Psychology of Cognition, 17, 881-899.

Péruch, P., Gaunet, F., Thinus-Blanc, C., Loomis, J. (2000) Understanding and learning virtual spaces. In R.M. Kitchin and S. Freundschuh (Eds.), Cognitive mapping: Past, present and future. London: Routledge, pp. 108-124.

Péruch, P., Lapin, E. (1993) Route knowledge in different spatial frames of reference. Acta Psychologica, 84, 253-269.

Presson, C.C., Hazelrigg, M.D. (1984) Building spatial representations through primary and secondary learning. Journal of Experimental Psychology: Learning, Memory, and Cognition, 10, 716-722.

Regian, J.W., Shebilske, W.L., Monk, J.M. (1992) Virtual reality: An instructional medium for visual-spatial tasks. Journal of Communication, 42, 136-149.

Richardson, A.E., Montello, D.R., Hegarty, M. (1999) Spatial knowledge acquisition from maps, and from navigation in real and virtual environments. Memory & Cognition, 27, 741-750.

Rose, F.D., Attree, E.A., Brooks, B.M., Parslow, D.M., Penn, P.R., Ambihaipahan, N. (1998) Transfer of training from virtual to real environments. In P.M. Sharkey, F.D. Rose, and J.I. Lindström (Eds.), Proceedings of the 2d European Conference on Disability, Virtual Reality and Associated Technologies. Skövde, Sweden. ISBN: 0-7049-1141-8.

Rossano, M.J., Moak, J. (1998) Spatial representations acquired from computer models: Cognitive load, orientation specificity and the acquisition of survey knowledge. British Journal of Psychology, 89, 481-497.

Ruddle, R.A., Payne, S.J., Jones, D.M. (1997) Navigating buildings in desk-top virtual environments: Experimental investigations using extended navigational experience. Journal of Experimental Psychology: Applied, 3, 143-159.

Ruddle, R.A., Payne, S.J., Jones, D.M. (1999) Navigating large-scale virtual environments: What differences occur between helmet-mounted and desk-top displays? Presence, 8, 157-168.

Siegel, A.W., White, S.H. (1975) The development of spatial representations of large-scale environments. In Reese, H.W. (Ed.), Advances in Child Development and Behavior, Vol. 10. New York: Academic Press, pp. 9-55.

Steck, S.D., Mallot, H.A. (1998) The role of global and local landmarks in virtual environment navigation. Max-Planck-Institute for Biological Cybernetics, Technical Report n°63.

Thinus-Blanc, C., Gaunet, F. (1999) The organizing function of spatial representations. Spatial processing in animal and man. In Golledge, R.G. (Ed.), Wayfinding: Cognitive mapping and spatial behavior. Baltimore, MD: John Hopkins, pp. 294-307.

Thorndyke, P.W., Hayes-Roth, B. (1982) Differences in spatial knowledge acquired from maps and navigation. Cognitive Psychology, 14, 560-589.

Tlauka, M., Wilson, P.N. (1994) The effects of landmarks on route-learning in a computer-simulated environment. Journal of Environmental Psychology, 14, 305-313.

Tlauka, M., Wilson, P.N. (1996) Orientation-free representations from navigation through a computer simulated environment. Environment and Behavior, 28, 647-664.

Tversky, B. (1981) Distortions in memory for maps. Cognitive Psychology, 13, 407-433.

Waller, D., Hunt, E., Knapp, D. (1998) The transfer of spatial knowledge in virtual environment training. Presence, 7, 129-143.

Wilson, P.N. (1997) Use of virtual reality computing in spatial learning research. In Foreman, N. and Gillett, R. (Eds.), Handbook of Spatial Research Paradigms and Methodologies. Vol. 1: Spatial Cognition in the Child and Adult. Hove, U.K.: Psychology Press, pp. 181-206.

Wilson, P.N., Foreman, N., Tlauka, M. (1997) Transfer of spatial information from a virtual to a real environment. Human Factors, 39, 526-531.

Witmer, B.G., Bailey, J.H., Knerr, B.W. (1996) Virtual spaces and real world places: Transfer of route knowledge. International Journal of Human-Computer Studies, 45, 412-428.

Coarse Qualitative Descriptions
in Robot Navigation

Rolf Müller[1], Thomas Röfer[1], Axel Lankenau[1],
Alexandra Musto[2], Klaus Stein[2], and Andreas Eisenkolb[3]

[1] Universität Bremen, Postfach 330440, 28334 Bremen, Germany
{rmueller,roefer,alone}@tzi.de
[2] Technische Universität München, 80290 München, Germany
{musto,steink}@in.tum.de
[3] Ludwig-Maximilians-Universität München, 80336 München, Germany
amadeus@imp.med.uni-muenchen.de

Abstract. This work is about the integration of the skills *robot control*, *landmark recognition*, and *qualitative reasoning* in a single autonomous mobile system. It deals with the transfer of coarse qualitative route descriptions usually given by humans into the domain of mobile robot navigation. An approach is proposed that enables the Bremen Autonomous Wheelchair to follow a route in a building, based on a description such as "Follow this corridor, take the second corridor branching off on the right-hand side and stop at its end." The landmark recognition uses a new method taken from the field of image processing for detecting significant places along a route.

1 Introduction

When designing human-computer interfaces for computer systems that solve configuration tasks (e.g. layout-manager), move in space on their own (e.g. semi-autonomous robots), or help humans to move in space (e.g. navigation systems), a good understanding of humans' spatial mental models is important.

It is well known [15] that in these spatial mental models the relations between the elements are coarse and include no metrical information. Typical spatial expressions are "next to", "left of", "east of", etc. Humans' spatial models typically do not only include no metric information, there also occur systematic distortions that influence judgements on distances and directions, e.g. the distance from a landmark to an ordinary building is judged smaller than the other way round, which leads to asymmetric distances [12].

Considering these findings, we see that computer systems dealing with motion in space should be able to understand coarse, *qualitative* relations and should be robust against errors humans make due to systematic distortions, e.g., they should not rely too much on metric information.

In the work described in this paper, qualitative route descriptions are used in an application from the robotics domain, i.e. controlling a semi-autonomous wheelchair along a route. In the field of spatial reasoning, qualitative relations

Ch. Freksa et al. (Eds.): Spatial Cognition II, LNAI 1849, pp. 265–276, 2000.
© Springer-Verlag Berlin Heidelberg 2000

are established between complex objects such as a *refrigerator* or a *person*, e. g. in [2]. In contrast, the items in the qualitative descriptions presented here are very simple, because the autonomous mobile system must be able to recognize them with its limited sensory equipment.

The intended scenario is as follows: In a hospital, a patient should visit a certain room, e. g. for having a medical examination. He or she is handicapped, so a wheelchair is used to travel to the examination room. Normally, the patient would be guided by a nurse. But, considering all the medical examination performed each day in a hospital, this costs a lot of the staff's time. Therefore, the hospital is equipped with intelligent power wheelchairs, enabling the nurse to instruct the wheelchair where to go. Then, the patient is automatically transported to the examination room. Currently, the experiments are carried out in an office building in the University of Bremen. Even though this building is accessible for wheelchairs, navigating there is more complex than in the hospital environment because the corridors and the doors are comparatively narrow.

The *Bremen Autonomous Wheelchair "Rolland"* serves as the experimental platform. It is based on the commercial power wheelchair *Genius 1.522* manufactured by the German company Meyra. The wheelchair is a non-holonomic vehicle that is driven by its front axle and steered by its rear axle. The human operator controls the system with a joystick. In addition, an external keyboard can be used by the service staff, e. g. to type in some instructions. The wheelchair is equipped with a standard PC (Pentium 233, 64 MB RAM) and a ring of sonar sensors to perceive its environment. Furthermore, the system is able to perform dead reckoning by measuring its speed and steering angle.

2   System Architecture

To navigate through an environment along a specified route requires a variety of skills: measuring locomotion (dead reckoning), perceiving the surroundings (obstacle and landmark detection), self-localization (mapping from reality to route description), planning (choosing an appropriate action in each situation), and moving as such.

In a technical system, the navigation skills can be implemented as asynchronously communicating hardware and software components that run in parallel. In the Bremen Autonomous Wheelchair, the communication is done via a real-time capable network (for more information cf. [5]). Since the implementation details do not matter here, Fig. 1 shows a schematic overview of the architecture and the information flow in the system described in this paper.

The wheelchair runs in a control loop of 32 ms cycles and provides three components relevant here: a *state monitor* that supplies information about the current state of the actuators, 27 *sonar sensors* that measure the distance to objects in the surroundings, and the *motor* which accepts driving commands.

In order to hide the specific properties of the hardware, an abstraction of a safe wheelchair had been introduced (cf. [5,9]). It is called *SAM* (short for Sensor/Actuator-Module). SAM runs in real-time, i. e. its main loop must not

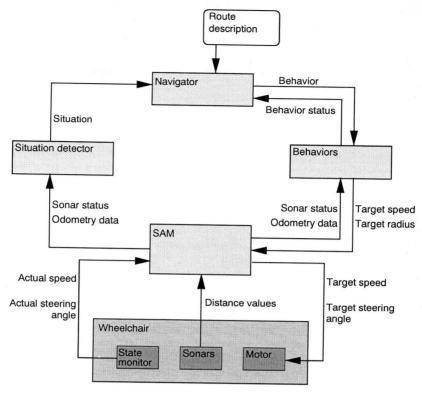

Fig. 1. System architecture (schematic overview; details in the text)

take longer than the 32ms frame time prescribed by the wheelchair. It processes the data produced by the wheelchair and the sensors, waits for driving commands in a (*speed, radius*)-format from other modules, and sends safe target motor commands to the vehicle. Furthermore, SAM provides the higher-level modules *Situation detector* and *Behaviors* with odometry data it derived from the actual speed and the actual steering angle by dead reckoning. In addition, it delivers the current status of the sonar sensors, i. e. the distance value measured by each sensor, the global position where the wheelchair was located when taking that specific measurement, and the absolute point in time of the measurement.

Concurrent to SAM, three high-level modules are responsible for the skills of environment perception and self-localization (*Situation detector*, cf. section 3), choosing adequate motion behaviors (*Navigator*, cf. section 4) and executing these behaviors (*Behaviors*, cf. [4]).

The situation detector extracts features from the sonar image that are mapped to certain landmark types, such as "CorridorRight". In the current implementation, this module produces a data record every 10 cm travel distance.

It contains the actual position of the wheelchair relative to its starting-point, and a 5-bit feature vector that indicates the detected landmark type.

This information is matched by the navigator with the initial route description which specifies the target track. According to the current situation, the navigator processes a behavior the execution of which allows the wheelchair to follow the target route.

This behavior (e. g. "FollowRightWall") is passed to the third high-level module. It simply converts the active behavior in a target motor command that consists of a speed and a radius component. This command is sent to SAM.

In the sequel, the situation detector and the navigator are described in detail.

3   Perceiving the Environment: Situation Detector

With its limited sensory equipment, the wheelchair can only perceive a small part of its environment. The data is analyzed to detect landmarks, which are used as points of reference during navigation.

In order to recognize them, the situation detector module (cf. section 2) processes the readings of the wheelchair's sonar sensors in several steps. At first, corridor walls are determined with a line detection algorithm. Then a coarse grid of relative positions in the wheelchair's immediate surroundings is searched for prominent properties, such as "there is a wall aligned with the corridor direction on the right side in front". Finally, typical patterns in the coarse grid are recognized yielding landmarks, e. g., "the corridor turns left".

3.1   Probabilistic Obstacle Map

The sonar sensors' measurements are accumulated in a probabilistic obstacle map. This is a local grid map covering an area of $4\,m \times 4\,m$. The center of the wheelchair's front axle is located in the center of the map. For each cell in this map, two values are kept track of: the number of measurements that covered this cell and the number of measurements that detected an obstacle in this cell. The ratio of the latter to the former is taken as the cell's probability to contain an obstacle (cf. Fig. 2b).

The map is shifted in the opposite direction of the wheelchair's movement in space, thus it moves in the same way as the real environment observed from the wheelchair.

3.2   Line Segment Detection

In the next step, walls are to be extracted from the probabilistic obstacle map. This is achieved by a method from image processing: edge detection using a Hough transform.

Firstly, the structure matrix [1] is determined. It contains information on the orientation and the contrast of an edge in the local environment of each map cell. This information can be regarded as an orientation vector which serves

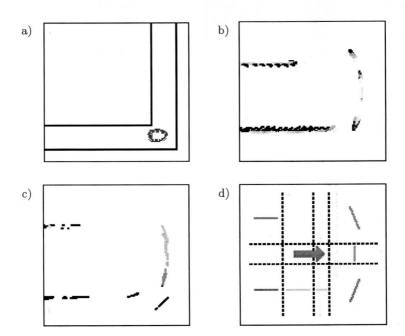

Fig. 2. Processing steps. a) In a simulation, the wheelchair is located in a corner of a corridor facing a wall. b) The probabilistic obstacle map. Darker entries indicate greater occupation probabilities of the map cells. c) The orientation map. The gray values encode the angles of the lines entered. Darker pixels mark clockwise increasing angles. The lightest shade of gray stands for a horizontal line. d) The orientation grid. The arrow indicates the corridor direction. The orientation of the line in each grid cell represents the main orientation angle in the corresponding area of the orientation map. The gray value of a line indicates the frequency of pixels of that orientation. A darker line means more pixels

for speeding up the Hough transform. The next subgoal is to find maxima of intensity in the Hough space. These correspond to straight lines in the obstacle map which in turn are likely to represent walls in the wheelchair's environment. To find out the exact coordinates of the wall candidates, both dimensions of the Hough space which represent the angle and the distance from the origin of the coordinate system are searched consecutively for relative extrema. A dynamic threshold value is used to reduce the influence of noise.

The result of this procedure is a set of coordinate pairs corresponding to lines of infinite length in the coordinate system of the probabilistic obstacle map. Each of these lines is traced through the obstacle map yielding line segments where the line covers entries in the map.

These solid line segments are inserted into another grid map. In this map, the line segments consist of pixels which encode the orientation angle of the line. The size and resolution of this orientation map are the same as those of the probabilistic obstacle map, and it is shifted according to the wheelchair's movements as well (cf. Fig. 2c).

3.3   Orientation Grid

As mentioned, the wheelchair currently operates in an office environment and will be used in a hospital in the future. Thus, the chief navigation task is to find routes along corridors. To accomplish this, the orientation of a corridor has to be determined first.

The walls of the corridor are expected to be the most frequent entries in the local obstacle map. Therefore, the orientation map is searched for the most prominent orientation angle. This is done by computing a histogram of the angles of the cells in the orientation map. The mean angle of the most frequented class yields the orientation of the corridor.

The orientation angle of the corridor is defined in an interval of length π, since an orientation, e. g., from north to south is equivalent to an orientation from south to north. To obtain a direction, the orientation angle is combined with the heading direction of the wheelchair. There are two directions which comply with the corridor orientation. The direction which differs less from the heading direction of the wheelchair is the direction of the corridor.

In the next step, the orientation map is divided into twelve areas relative to the position of the wheelchair, facing in the corridor direction. These categories of relative positions mark areas of interest for assessing the features of the wheelchair's surrounding. They make up a coarse grid in the orientation map (cf. Fig. 2d). In each of these areas, the main orientation is computed with a histogram in the same way as described above.

3.4   Landmark Detection

The orientation grid is searched for typical patterns that indicate prominent features of the wheelchair's surroundings. These landmark categories are: wall in front, corridor left, corridor right, door left and door right. A Boolean variable corresponds to each of these landmark categories, the state of which is determined according to the presence of the feature. For instance, a "wall in front" is detected if the main orientation angle of the two center grid cells in front of the wheelchair is perpendicular to the corridor direction, and if the sum of the numbers of the wall pixels of that orientation in these grid cells is greater than a threshold value.

Finally, the Boolean vector, the components of which represent the results of the detection of the five categories, is mapped to a specific type of landmark. In the example given in Fig. 2, the landmark category detection yields a "true" value for the categories "wall in front" and "corridor left". The combination of these is the landmark "LeftHandBend."

4   The Navigator

The navigator matches the landmark information computed by the situation detector with the initial route description which specifies the target track. Depending on the current situation, the navigator determines a behavior allowing the wheelchair to follow the route.

4.1   Coarse Route Descriptions

According to [13], humans are used to give route descriptions that can be segmented into pieces mainly belonging to four categories: starting-point, reorientation (direction), path/progression, and goal. However, it was also found that people often give additional information such as extra landmarks (not only at turning points), cardinal directions, and the shape of the path between landmarks. "This information, while not essential, may be important for keeping the traveler confidently on track" [13], p. 169.

Humans typically give route descriptions as a sequence of elementary pieces which consist at least of some of the following items:

- starting-point
- reorientation
- path/progression (additional landmarks, approximate distances, . . .)
- goal

In order to ease the communication of a human operator with the wheelchair, it should understand route descriptions that consist only of these elements, and thus enable the vehicle to find its way in a building based on such a description.

The starting-point is always the current position of the robot. The goal is the end of the route. A route description is specified by a sequence of tuples of the following kind:

$$< [\ \{ \ controlmarks \ \} \ router \ ] \ reorientation \ >$$

A *reorientation* is some directional instruction that humans often use in route descriptions [14], such as "TurnLeft", "EnterRightDoor", or "FollowCorridor". A *router* is a landmark where a directional change can take place. The last router is the goal of the route. *Controlmarks* support following routes over longer distances without directional changes; they are especially useful to describe locations where no turn should take place. Depending on the situation, landmarks found by the situation detector are interpreted as controlmarks and routers, respectively.

The coarse qualitative route description A

< RightHandBend TurnRight >
< CorridorRight CorridorLeft TurnLeft >
< CorridorRight DeadEnd Stop >

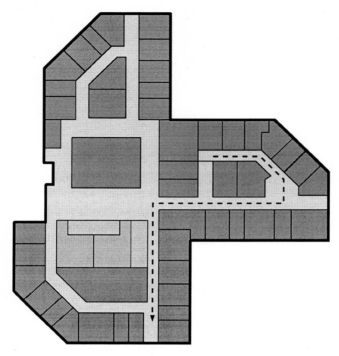

Fig. 3. Plan of the second floor of the MZH-building of Bremen University and the route A (dashed line, approx. 65 m in reality)

corresponds to a route depicted as a dashed line in Fig. 3. However, the wheelchair is not able to directly perform operations such as "TurnRight" because it cannot determine how far it has to turn. Instead, basic behaviors such as "FollowRightWall" are employed. They are started before arriving at the router and may end after it has successfully been passed.

In a first step, the route description is converted into a representation that takes such demands into account:

< FollowRightWall RightHandBend>
< FollowLeftWall CorridorRight CorridorLeft >
< FollowLeftWall CorridorRight DeadEnd >
< Stop >

This representation does not yet prevent the wheelchair from turning into a corridor that is part of a controlmark and therefore should not be entered. Instead, this is achieved when following the route.

4.2   Generation of Driving Commands

The navigator generates driving commands, i. e. it processes the route representation and derives basic behaviors that are adequate in the specific situations. The basic navigation algorithm works as follows: the navigator always selects the first elementary piece of the route representation, the tuple T. Depending on the contents of T, one of the following three cases can occur:

T consists only of a reorientation. Then, the behavior must have an intrinsic end, and it is able to detect when it is has reached this state, as e. g. "Stop" or "TurnRound". The navigator waits until the behavior module announces that this action has been terminated by setting the behavior status accordingly. The navigator will then remove T and switch to the next tuple.

There are controlmarks in T. These controlmarks have to be straightly passed in order to reach the router where a reorientation has to take place. Therefore, a default behavior is associated with each controlmark type that is necessary to avoid entering, e. g., a branching corridor which is part of the mark. After a controlmark has successfully been passed, it is removed from T and the navigator continues with the rest of the current tuple.

There are no controlmarks or they all have successfully been passed, i. e., they were removed from T. Then, it has to be searched for the router. The major difference between a controlmark and a router is that the wheelchair is not prevented from turning off at the router, so if the router is, e. g., a "Crossing" and the behavior is "FollowLeftWall", the wheelchair is allowed to—and in fact should—turn into the left branch of the crossing. Again, if the router has successfully been passed, T is deleted from the route representation and the next tuple becomes the actual one.

When the route description is empty, the navigator stops the wheelchair, and it is assumed that the goal has been reached.

5   Experimental Results

Experiments with the wheelchair robot prove that the approach described in this article does work in practice. In Fig. 4 a successful trial to follow route A (for details cf. above) is depicted.

However, relying on sonar sensors for perceiving the environment has its drawbacks. As the angular resolution of these sensors is relatively weak and their distance information is prone to be erroneous (cf. [10]), extracting features of the environment reliably from sonar data was not always successful. For instance, an open door of an office was occasionally interpreted as a branching corridor by the situation detector. Moreover, a closed door could hardly be distinguished

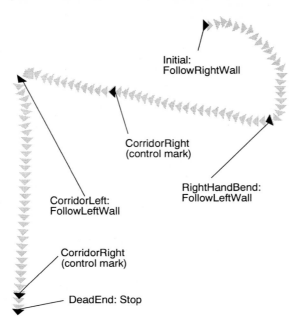

Fig. 4. Path of the wheelchair successfully following route A. The triangles indicate the position and orientation of the robot in intervals of about 0.25 s. The distortions of the depicted path, particularly the turns not appearing as right angles, are due to the non-ideal dead reckoning system of the wheelchair. For a map of the setting of this experiment cf. Fig. 3

from a wall. Therefore, doors have not been used as controlmarks or routers for these experiments.

6   Conclusion and Outlook

In order to improve the acceptance of service and rehabilitation robots such as the Bremen Autonomous Wheelchair, the problem of human-computer interaction has to be tackled. Especially in the context of elder and handicapped users, these robots have to satisfy some conditions: Firstly, they should be *safe* in the sense that they work as they are intended to do and do no harm to people or objects in their environment. Secondly, they should only *assist* the human operator and provide skills he or she lost due to age, illness or impairments, but they should not replace the user's remaining capabilities. And thirdly, the robots should be easy to control.

This paper deals with the latter in that it presents an approach to feed a mobile robot with a route description that makes use of coarse and qualitative expressions such as "left of". This approach integrates the topics *control of mo-*

bile robots, *landmark recognition* and *qualitative reasoning* in a single service robotics application.

While traveling, the wheelchair perceives its environment by sonar sensors and extracts information about special landmarks, the *controlmarks* and the *routers*. This is done by an algorithm based on line detection which is borrowed from the field of image processing. In accordance with the current situation, the robot chooses one of the available basic behaviors, e. g. "FollowCorridor".

The first experimental results show that the approach presented here is a promising method to guide mobile systems through buildings.

In order to improve the robustness of the algorithm, the recorded odometry data should be generalized (cf. [6,11] and [8] for incremental generalization of routes) and matched with the route descriptions augmented by qualitative distances.

When using qualitative distances in route descriptions, the performance of the landmark detection can significantly be increased, because the likelihood of the existence of a particular landmark varies with its position along the route. By taking this fact into account, the misinterpretations of sonar images can be minimized.

The integration of the segmentation and classification algorithm presented in [7] will result in a further improvement by extracting motion shapes from the route descriptions. The algorithm exploits the fact that the shape of the path between two landmarks can be considered as a landmark itself, e. g. in a description such as "The street *makes a sharp turn to the left*, which you follow. After this turn, take the next street on the right." The wheelchair can make use of this additional information without further sensor equipment.

Future work will also deal with supplying route descriptions in natural language rather than in the formal language used in this article.

Acknowledgements

The authors are supported by the Deutsche Forschungsgemeinschaft (DFG) within the Priority Program "Spatial Cognition".

References

1. J. Bigün and G. H. Granlund. Optimal orientation detection of linear symmetry. In *Proc. of the First International Conference on Computer Vision*, pages 433–438. IEEE Computer Society Press, 1987.
2. B. Claus, K. Eyferth, C. Gips, R. Hörnig, U. Schmid, S. Wiebrock, and F. Wysotzki. Reference frames for spatial inference in text understanding. In *Spatial Cognition - An interdisciplinary approach to representing and processing spatial knowledge*, number 1404 in Lecture Notes in Artificial Intelligence. Springer, 1998.
3. C. Freksa and D. M. Mark, editors. *Spatial Information Theory. Cognitive and Computational Foundations of Geographic Information Science. International Conference COSIT'99*, volume 1661 of *Lecture Notes in Computer Science*, Berlin,Heidelberg,New York, August 1999. Springer.

4. B. Krieg-Brückner, Th. Röfer, H.-O. Carmesin, and R. Müller. A taxonomy of spatial knowledge and its application to the Bremen Autonomous Wheelchair. In Ch. Freksa, Ch. Habel, and K. F. Wender, editors, *Spatial Cognition*, volume 1404 of *Lecture Notes in Artificial Intelligence*, pages 373–397, Berlin, Heidelberg, New York, 1998. Springer.

5. A. Lankenau, O. Meyer, and B. Krieg-Brückner. Safety in robotics: The Bremen Autonomous Wheelchair. In *Proceedings of AMC'98, 5th Int. Workshop on Advanced Motion Control*, pages 524–529, Coimbra, Portugal, 1998.

6. A. Musto, K. Stein, A. Eisenkolb, and Th. Röfer. Qualitative and quantitative representations of locomotion and their application in robot navigation. In *Proceedings of the 16th International Joint Conference on Artificial Intelligence (IJCAI-99)*, August 1999.

7. A. Musto, K. Stein, A. Eisenkolb, K. Schill, and W. Brauer. Generalization, segmentation and classification of qualitative motion data. In H. Prade, editor, *Proceedings of the 13th European Conference on Artificial Intelligence (ECAI-98)*, pages 180–184. John Wiley & Sons, 1998.

8. A. Musto, K. Stein, A. Eisenkolb, K. Schill, Th. Röfer, and W. Brauer. From motion observation to qualitative motion representation. In C. Freksa, C. Habel, and K. Wender, editors, *Spatial Cognition II*, Lecture Notes in Artificial Intelligence. Springer, 2000.

9. Th. Röfer and A. Lankenau. Architecture and applications of the Bremen Autonomous Wheelchair. In P. P. Wang, editor, *Proc. of the Fourth Joint Conference on Information Systems*, volume 1, pages 365–368. Association for Intelligent Machinery, 1998.

10. Th. Röfer and A. Lankenau. Ensuring safe obstacle avoidance in a shared-control system. In J. M. Fuertes, editor, *Proc. of the 7th International Conference on Emergent Technologies and Factory Automation*, pages 1405–1414, 1999.

11. Th. Röfer. Route navigation using motion analysis. In Freksa and Mark [3], pages 21–36.

12. E.K. Sadalla, W.J. Burroughs, and L.J Staplin. Reference points in spatial cognition. *Journal of Experimental Psychology: Human Learning and Memory*, 6:516–528, 1980.

13. B. Tversky and P. U. Lee. How space structures language. In C. Freksa, C. Habel, and K.-F. Wender, editors, *Spatial Cognition. An Interdisciplinary Approach to Representing and Processing Spatial Knowledge*, volume 1404 of *Lecture Notes in Artificial Intelligence*. Springer, 1998.

14. B. Tversky and P. U. Lee. Pictorial and verbal tools for conveying routes. In Freksa and Mark [3], pages 51–64.

15. B. Tversky. Cognitive maps, cognitive collages, and spatial mental models. In Andrew U. Frank and Irene Campari, editors, *Spatial Information Theory. A Theoretical Basis for GIS. European Conference, COSIT'93*, volume 716 of *Lecture Notes in Computer Science*, pages 14–24, Berlin, Heidelberg, New York, 1993. Springer.

Oblique Angled Intersections and Barriers: Navigating through a Virtual Maze *

Gabriele Janzen[1], Theo Herrmann[1], Steffi Katz[1] and Karin Schweizer[2]

[1]Department of Psychology, University of Mannheim, D-68131 Mannheim
(janzen, fprg, katz)@psychologie.uni-mannheim.de
[2] University of the Federal Armed Forces at Munich
(karin.schweizer@unibw-muenchen.de)

Abstract. The configuration of a spatial layout has a substantial effect on the acquisition and the representation of the environment. In four experiments, we investigated navigation difficulties arising at oblique angled intersections. In the first three studies we investigated specific arrow-fork configurations. In dependence on the branch subjects use to enter the intersection different decision latencies and numbers of errors arise. If subjects see the intersection as a fork, it is more difficult to find the correct way as if it is seen as an arrow. In a fourth study we investigated different heuristics people use while making a detour around a barrier. Detour behaviour varies with the perspective. If subjects learn and navigate through the maze in a field perspective they use a heuristic of preferring right angled paths. If they have a view from above and acquire their knowledge in an observer perspective they use oblique angled paths more often.

1 Introduction

In everyday life people have to deal with tasks such as reaching a certain destination, finding one's way back, finding a shortcut or making a detour around a barrier. All of these tasks require spatial knowledge. There are a lot of different ways how people can acquire such knowledge about their spatial surroundings (for an overview see Golledge, 1999a; Herrmann & Grabowski, 1994; Herrmann & Schweizer, 1998). For example, they are able to read a map, ask somebody for a description of the way, navigate through the environment on different routes or use land- or waymarks to orientate themselves (e.g. Buhl, Katz, Schweizer & Herrmann, 2000; Daniel & Denis, 1998; Habel, 1988; Rothkegel, Wender & Schumacher, 1998). Different tasks can require different aspects of spatial knowledge (e.g. Mecklenbräuker, Wippich, Wagener & Saathoff, 1998). For most of the wayfinding tasks route knowledge is sufficient. For tasks where people have to find novel paths, for instance making a detour, a more elaborated knowledge such as survey knowledge is needed (Herrmann, Schweizer, Janzen & Katz, 1998).

---

* This work was supported by a grant from the Deutsche Forschungsgemeinschaft (DFG) in the framework of the Spatial Cognition Priority Program (He 270/19-2).

Ch. Freksa et al. (Eds.): Spatial Cognition II, LNAI 1849, pp. 277-294, 2000.

Regarding the development of survey knowledge, the stage theory has been the most influential theory during the last decades (Siegel & White 1975). People who get to know a new environment, first acquire a stage of landmark knowledge where a few unconnected landmarks are stored in memory. In the following, we distinguish between landmarks which are objects seen from far away and waymarks which are objects only seen along a route (see also Krieg-Brückner, Röfer, Carmesin & Müller, 1998, who distinguish between landmarks and routemarks). When people learn more about their surroundings, they acquire a stage of route knowledge. People mentally represent not only land- and waymarks but also a connecting single familiar route (for a discussion of the term mental representation see Herrmann, 1993). After learning other routes, people are able to integrate their route knowledge and reach a stage of survey knowledge. Now they are able to navigate on novel paths, find shortcuts and make a detour (e.g. Appleyard, 1970; Chown, Kaplan & Kortenkamp, 1995; Downs & Stea, 1977; Herrmann, et al., 1998; Levine, Jankovic & Palij, 1982; May & Klatzky, in press; Schumacher, Wender, & Rothkegel, this volume; Thorndyke & Hayes-Roth, 1982).

A common opinion considers route and survey knowledge as the development of directed relations between decision points along a route (see Schweizer, Herrmann, Janzen & Katz, 1998; Werner, Krieg-Brückner & Herrmann, this volume). Newly published results suggest that people can achieve the stage of survey knowledge without passing through the stage of route knowledge first. They integrate different aspects of the route like turns directly into a map-like survey knowledge (e.g., Aginsky, Harris, Rensink & Beusmans, 1997; Montello, 1998; Rothkegel et al., 1998). Aginsky et al. (1997) assume that people use different strategies while learning a route. The visually-dominated strategy is comparable with the acquisition of route knowledge and means that decision points along a route are stored in memory. However, in the spatially-dominated strategy people integrate their knowledge right from the beginning into a map-like knowledge without having landmark or route knowledge first.

Herrmann et al. (1998) proposed that the visual perspective during knowledge acquisition has an effect on the stages of knowledge and enables people to carry out different performances. They distinguish between a field perspective and an observer perspective. People can either have a field perspective taking the viewpoint of a person who is directly located in the scene or an observer perspective taking a viewpoint from above comparable to having a bird's-eye view. With this distinction they refer to results of Nigro and Neisser (1983) who found that subjects remember autobiographic events from different perspectives (see also Frank & Gilovich, 1989). Herrmann et al. (1998) assume that the perspective and the stage of spatial knowledge can be independent of each other. People can represent route knowledge in a field (F) or in an observer (O) perspective dependent on the perspective during knowledge acquisition. They can also use a strategy of mentally recoding or transforming the perspective. For people who learn an environment in a field perspective it can be useful to mentally recode their perspective from a field into an observer perspective (see figure 1). Different performances are possible in dependence of the stage of spatial knowledge and the perspective. For instance, a good estimation of Euclidean distances is expected when people are at a stage of survey knowledge which is represented in an observer perspective. In contrast, people having survey knowledge represented in a field perspective should estimate Euclidean distances more poorly. Figure 1 depicts the theoretical assumptions of Herrmann et al. (1998).

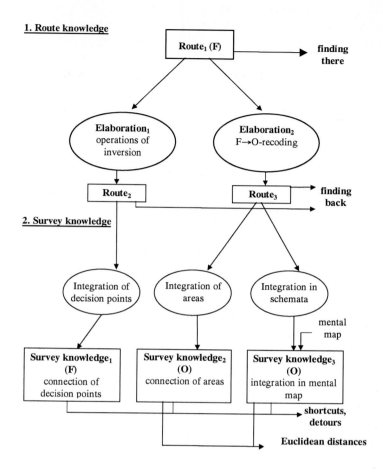

Fig. 1. Theoretical assumptions of Herrmann et al. (1998)

Besides the perspective during knowledge acquisition, the configuration of the environment is another important factor which influences the stage of spatial knowledge. The configuration of spatial layouts determines to a great extent how easily knowledge about an environment can be acquired. Besides the influential work of Tolman in 1948 on cognitive maps, geographers and planners like Lynch (1960) and Appleyard (1973) had a great impact on research in the field of spatial cognition. With Downs and Stea (1973, 1977), a geographer and a psychologist worked together for the first time. In his book "The image of the city" Lynch (1960) analysed different configurational factors and their perception.

In the following section we discuss different configurational parameters and describe experiments in detail we carried out to investigate wayfinding performance at different intersections along a route. In the third section, we focus on strategies people use to find a detour around a barrier. We carried out an experiment to investigate whether detour heuristics differ in dependence of the perspective (field or observer perspective) people have during learning and navigating.

2 Configurational Parameters

Lynch (1960) distinguished different elements of a city: paths, lines, areas, connecting points and landmarks. Paths can meet at intersections of different complexity. Lynch defined an intersection as a point of decision. It therefore becomes a substantial component of a city (for a general view of decision points see Janzen, 1999). Paths and intersections can be divided into different levels of difficulty depending on the angles in which the paths meet (see Casakin, Barkowsky, Klippel & Freksa, this volume). Sketch maps from Lynch's subjects showed that right angled intersections are the easiest to draw whereas intersections with oblique angles and those with more than four branches are difficult. The most difficult intersections are those where different branches leave the intersection in small angles. Lynch (1960) assumed that cities with regular street grids are easier to comprehend than those with an irregular structure and quickly lead to the representation of survey knowledge. Likewise, Thorndyke and Hayes-Roth (1982) proposed that the regularity of the environment plays an important role in knowledge acquisition. In line with these assumptions, Evans (1980) found that straightening of curved paths, squaring of oblique intersections and aligning of nonparallel streets are common cognitive biases (for an impact of the configurational structure, see already the Gestalt psychologists, for instance, the Gestalt principles of Prägnanz and good continuation, e.g., Koffka, 1935; Wertheimer, 1922, 1923). Tversky (1981) also found that students who were told to draw a map of their campus turned 60° angles in the environment into 90° angles on the map (see also Gillner & Mallot, 1998).

Sadalla and Montello (1989) presented four arguments supporting the suggestion that orthogonal reference axes are a fundamental component of egocentric orientation which could explain the preference of 90° angles. First, there is the structure of the human body with a clear front and back (e.g. Franklin, Tversky & Coon, 1992). Secondly, the earth's gravitational field with a vertical and a horizontal dimension has an influence. Thirdly, it is easier to verbally label orthogonal changes in direction than oblique turns. Fourthly, the "carpentered world" hypothesis (Allport & Pettigrew, 1957; Segall, Campbell & Herskovits, 1966) suggests that people who live in perpendicular environments develop schemata for right angles.

Sadalla and Montello (1989) carried out an experiment where subjects walked pathways each with an angular turn of a different size. After each path, subjects had to estimate the angle of each turn and the direction to the starting point. Their results confirm the model of orthogonal axes. Error scores are at maximum between the two orthogonal axes and decrease with angles approaching 90° and 180°.

Schweizer, Herrmann, Katz, Janzen & Trendler (1999) investigated the possibility to take sidelong glances from a route. The opportunity to see further parts of the spatial layout should result in better Euclidean distance estimations than an environment without the possibility to get oriented by sidelong glances. They designed two different spatial layouts. One was a virtual maze with a corridor where walls did not allow to see other parts of the environment. The other environment was a village with open fields where subjects had the possibility to get orientated by seeing further parts of the road, for example houses which are located at a road parallel to the one subjects were moving on.

Subjects acquired knowledge about the environment through film sequences showing the objects along the route sequentially. The films of both layouts show the route in a field perspective. Afterwards, subjects had to estimate distances between two objects in relation to a comparative distance. The instruction did not specify what kind of distance subjects have to estimate. Thus, the subjects were free to judge either Euclidean or route distances. The analyses of the distance estimations for object-pairs in which the Euclidean and the route distance were different showed that the estimations of subjects in both experiments were in-between the Euclidean and the route distance of the object pairs. In particular, the tendency to estimate closer to the Euclidean distance is stronger if the configuration of the environment allows sidelong glances than if there is no possibility to see other parts of the layout. In the layout with the corridor system the estimations of the subjects were closer to the route distance.

The results confirm the assumption of a strong influence of the environmental configuration. All subjects learned the route in a field perspective, i.e. in both experiments they were neither able to acquire map-like knowledge nor to have a general outlook over larger parts of the environment. Even slight differences between the layouts such as the possibility to take sidelong glances at only a few parts along the route can affect the representation of distances between objects.

The studies depicted above show that there is no general factor of complexity or difficulty applying to all different environments. Not only the angles of turns along a route but also the possibility to take different viewpoints influence the representation of the spatial configuration. It is therefore necessary to systematically vary the perspective (field and observer perspective) and the viewpoints at a specific section of the environment. In the experiments described in the following, subjects navigate through a virtual maze and become acquainted with arrow-fork-intersections under different perspectives.

Experiments with Oblique Angled Intersections

In all of the experiments subjects learned a virtual maze with right and oblique angled intersections. Of main interest was the configuration of a specific oblique angled intersection which subjects entered from different sides (for a detailed description of the experiments see Katz, Janzen, Schweizer & Herrmann, 1999). Figure 2 shows such an oblique angled intersection.

If this intersection is entered from branch A it looks like a fork (fork viewpoint). On the other hand if it is entered from branch C or from branch D it looks like an arrow (arrow viewpoint). We did not differentiate between the branches C and D. (If one enters the intersection from branch B and goes straight ahead to branch A or in the reverse direction from A to B no rotation is required. This may be a special case, we did not test so far.) We assume that it is easier to make a decision about the correct way if the intersection looks like an arrow than like a fork. Easier means that the decision about the correct way should be faster and with less errors. A second commonly hold assumption refers to the ability of finding the way back. It should be easier to find the way to the goal location than to find the way back.

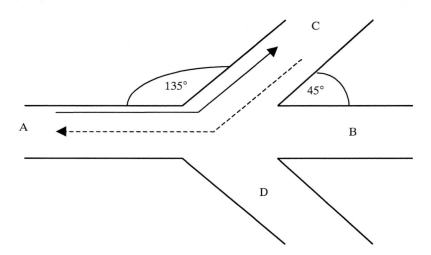

Fig. 2. Arrow-Fork-Configuration

With three experiments we tested whether the branch from which the arrow-fork-intersection is entered has an effect on the decision latencies and the accuracy of finding the correct path. Figure 3 shows the virtual maze subjects learned in all three experiments through film sequences presented with a head mounted display (HMD, Virtual Research V6, field of view: 60° angle, 60 Hz refresh rate, screen resolution: 480 x 640 pixels, for the use of virtual environments see also Janzen, Katz, Schweizer & Herrmann, in press).

The HMD was equipped with a sensor which registered the head movement. Sensor signals were transmitted to a tracking system (Polhemus 3Space Insidetrak). Subjects' head movements caused a corresponding change of the view in the virtual labyrinth. The virtual maze was constructed with the computer graphic software "Superscape VRT". The film sequences were generated with average frame rates of 14 frames/s. In real world dimensions, the environment had a length of 218 meters and was 114 meters wide in relation to the simulated eye level of an observer from 1.70 meters. The film sequences for both field and observer perspective experiments last for 2.45 minutes.

Before seeing the film sequence, in a practice phase subjects learned to move with the joystick in a virtual petrol station. After seeing the film sequence, in a learning phase subjects navigated through the maze with the help of the joystick. They first had to find the same way they saw in the film sequence, then they should find the way back to the starting position. In order to measure the time subjects needed to decide at an intersection which way is the correct way to go invisible sensors were placed in the virtual environment. The decision latencies were measured with the sensors implemented at each branch registering when the intersection is entered and left.

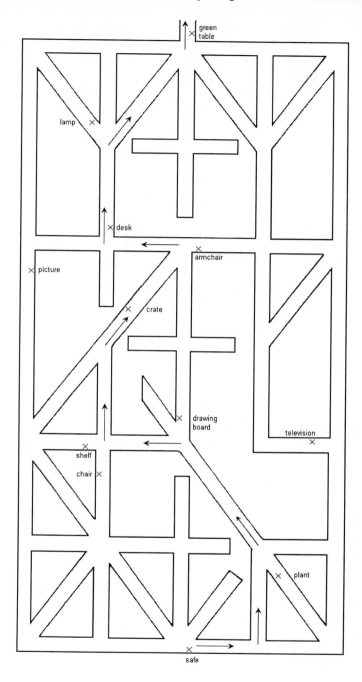

Fig. 3. The virtual maze

Experiment 1. In this experiment 20 subjects saw the film sequence of the maze in a field perspective. They also navigated through the maze in field perspective. The sensor recorded decision latencies were corrected for errors and outliers. Figure 4 presents the decision latencies for fork and arrow viewpoints on subjects' way to the goal location and on their way back.

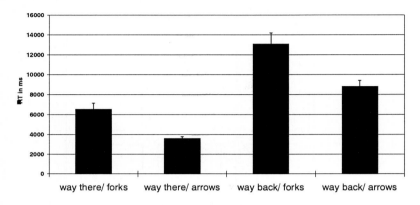

Fig. 4. Decision latencies at the fork/arrow intersections separated for the way there and back in experiment 1 (means and standard error)

Analyses show that there is a considerable time difference between having a fork or an arrow viewpoint. Additionally, the way to the destination was covered faster than the way back. The error rate (i. e. choosing a wrong branch) was higher with a fork viewpoint than with an arrow viewpoint and was also higher on the way back than on the way there. These results can be explained by two competing hypotheses. The *assumption of visible alternatives* explains time delays between arrow and fork viewpoints by a larger number of clearly visible alternatives of branches when having a fork viewpoint. It is possible that it is more complicated to decide which way to go if three alternatives of branches are visible without moving the head. When having an arrow viewpoint, only one branch is clearly visible allowing a decision which is faster and contains less errors.

The second assumption is the *assumption of a small rotational angle*. This assumption does not depend on the visual information after entering the intersection but on the number of alternatives of choosing a small body rotation. When having a fork viewpoint, three alternatives of branches with rotations smaller than 90° are possible. On the other hand, when entering the intersection with an arrow view only for one branch a small body rotation is required. For all other branches rotations of 90° or larger have to be made. It could be easier to manage the intersection having an arrow view. There is only one branch requiring a small body rotation and therefore no decision has to be made about minimising the effort. With a fork viewpoint all possible branches cost about the same rotational effort and therefore a decision demands a greater amount of time.

In order to test these two competing hypotheses we carried out two other experiments. If a difference between the decision latencies of an arrow versus a fork viewpoint is due to the greater number of visible alternatives, this time difference should disappear: either when subjects know the configuration of the intersection before they see the film sequence or when they have an observer perspective during navigating in which the visibility of branches is the same with an arrow and a fork viewpoint.

Experiment 2. In this experiment 20 subjects saw a map of the layout before they were presented with the film sequence. They had time to study the map for 2 minutes. Afterwards, they performed the same navigation tasks as in experiment 1. The decision latencies are shown in figure 5.

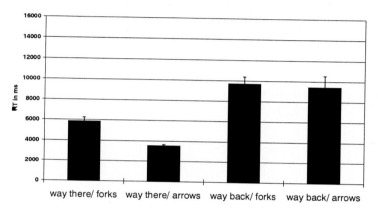

Fig. 5. Decision latencies at the fork/arrow intersections separated for the way there and back in experiment 2 (means and standard error)

The results show the same decision latency pattern on the way there as in experiment 1. On the way back there is no difference between arrow and fork viewpoints. Subjects seem to profit from the map only on their way back. The decision latency pattern does not distinguish between the two hypotheses. The error pattern is the same as in experiment 1.

Experiment 3. In this experiment, we varied the perspective. Subjects saw the environment in an observer perspective (for an example of an intersection in field and in observer perspective see figure 7). The procedure is the same as in experiment 1. Figure 6 shows the decision latencies.

The results show a decision latency pattern similar to the one observed in experiment 1. The analysis of the decision latencies confirms the assumption of a small rotational angle but does not confirm the assumption of visible alternatives. Decision latencies are shorter when subjects have an arrow viewpoint than a fork viewpoint when entering the intersections. Subjects are also faster and make less errors on the way there than on their way back in all three experiments. The error score was again higher with a fork viewpoint compared to an arrow viewpoint.

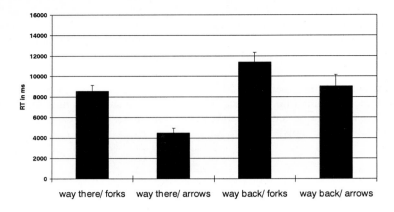

Fig. 6. Decision latencies at the fork/arrow intersections separated for the way there and back in experiment 3 (means and standard error)

Discussion

In all three experiments the decision latency pattern is basically the same. Except in the map experiment (experiment 2) there is no significant difference between decision latencies for fork and arrow viewpoints on subjects' way back. The results do not correspond with the assumption of visible alternatives. Neither the presentation of a map nor the observer perspective reduced the time differences between fork and arrow viewpoints when entering the intersection. The perception of branches does not have an influence on the decision on the correct way.

The results may also be interpreted by an *assumption of general orientation*. This assumption refers to the overall direction of the destination when choosing a branch. Experiments with animal navigation as well as human navigation and the models of path integration show that organisms are capable of integrating rotational and distance information during navigation to compute a homing vector which indicates the direction to the end and starting point of the route (e.g. May & Klatzky, in press; Mittelstaedt & Mittelstaedt, 1982; Wehner & Menzel, 1990; Werner, Krieg-Brückner & Herrmann, this volume). The learned way at intersections when entering from the arrow side always leads to the direction of the destination whereas at one intersection (intersection with drawing board) when having a fork viewpoint the correct branch leads away from the end of the route. This difference in the direction of the destination could explain the different decision latencies.

In order to test this assumption we carried out another experiment. The procedure was the same as in the first experiment. To control the general direction towards the end we had to change the route. Now, when choosing the correct fork branch at the intersection the destination was exactly in the opposite direction. Therefore, if subjects represent a general direction to the destination decision latencies should increase. The results depict that there is no difference between the decision latencies for the fork intersection in both experiments. Therefore, the results also confirm the assumption of a small rotational angle.

3 Detour Heuristics

Besides the configuration of an environment, heuristics of coping with navigational problems play an important role for wayfinding abilities. Navigation through an unknown environment is a very complex behaviour which requires plenty of heterogeneous skills. It is therefore likely that it can easily be affected by spatial navigation strategies or heuristics reducing the difficulties.

Finding a new way, a novel shortcut or a detour around a barrier are abilities which usually demand a more elaborated knowledge about the spatial layout than route knowledge. For Tolman (1948) these abilities stated the main criteria for the existence of cognitive maps (for a detailed discussion of the term cognitive map see Golledge, 1999b; Kitchin, 1994). In our daily life people often have to deal with similar tasks without having survey knowledge. Heuristics can therefore be helpful in solving a detour problem.

Maguire, Burgess, Donnett, Frackowiak, Frith, and O'Keefe (1998) carried out a positron emission tomography (PET) study. Their subjects had to navigate through a complex virtual environment and to perform different dynamic and static tasks. During a detour finding task they found activation in an area (left frontal cortex) which is usually associated with general planning and decision making (see also Maguire, Burgess & O'Keefe, 1999). This result underlines that detour finding is a complex task where no automatic coping strategies exist.

Because of different environmental factors it is plausible that there is no general solution strategy for a detour around a barrier. The most general heuristic is probably the heuristic to return to the learned route after surrounding the obstacle. A more specific detour heuristic can be applied if the environment allows to use branches with 90° angles to return to the familiar route. In a city with street grids there is no other possibility but even when there are other routes which may be even shorter people still show this heuristic behaviour. Evidence in supporting this heuristic is the tendency to represent 90° angles. Tversky (1981) found that students who should draw a map from their campus turned 60° angles in the real environment into 90° angles on the map (e.g. Gillner, 1997; Gillner & Mallot, 1998).

Another possible heuristic is to take the shortest way back to the former route even if it is necessary to use an oblique angled path. It is very likely that the heuristic used for a detour also depends on the visual perspective. In one experiment, we tested what kind of detour heuristics subjects use if there are right as well as oblique angled paths to choose for a detour. If a general *right angle strategy* for making a detour with turnoffs in 90° angles exists, it should be used independent from the perspective subjects have. Alternatively, detour behaviour can depend on a representation of the familiar route including the learned perspective. A different detour strategy may then be used more frequently in an observer perspective than in a field perspective.

Experiment about Detour Heuristics

In an experiment (Janzen, Schade, Katz & Herrmann, 1999) we investigated the effect of different visual perspectives on detour heuristics in a virtual maze. We varied the perspective during knowledge acquisition and navigation. Subjects either had a field perspective or an observer perspective (see introduction). In a third condition they should mentally transform their perspective from a field into an observer perspective. Subjects saw the film sequence in field perspective and were asked to mentally imagine the route as seen in an observer perspective. A picture of an object was shown in observer perspective before the film sequence to help subjects in trans- forming the perspective.

If general planning behaviour is the basis for spatial navigation tasks path finding behaviour should not be affected by visual characteristics. We tested whether general navigation heuristics like the preference of using right angled paths are used for detour finding. If a general heuristic is used detours should be the same in a field and in an observer perspective. The detour heuristic should also be unaffected by a mental transformation task like the recoding from a field into an observer perspective. On the other hand, finding a novel path around a barrier can depend on the perspective during knowledge acquisition. Then we expect different detour finding abilities under a field and an observer perspective. We assume that subjects in the observer condition do not rely on default strategies as subjects in a field perspective will do. Although subjects in an observer perspective do not have a survey view over a greater area of the route in our study their perspective from above should enable them to integrate different paths and to find other more complex detours than subjects in a field perspective. Figure 7 shows an intersection of the route in field and in observer perspective.

a   b

Fig. 7. Intersection in field (a) and observer perspective (b)

We realized three perspective conditions (60 subjects): a field perspective condition, an observer perspective condition and a mental recoding condition. Subjects learned a route through a virtual maze presented with a head mounted display. Subjects in the mental recoding condition saw the film sequence in field perspective and were told to mentally recode the visual input into an observer perspective. The experimenter told the subjects that this mental strategy is known to be a good technique to learn the route and showed a picture of the starting point in observer perspective to them. The picture was shown to help subjects how they should mentally imagine the route. The route is shown in figure 3.

After seeing the film sequence three times, subjects navigated through the maze up to a barrier which blocked the former path. We analysed the detour strategies subjects used to find a novel path around the obstacle. Figure 8 shows the part out of the route where the barrier is located and the possible routes at the next intersection.

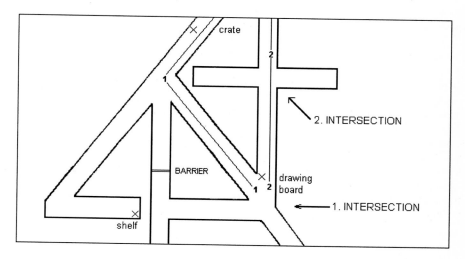

Fig. 8. Part of the map with detours around the barrier

Subjects had two different choices at this intersection (intersection 1). One was a right angled path and the other one an oblique angled path. Table 1 lists the frequencies for both paths in all three perspective conditions.

Table 1. Frequencies of chosen detour

	Oblique Angled Path 1	Right Angled Path 2
Field-Perspective	1	19
Observer-Perspective	9	11
Field-Observer Recoding	6	14

The results (χ^2-tests) show that in all three conditions the right angle heuristic is the most frequently chosen. But the chosen detours differ significantly in all three conditions. If we compare the field and the observer perspective condition, a significantly larger number of subjects in the observer condition chose the oblique angled path. In contrast almost all subjects in the field perspective condition chose the right angled heuristic.

Further analysis confirms the assumption of a right angled heuristic chosen by subjects in a field perspective. We analysed the subjects' behaviour after choosing the right angled path. The following intersection (intersection 2) seen in figure 8 consists of only right angles and has two dead ends. All subjects were unable to see the dead ends before making a turn. We determined the number of subjects who wanted to make a right angled turn at this intersection. 58% of subjects in the field perspective condition and 64% of subjects in the mental recoding condition tried to make a right angled turn at this intersection in the direction of the former route. But only 27% of subjects in the observer perspective condition tried to walk into this wrong path.

In a next step, sensor recorded decision latencies at the intersections were analysed. All times were corrected for a motor component of joystick movement. Analyses of the corrected decision latencies at each intersection do not show any significant differences between the three perspective conditions. Likewise, the time needed for completing the route corrected by the route length is not significantly different. This result indicates that a particular heuristic is not chosen because of a time advantage. The preference for an oblique or a right angle (path 1 and path 2) does not exist because of a time advantage in the decision process or in moving around a specific angle. The different detour heuristics are due to the visual perspective during knowledge acquisition and navigation and the resulting different representation of angles (i.e., subjects who have an observer perspective have a more accurate representation of oblique angles) rather than to a heuristic to choose the least time consuming path.

Discussion

We can draw the conclusion that most subjects in all of the three conditions used a heuristic of returning to the learned route after the barrier. We can conclude that a heuristic of using right angles is the most common. Another experiment with a more simple path also showed that subjects choose a right angled heuristic to make a detour. This behaviour is not the only heuristic for a detour. Detour behaviour varies with the perspective during knowledge acquisition and navigation. In a field perspective subjects preferred a heuristic of right angles whereas subjects in observer perspective also used oblique paths leading to the former route more quickly. The subjects in the mental recoding condition behaved in between the field and the observer perspective condition at the first intersection but those who once decided to choose a right angle heuristic used it for the entire detour as behaviour at the dead end intersection shows.

Besides the influence of different perspectives detour behaviour also depends on spatial aspects of the environment. A virtual labyrinth like the one we used can lead to different detours than an open field environment. Results from Schweizer et al. (1999) show differences in distance estimations from a "closed" environment versus an open field. It is also possible that different detour heuristics result when further aspects of the environment like the known path can be clearly seen from the position of the barrier.

4 Conclusion

Our studies demonstrate that first the configuration of an environment and secondly the perspective people have while learning and navigating through a maze have an essential effect on the wayfinding abilities. Likewise, the representation of a spatial layout depends on characteristics of an environmental configuration. In the stage theory of a development from route to survey knowledge it is usually assumed that the spatial environment is simply represented by directed object relations (for the representation of directions between objects along a route see Herrmann, Buhl & Schweizer, 1995; Schweizer, 1997; Schweizer & Janzen, 1996). The present results, however, show that configurational parameters and people's perspective during knowledge acquisition and navigation should be additionally taken into consideration.

Acknowledgements

We wish to thank Sabine Schumacher and two anonymous reviewers for helpful comments on an earlier version of this manuscript. We also like to thank Marion Schade and Christian Westermeier for their assistance in data collection.

References

Aginsky, V., Harris, C., Rensink, R., & Beusmans, J. (1997). Two strategies for learning a route in a driving simulator. *Journal of Environmental Psychology, 17*, 317-331.

Allport, G., & Pettigrew, T. (1957). Cultural influence on the perception of movement: The trapezoidal illusion among the Zulus. *Journal of Abnormal and Social Psychology, 55*, 104-113.

Appleyard, D. (1970). Styles and methods of structuring a city. *Environment and Behavior, 2*, 100-116.

Appleyard, D. (1973). Notes on urban perception and knowledge. In R. M. Downs & D. Stea (Eds.), *Image and environment. Cognitive mapping and spatial behavior* (pp. 109-114). Chicago, IL: Aldine.

Buhl, H. M., Katz, S., Schweizer, K., & Herrmann, Th. (2000). Einflüsse des Wissenserwerbs auf die Linearisierung beim Sprechen über räumliche Anordnungen. *Zeitschrift für Experimentelle Psychologie, 47*, 17-33.

Casakin, H., Barkowsky, T., Klippel, A., & Freksa, Ch. (in press). Schematic maps as wayfinding aids. In Ch. Freksa, W. Brauer, Ch. Habel, & K. F. Wender (Eds.), *Spatial Cognition II (working title)*. Berlin: Springer.

Chown, E., Kaplan, S., & Kortenkamp, D. (1995). Prototypes, location and associative networks (PLAN): Towards a unified theory of cognitive mapping. *Cognitive Science, 19*, 1-51.

Daniel, M. P., & Denis, M. (1998). Spatial descriptions as navigational aids: A cognitive analysis of route directions. *Kognitionswissenschaft, 7*, 45-52.

Downs, R. M., & Stea, D. (1973). Cognitive maps and spatial behavior: Process and products. In R. M. Downs & D. S. Stea (Eds.), *Image and environment. Cognitive mapping and spatial behavior* (pp. 8-26). Chicago: Aldine.

Downs, R. M., & Stea, D. (1977). *Maps in minds: Reflections on cognitive mapping*. New York: Harper & Row.

Evans, G. W. (1980). Environmental cognition. *Psychological Bulletin, 88*, 259-287.

Frank, M. G., & Gilovich, T. (1989). Effect of memory perspective on retrospective causal attributions. *Journal of Personality and Social Psychology, 57*, 399-403.

Franklin, N., Tversky, B., & Coon, V. (1992). Switching points of views in spatial mental models. *Memory & Cognition, 20*, 507-518.

Gillner, S. (1997). *Untersuchungen zur bildbasierten Navigationsleistung in virtuellen Welten.* Unveröffentlichte Dissertation, Eberhard-Karls-Universität, Tübingen.

Gillner, S., & Mallot, H. A. (1998). Navigation and acquisition of spatial knowledge in a virtual maze. *Journal of Cognitive Neuroscience, 10*, 445-463.

Golledge, R. G. (Ed.). (1999a). *Wayfinding behavior: Cognitive mapping and other spatial processes.* Baltimore: The John Hopkins University Press.

Golledge, R. G. (1999b). Human wayfinding and cognitive maps. In R. G. Golledge (Ed.) *Wayfinding behavior: Cognitive mapping and other spatial processes* (pp. 1-45). Baltimore: The John Hopkins University Press.

Habel, Ch. (1988). Prozedurale Aspekte der Wegplanung und Wegbeschreibung. In H. Schnelle & G. Rickheit (Hrsg.), *Sprache in Mensch und Computer* (S. 107-133). Wiesbaden: Westdeutscher Verlag.

Herrmann, Th. (1993). Mentale Repräsentation - ein erläuterungswürdiger Begriff. In J. Engelkamp & Th. Pechmann (Hrsg.), *Mentale Repräsentation* (S. 17-30). Bern: Huber.

Herrmann, Th., & Grabowski, J. (1994). *Sprechen - Psychologie der Sprachproduktion.* Heidelberg. Spektrum Akademischer Verlag.

Herrmann, Th., Buhl H. M., & Schweizer, K. (1995). Zur blickpunktbezogenen Wissensrepräsentation: der Richtungseffekt. *Zeitschrift für Psychologie, 203*, 1-23.

Herrmann, Th., & Schweizer, K. (1998). Sprechen über Raum. Sprachliches Lokalisieren und seine kognitiven Grundlagen. Bern: Huber.

Herrmann, Th., Schweizer, K., Janzen, G., & Katz, S. (1998). Routen- und Überblickswissen - konzeptuelle Überlegungen. *Kognitionswissenschaft, 7*, 145-159.

Janzen, G. (1999). *Die Organisation räumlichen Wissens. Untersuchungen zur Orts- und Richtungsrepräsentation.* Unveröffentl. Dissertation. Universität Mannheim.

Janzen, G., Schade, M., Katz, S., & Herrmann, Th. (1999). *Detour behaviour depends on the visual perspective during learning and navigating in a virtual maze* (Arbeiten des Mannheimer Teilprojekts im DFG-Schwerpunktprogramm "Raumkognition" Bericht Nr. 4). Mannheim: Universität Mannheim, Lehrstuhl Psychologie III.

Janzen, G., Katz, S., Schweizer, K., & Herrmann, Th. (in press). Experimentalpsychologische Untersuchungen zum Wegfindeverhalten in einem virtuellen Labyrinth (Themenheft "Intelligente virtuelle Umgebungen", herausgegeben von B. Jung, & J.-T. Milde). *KI Zeitschrift Künstliche Intelligenz.*

Katz, S., Janzen, G., Schweizer, K., & Herrmann, Th. (1999). *Hin- und Zurückfinden beim Routenlernen: Der Einfluß schiefwinkliger Kreuzungskonfigurationen auf die Navigationsleistung* (Arbeiten des Mannheimer Teilprojekts im DFG- Schwerpunktprogramm "Raumkognition" Bericht Nr. 5). Mannheim: Universität Mannheim, Lehrstuhl Psychologie III.

Kitchin, R. M. (1994). Cognitive maps: What are they and why study them? *Journal of Environmental Psychology, 14*, 1-19.

Koffka, K. (1935). *Principles of Gestalt psychology.* London: Kegan Paul.

Krieg-Brückner, B., Röfer, T., Carmesin, H.-O., Müller, R. (1998). A taxonomy of spatial knowledge for navigation and its application to the Bremen Autonomous Wheelchair.In Ch. Freksa, Ch. Habel, & K. F. Wender (Eds.), *Spatial cognition: An interdisciplinary approach to representing and processing spatial knowledge* (pp. 373-397). Berlin: Springer.

Levine, M., Jankovic, I. N., & Palij, M. (1982). Principles of spatial problem solving. *Journal of Experimental Psychology: General, 111*, 157-175.

Lynch, K. (1960). *The image of the city.* Cambridge: The Technology Press & Harvard University Press.

Maguire, E. A., Burgess, N., Donnett, J. G., Frackowiak, R. S. J., Frith, C. D., & O'Keefe, J. (1998). Knowing where and getting there: A human navigation network. *Science, 280*, 921-924.

Maguire, E. A., Burgess, N., & O 'Keefe, J. (1999). Human spatial navigation: Cognitive maps, sexual dimorphism, and neural substrates. *Current Opinion in Neurobiology, 9*, 171-177.

May, M., & Klatzky, R. L. (in press). Path integration while ignoring irrelevant movement. *Journal of Experimental Psychology: Learning, Memory, & Cognition.*

Mecklenbräuker, S., Wippich, W., Wagener, M., & Saathoff, J. E. (1998). Spatial information and actions. In Ch. Freksa, Ch. Habel, & K. F. Wender (Eds.), *Spatial cognition: An interdisciplinary approach to representing and processing spatial knowledge* (pp. 39-61). Berlin: Springer.

Mittelstaedt, H., & Mittelstaedt, M. L. (1982). Homing by path integration. In F. Papi, & H. G. Wallraff (Eds.), *Avian navigation* (pp. 290-297). New York: Springer.

Montello, D. R. (1998). A new framework for understanding the acquisition of spatial knowledge in large-scale environments. In M. Egenhofer, & R. G. Golledge (Eds.). *Spatial and temporal reasoning in geographic information systems* (pp. 143-154). Oxford: Oxford University Press.

Nigro, G., & Neisser, U. (1983). Point of view in personal memories. *Cognitive Psychology, 15*, 467-482.

Rothkegel, R., Wender, K. F., & Schumacher, S. (1998). Judging spatial relations from memory. In Ch. Freksa, Ch. Habel, & K. F. Wender (Eds.), *Spatial cognition: An interdisciplinary approach to representing and processing spatial knowledge* (pp. 79-106). Berlin: Springer.

Sadalla, E. K., & Montello, D. R. (1989). Remembering changes in direction. *Environment and Behavior, 21*, 346-363.

Schumacher, S., Wender, K. F., & Rothkegel, R. (in press). Influences of context on memory for routes. In Ch. Freksa, W. Brauer, Ch. Habel, & K. F. Wender (Eds.), *Spatial Cognition II (working title).* Berlin: Springer.

Schweizer, K. (1997). *Räumliche oder zeitliche Wissensorganisation? Zur mentalen Repräsentation der Blickpunktsequenz bei räumlichen Anordnungen.* Lengerich: Pabst Science Publishers.

Schweizer K., & Janzen, G. (1996). Zum Einfluß der Erwerbssituation auf die Raumkognition: Mentale Repräsentation der Blickpunktsequenz bei räumlichen Anordnungen. *Sprache & Kognition, 15*, 217-233.

Schweizer, K., Herrmann, Th., Janzen, G., & Katz, S. (1998). The route direction effect and its constraints. In Ch. Freksa, Ch. Habel, & K. F. Wender (Eds.), *Spatial cognition: An interdisciplinary approach to representing and processing spatial knowledge* (pp. 19-38). Berlin: Springer.

Schweizer, K., Herrmann, Th., Katz, S., Janzen, G., & Trendler, G. (1999). *Überblick und Seitenblick. Untersuchungen zum Distanzschätzen.* (Arbeiten des Mannheimer Teilprojekts im DFG- Schwerpunktprogramm "Raumkognition" Bericht Nr. 3). Mannheim: Universität Mannheim, Lehrstuhl Psychologie III.

Segall, M. H., Campell, D. T., & Herskovits, M. J. (1966). *The influence of culture on visual perception.* Indianapolis: Bobbs-Merrill.

Siegel, A. W., & White, S. H. (1975). The development of spatial representations of large-scale environments. In H. R. Reese (Ed.), *Advances in Child Development and Behavior* (Vol. 10, pp. 10-55). New York: Academic Press.

Thorndyke, P. W., & Hayes-Roth, B. (1982). Differences in spatial knowledge acquired from maps and navigation. *Cognitive Psychology, 14*, 560-589.

Tolman, E. C. (1948). Cognitive maps in rats and men. *Psychological Review, 55*, 189-208.

Tversky, B. (1981). Distortions in memory for maps. *Cognitive Psychology, 13*, 407-433.

Wehner, R., & Menzel, R. (1990). Do insects have cognitive maps? In W. Cowan, E. Shooter, C. Stevens, & R. Thompson (Eds.), *Annual review of neuroscience* (Vol. 13, pp. 403-414). Palo Alto, CA: Annual Reviews.

Werner, S., Krieg-Brückner, B, & Herrmann, Th. (in press). Modelling navigational knowledge by route graphs. In Ch. Freksa, W. Brauer, Ch. Habel, & K. F. Wender (Eds.), *Spatial Cognition II (working title).* Berlin: Springer.

Wertheimer, M. (1922). Untersuchungen zur Lehre von der Gestalt I. *Psychologische Forschung, 1*, 47-58.
Wertheimer, M. (1923). Untersuchungen zur Lehre von der Gestalt II. *Psychologische Forschung, 4*, 301-350.

Modelling Navigational Knowledge
by Route Graphs

Steffen Werner[1], Bernd Krieg-Brückner[2], and Theo Herrmann[3]

[1] Institute of Psychology, University of Göttingen, Gosslerstr. 14, D-37073 Göttingen
swerner@uni-goettingen.de, www.uni-goettingen.de/~sppraum
[2] Bremen Institute of Safe Systems, University of Bremen, PBox 330440, D-28334 Bremen
bkb@Informatik.Uni-Bremen.DE, www.uni-bremen.de/~sppraum
[3] Institute of Psychology, University of Mannheim, Schloss–E0, D-68131 Mannheim
fprg@rumms.uni-mannheim.de

Abstract. Navigation has always been an interdisciplinary topic of research, because mobile agents of different types are inevitably faced with similar navigational problems. Therefore, human navigation can readily be compared to navigation in other biological organisms or in artificial mobile agents like autonomous robots. One such navigational strategy, route-based navigation, in which an agent moves from one location to another by following a particular route, is the focus of this paper. Drawing on the research from cognitive psychology and linguistics, biology, and robotics, we present a simple, abstract formalism to express the key concepts of route-based navigation in a common scientific language. Starting with the distinction of places and route segments, we develop the notion of a route graph, which can serve as the basis for complex navigational knowledge. Implications and constraints of the model are discussed along the way, together with examples of different instantiations of parts of the model in different mobile agents. By providing this common conceptual framework, we hope to advance the interdisciplinary discussion of spatial navigation.

1 Introduction

Our daily life is filled with activities travelling from one location to another. Every morning, we walk from the bedroom to the bathroom, from there to the kitchen, etc. Later, we drive from home to work or take the bus into town. We stroll through the city we live in or drive to places far away to visit friends. All these activities share one important component: navigation in a spatial environment.

The navigational abilities of humans, and of other organisms, always have been an interdisciplinary topic of research. This is due to the fact that problems of navigation, in their general form, similarly apply to very different mobile agents in very different environments. The desert ant, foraging for food under the desert sun, has to return home after it has found a dead insect (Wehner & Menzel, 1990) as does the hamster

Ch. Freksa et al. (Eds.): Spatial Cognition II, LNAI 1849, pp. 295-316, 2000.
© Springer-Verlag Berlin Heidelberg 2000

after pouching food (Etienne, Maurer, Georgakopoulos, & Griffin, 1999). Pigeons, like other birds, take extensive flights away from their nest but still return home safely, while migrating birds fly for thousands of miles to reach their winter quarters (Wiltschkow & Wiltschkow, 1999).

Human navigation ranges from our daily way to work to global travel. Unlike most other organisms, humans have also developed navigational tools to complement or enhance their navigational abilities. Maps, compasses, sextants, and global positioning devices (GPS) help to locate oneself in an environment or to plan one's travel. By using language, navigational information can be communicated to others, e.g. by giving route directions. Using remotely controlled objects, the space in which humans have to navigate has been greatly extended. Navigation today might entail directing minuscule medical instruments within the human body or controlling a remote vehicle on a different planet.

Artificial autonomous mobile agents, such as robots, also have to deal with similar navigational problems. Although relying on a host of different sensory and motor mechanisms, the abstract tasks of planning a route and executing the right actions along the way are nearly the same (e.g. Werner *et al.*, 1997). It is thus not surprising that in this field, disciplines such as biology, psychology, linguistics, artificial intelligence, and geography jointly contribute to our understanding of the basic processes underlying spatial navigation.

In this paper, we wish to outline a simple model to describe how mobile agents can acquire spatial knowledge to successfully navigate in an environment. Our approach is aimed at bringing together the basic ideas and models from different disciplines and providing a common platform to express concepts relevant for navigation and related types of spatial knowledge. Our main interest lies in so-called *route-based navigation* because it has proven to be an important navigational method for humans and other organisms, and in addition constitutes a common method employed in robot navigation. Of course this approach is not new, and many aspects of the proposed model can be found in a number of predecessors. Before we will describe the model, a brief overview of similar models from different disciplines will therefore set the stage.

2   Route-Based Navigational Knowledge for Different Agents

The number of particular navigational strategies employed by different agents is quite large. However, only a few basic kinds of navigational strategies can be distinguished at an abstract level (e.g., Allen, 1999). Agents can follow a marked trail, as in the case of Theseus following Ariadne's thread to the exit of the labyrinth, or use a gradient ascent or descent strategy, for example when moving in the direction increasing the intensity of a specific smell or following the magnetic field of the earth (Wiltschkow & Wiltschkow, 1999). Alternatively, agents can compute their position in relation to a starting point or goal location by integrating all their movements in space as in path-integration (Loomis *et al.*, 1999). In this case, the direction and distance of the home-base or the location of the goal are continually updated. Some other agent might be able to compute its position in terms of a global co-ordinate system, allowing the computation of the direction and distance of any goal specified in the same reference system. Other navigational means might entail systematic search strategies, such as a grid search to find a shipwreck in the ocean or meandering to locate one's nest for desert ants (Wehner *et al.*, 1996).

In this paper we will mainly deal with route-based navigation, in which an agent moves from one location to another by following a particular route, i.e. a sequence of different places and means to get from one place to the next (an exact terminology will be developed later). Other strategies, such as the ones mentioned above, will be discussed only as they relate to route-based navigation. This, of course, does not imply that these other strategies are in any way inferior to route-based methods as general navigational strategies. However, route-based navigation plays a dominant role in human and robot navigation (and probably in animal navigation, too) and is thus the main focus of this paper.

2.1   Route-Based Navigation in Humans

Routes are a concept commonly encountered when dealing with human spatial navigation. When asking another person for directions to a goal, the usual response consists of a route description, sometimes accompanied by a sketch-map (Habel, 1988; Tversky & Lee, 1998). When planning a trip from one point to another, a number of intermediate points are usually identified and the total trip is broken down into a number of different segments, the endpoint of one segment also serving as the start of the next. The routes we travel are often physically defined in our environment, e.g. by the flow of a river when travelling by boat, or by the highway system, determining the sequence of cities and exits when driving in a particular direction.

At least since the classic paper by Siegel and White (1975), drawing heavily on earlier work by Shemyakin (1962), the distinction of landmark knowledge, route knowledge, and survey knowledge has played a prominent role in cognitive psychology. In their view, route knowledge is an integral part in the acquisition of spatial information. Route knowledge consists of a sequence of decisions, which are triggered by the perception of landmarks identifying a particular location (for the definition of landmarks see for example Lynch, 1960)[1]. Simple learning paradigms, such as paired-associate learning, can account for route learning. Learning is thus "organised around the nodes of the decision system, the landmarks" (p. 29), while learning between landmarks is incidental and largely irrelevant. With more experience, however, information about different routes is integrated into a network-like structure and distance information is added. This integration of spatial knowledge ultimately leads to survey knowledge, which contains configurational information, such as regions or the geometrical shape of the layout of landmarks.

The model of Siegel and White (1975) has been criticised for its rigid developmental sequence of spatial knowledge acquisition, both for adult spatial cognition and

---

[1] We will later distinguish different uses of landmarks, as global landmarks or local routemarks, see Section 3.1.2.

children's development of spatial knowledge (e.g., McDonald & Pellegrino, 1993; Montello, 1998). However, the concept of route knowledge has stimulated a number of research questions.

One way to investigate route knowledge is to conduct linguistic analyses of route descriptions. In one study, for example, Habel (1988) asked 73 participants to describe a route through the town of Trier. As his analyses show, more than two-thirds of the descriptions used landmarks together with a directional change at that landmark to describe the route. Results from Tversky & Lee (1998) suggest that similar elements are used when giving a verbal route description and sketching a map depicting the route. This indicates the important role of landmark and directional-change information for route navigation.

Route descriptions necessarily consist of sequences of verbal statements, e.g., about landmarks and direction changes. However, route knowledge seems to be bound to a sequential order of places or objects even outside the realm of verbal descriptions. As Herrmann et al. (1995) were able to show, retrieval of information about objects along a route is easier in the direction of the route than in the opposite direction. If an object is briefly presented as a cue it is thus easier to recognise the name of the subsequent object than the preceding object on the route. This so-called *route direction effect* does not occur if the objects are just *displayed* in the same sequence as when travelling the path or when the optical flow is reversed while learning the route (Herrmann et al., 1995; Schweizer et al., 1998). These results indicate that spatial relations are encoded in a direction-specific way, and that the relation between two objects or places A and B is not necessarily the same as the relation between B and A.

In light of the distinction between route and survey knowledge, the question of distance estimation between two points on a route has also been addressed in a number of publications (Rothkegel, Wender, & Schumacher, 1998; Sadalla, Staplin, & Burroughs, 1979; Thorndyke & Hayes-Roth, 1982). Two different distance estimations are usually compared: route-distance vs. Euclidean distance estimation. In their study, Rothkegel et al. analysed the time it took participants to estimate the route-distance between different objects. Their results suggest that route-distance estimation times increase with increasing number of intervening objects along the route, while the length of the path itself only marginally determines the estimation time. They interpret their results as indicating that only route-distances between neighbouring objects (or places) are explicitly stored in memory. In contrast, route-distances encompassing more than two objects are mentally computed by adding the different legs from one object to another along the route.

In contrast, Rothkegel et al. were able to show in a different experiment that the required time for estimates of the Euclidean distance between two objects learned on a route increases with increasing Euclidean distance. In this case, the estimation time was largely independent of the route-distance between the two objects. Rothkegel et al. see this potential difference between the estimation times for route and Euclidean distance as one way to distinguish between route and survey knowledge. Route-distances are computed by integrating the length of individual legs along the route and are thus affected by the number of intervening objects. Survey knowledge should be largely independent of the route(s) experienced within an area and Euclidean distance estimations should therefore not be affected by the number of objects between the target pair. Assuming a mental scan process to estimate Euclidean distances in survey knowledge, an increase in the required time with increasing Euclidean distances fits well with this notion of survey knowledge.

In summary, the concept of route knowledge for human navigation has led to a significant body of research. Routes are commonly seen as a sequence of places which act as decision points for further travel. Route knowledge is direction-specific. Route knowledge in human memory also seems to consist of local information, e.g. the distance between a place and a successor, whereas integrated spatial knowledge of configurations or arbitrary Euclidean distances is attributed to survey knowledge.

In the following section we will briefly outline the role of route-based navigation in animals and how it compares to research on humans.

2.2   Route-Based Navigation in Animals

In the vast literature on animal navigation and spatial competences of different organisms, the role of route-based navigation has been a topic of research in many different species (for an overview see Able, 1996). Research in this area focuses on two main areas. First, the spatial behaviours of foraging animals, such as rodents or insects, enable researchers to focus on navigational abilities at time scales of minutes to hours, covering long distances with respect to the organism's body size. Second, migrating behaviour in long-distance migrating birds or sea animals, such as sea-turtles or fish, allows researchers to investigate navigational skills at a much larger scale of both time and space. In this short overview we will mainly focus on the first part, excluding migration for reasons of brevity.

As in the discussion on human spatial cognition, the issue whether navigation relies on a spatial cognitive map or whether it is based on different principles has also been addressed in the animal literature. A prominent model for the development of such a spatial cognitive map, explicitly representing the spatial relations between a large number of locations within the same reference system, has been put forward by Poucet (1993). The basic units in his hierarchical model are place representations. A place is defined by a collection of inter-linked views that an animal experiences through movement, especially rotations, within a small part of space. This set of views allows an animal to recognise the same place from different vantage points, e.g., when approaching a place from different directions. Connections between two places are encoded in a local reference system in terms of the direction and distance from one place to another. These multiple local reference systems make it impossible to compare directional information from one place directly with directional information from another place. The set of places within an animal's environment, which are connected by this vector information, constitutes a local chart. It is important to note that not all connections between places have to be available, nor does knowledge of the direction from A to B imply that the reverse direction from B to A is known. Besides this multiple-point reference system connecting some of the places, a second, so-called topological network yields information about the connectivity and order of the places. In Poucet's model both kinds of information, vectorial and topological, work together to enable an organism to find novel routes and shortcuts. The final step in the model consists of computing an overall reference direction. This global, environment-centred reference direction is assumed to correspond to the preferred axes of movement. It allows an organism to integrate the directional information from different local reference systems with respect to a common reference system, enabling the computation of previously unknown spatial relations and thus enhancing an organism's navigational precision and flexibility.

The model of Poucet (1993) captures a number of points relevant to route-based navigation. First, the role of places for the integration of navigational knowledge is emphasized. Second, the model clearly distinguishes between a network of connected places with multiple location-dependent reference systems and topological information on the one hand, and spatial information integrated within a common, global reference system on the other. Route-following behaviour and the finding of new routes is assigned to the lower level of integration of spatial information. Third, the gradual construction of an increasing number of connections between places, which takes place through active spatial exploration by the animal, can result in highly integrated spatial cognitive maps.

One goal of Poucet's (1993) model is the analysis of how integrated spatial knowledge might develop out of locally available information, mainly focussing on mammalian spatial navigation. However, for many species it can be demonstrated that no such maps are necessary to explain their navigational skills. As Wehner et al. (1996) point out, an animal's brain did not necessarily "evolve ... to reconstruct a full representation of the three-dimensional world, but to find particular solutions to particular problems within that world" (p. 138). They argue against a cognitive map metaphor and propose instead that a close analysis of the computational strategies used by a particular brain in a particular situation might prove more useful for our understanding of spatial behaviour.

In their elaborate model of spatial navigation in the desert ant (Cataglyphis), Wehner and colleagues were able to identify a number of important components of the ant's navigational techniques (Wehner & Menzel, 1990; Wehner et al., 1996). The most striking ability of the desert ant rests in its capacity to use path-integration to integrate all directional changes and distances travelled into a homing vector, representing the exact direction and distance of the nest entrance (from which its travel originated). To avoid cumulation of directional errors, the ant is equipped with a simple but efficient means of computing a global reference direction from an innate skylight compass. This enables the ant to correctly identify the direction it is travelling at each point in time. The importance of the homing vector becomes evident when the ant is transferred to a different environment before it is able to return to the nest. In this case, the ant ignores the repositioning and travels exactly to the location where its nest would have been if it were still in the old environment, clearly indicating that the homing vector drives the ant's spatial behaviour. More precisely, the behaviour is governed by two vectors: the homing vector and the target vector, both re-adjusted by path-integration at any moment in time. For a honeybee, the target vector may have been communicated prior to the foraging flight by a scout sister using the honeybee language (von Frisch, 1967).

Besides the dominant role of path-integration for *Cataglyphis*-navigation, the use of *landmark information* has also been demonstrated. One piece of evidence rests on the fact that individual desert ants show consistent trajectories from the nests to the feeding site, presumably by using landmark information on the way to guide their course. In addition, if an ant has reached the nest entrance, thus resetting its homing vector to zero, and is subsequently passively transported and released at the former feeding point, it returns to the nest along the same route as before (Wehner *et al.*, 1996). This indicates that features of the environment have been learned and can be used by the ant to find its way back home. There is also evidence that insects follow routes, e.g. a honeybee systematically flying from one food source to the next following the same route as on the previous flight.

One mechanism by which insects, such as desert ants or bees, can use landmark information has been proposed by Cartwright and Collett (1983). In their snapshot model, a 360° view of the environment is encoded by the animal at a particular location, e.g., the nest entrance or a salient part of a route. Using simple techniques, this information can then be used to steer towards the original position from which the snapshot was taken whenever the animal is close to it. By physically turning the landmark configuration it can be demonstrated that these snapshots are encoded with respect to a particular direction. If the configuration is turned, the animals are less likely to recognise it and thus don't use it for navigation (Judd, Dale, & Collett, 1999; Wehner *et al.*, 1996). Honeybees, for example, make a few circular flights around their nest before leaving to learn this landmark configuration; they do so in a well-defined fixed orientation, the same as when returning, to avoid the influence of rotation on the 360° view. Interestingly, a landmark strategy is not always used by the animal. The desert ant, for example, only reacts to the visual landmark configuration around the nest entrance when its homing vector indicates that it should be close to it. Expectancies, therefore, seem to play an important role in landmark usage. This simple mechanism to identify a previously visited location can also be coupled with additional navigational information. However, as Wehner *et al.* (1996) show, such route following is less accurate and slower when the homing vector is missing than for the original route home including the homing vector.

Similar processes like those proposed above have also been investigated in humans (Gillner & Mallot, 1998). Gillner and Mallot's view-graph model assumes that particular views are represented in a network of views together with the action required to get from one view to another (see also section 2.3).

In conclusion, research on route knowledge in non-human organisms has identified a number of detailed mechanisms how landmarks can be used to guide an organism's travel. Although there are differences in the importance and exact role of route knowledge for each species, the main topics are the same. Landmarks or places have to be recognised, expectancies for certain places can be built up, and navigational decisions have to be triggered at the correct point of the route. In addition, research on detailed navigational mechanisms in different animals, in particular non-vertebrates, provides an excellent means to explain complex human spatial behaviour by comparing it to simple mechanisms that can be found in other species.

In the last section of this chapter we will now focus on the navigational strategies employed in artificial mobile agents, e.g. autonomous robots.

2.3   Route-Based Navigation in Robots

The set of navigational strategies found in artificial moving agents mirrors the complexity of navigational strategies employed by biological organisms. In many instances the close resemblance between biological and artificial navigation is not a mere coincidence. Artificial navigation often mimics evolutionary proven strategies in an attempt to build robust technologies. In some instances, artificial agents are even specifically designed to test biological models of navigation, e.g. by Möller *et al.* (1998).

There are, however, a number of important differences between biological and artificial agents. First, modern technology makes a wide range of very accurate sensors possible, such as laser range finders, radar and ultrasonic sensors, global positioning systems, etc., providing information about the environment or the position of an agent. Unlike biological organisms, who might possess a few highly developed sensory systems for particular sources of information (e.g., the innate compass for the desert ant), the designer of a robot system is free to use an arbitrary combination of different sensors. In addition, robots sometimes are equipped from the start with a precise representation of their spatial environment (e.g., a map). In many cases, the environment is also specifically designed to match the robots navigational or sensory abilities, e.g. by placing markers at important points of a route or using well demarcated paths along which the robot travels. The basic tasks of spatial navigation, however, remain the same for both robots and biological organisms. According to Trullier *et al.* (1997), four different, broad categories to classify natural and artificial navigation can be distinguished: guidance, place recognition – triggered response, topological navigation, and metrical navigation.

Guidance is mainly concerned with directly leading an agent by external cues – either by following a particular gradient or moving to match the current sensory image with a stored image of the target or of the surroundings. In all these cases, the agent tries to locally maximise a predefined criterion without knowledge of spatial relations in the environment or about its own position. A robot could, for example, try to follow a wall of a corridor by keeping a constant distance to it (Krieg-Brückner *et al.*, 1998). For vehicle control, the guidance system could be designed to steer the vehicle parallel to the side of the road. A different robot could try to increase the match of its current visual image of the surroundings with a target image. This latter approach is directly modelled after the snapshot model of insect navigation, as discussed above (Röfer, 1998; Franz *et al.*, 1997).

Place Recognition – Triggered Response. For place recognition based strategies, complex spatial behaviours are triggered at distinct points in space. Once the correct place is recognised, the associated action (e.g., movement in a particular direction or guided behaviour) will lead to complex trajectories. The main problem of this strategy obviously consists in the correct identification of a place. One approach uses the perceptual input (the current state of the sensors of a robot) as the defining characteristic of a place (e.g., Schölkopf & Mallott, 1995). Another method is to detect landmarks (Krieg-Brückner et al., 1998). Of course both might lead to problems if the same sensory state or landmark can be evoked at different spatial locations (*perceptual aliasing*; Duckett & Nehmzow, 1997). A second problem is sensory noise. Even with highly accurate sensors it is unlikely that a second visit to the identical location will yield exactly the same sensory state.

Topological Navigation describes navigation based on topological networks and is thus a more flexible extension of place-triggered navigation. The basic elements of this type of network are places and the connections between these places. In this case the agent possesses knowledge how to get from one place to a second place which is connected to it. The agent's spatial knowledge is confined, however, to the topological network. New places not included in the network, or new connections between two places cannot be found by navigating on the basis of a topological network. One example for a topological network is Schölkopf and Mallott's (1995) view graph model. In this type of graph, each view (i.e. a location coupled with a viewing direction) is connected to a second view by an action. The view graph thus consists of all transitions from one view to the next, and the actions to achieve these transitions (Franz et al., 1997). As before, no further spatial knowledge about the directions or distances between different places is necessary for navigation.

Metrical Navigation. Unlike the last two approaches, which divide space in a small number of distinct places and the space in between, metrical navigation does not require such a distinction in principle. The metric most frequently used is Euclidean, thus distances and angles are well defined and can be used to drive spatial navigation. Pre-existing maps, which specify the metrical relations between objects in the environment of the agent, are often supplied directly or are autonomously constructed by triangulation and integration of sensory information. A coarse version of metrical navigation can be seen in world representations such as *occupancy grids*. Unlike navigation based on topological networks, the agent can determine its position in space and its spatial relation to each other object within the same metrical space for each position. However, knowledge of its position does not, as in topological navigation, automatically trigger the correct action. Instead, the action usually has to be computed for the current situation.

The Spatial Semantic Hierarchy Model. Kuiper's (1998) Spatial Semantic Hierarchy (SSH) model nicely illustrates the interplay of the different navigational strategies listed above. At the lowest level, *sensorimotor systems* of an agent provide information about its current state, which can be used at the *control level* to reach so-called *distinctive states.* A distinctive state is characterised by a local maximum of some *distinctiveness measure* which is computed from the change of the agent's sensori-motor states, but also more complex operations as detecting a landmark (Krieg-Brückner *et al.*, 1998) or recognising a kink in the robot's trajectory (Musto *et al.*, 1999; Röfer, 1999) can be employed to generate such a state. The measure thus provides the agent with a discrete set of distinctive states which can be reached by simple guidance techniques, such as gradient hill-climbing or following a particular trajectory from one distinct state to the next. At the causal level, these distinct states are treated as *places*, from which another place can be reached through a particular *action*. Combining different associations between two places through an action, the agent can build a *topological* representation, consisting of *places*, *paths*, and *regions*. At this level, route-finding problems can be solved. Finally, at a *metrical level*, the topological representation is enhanced with metrical properties, such as distance or direction between different places. This hierarchical strategy has been successfully employed in different simulated and physical robots (Kuipers & Buyn, 1991). However, it does not seem clear that the metrical level is necessarily acquired last or acquired at all. As Gutmann and Nebel (1997) have shown, knowledge at the topological level might be more important and more difficult to acquire for navigational tasks than a metrical representation (for a real-life example of the acquisition of such a topological map see Fu, Hammond & Swain, 1996).

Route based navigation can be found at different levels in the SSH model of Kuipers (1998). Starting with the causal level, different places are connected through spatial actions that the agent can perform. At the topological level, the flexibility of continuing from one place to another is greatly enhanced. The level of metrical spatial information, on the other hand, only enhances existing knowledge about spatial relations between different places, without adding significantly to route following behaviours. Eventually, however, metrical information can be used to expand topological networks by exploring new connections or defining new places.

3   The Model

In the following section, we will present a simple model describing key elements for route based navigation as part of an agents' general spatial knowledge. Drawing on the topics discussed above for human, animal, and artificial navigation, we will present basic concepts relevant for route based navigation. Our main goal consists of sketching a potential formalism expressing the key notions of route based navigation without restricting the formalism to a particular implementation, agent, domain, or discipline. Therefore, instead of focussing on a particular implementation or a specialised purpose, we will develop the general concepts step-by-step and discuss the implications of different possible assumptions at each stage, leaving open many details of the model for the reader to fill in. We will emphasise the core assumptions and the implications that they carry at each stage. This attempt of a flexible and open model is mirrored by our use of the term "agent" to describe any kind of autonomous mobile entity, whether it is a biological organism, a machine, or a virtual entity existing only in a simulation. It is our experience that the introduction of abstract concepts in this way aids the communication between disciplins, both at the empirical and theoretical level. In addition, new empirical questions will arise from a better understanding of the relevant concepts for route-navigation.

3.1 Single Routes

A **Route** is a concatenation of directed Route Segments from one Place to another. As we will see below, a Place is a tactical decision point where to continue, i.e. about the choice of the next Route Segment. This dichotomy of Places and Route Segments is consistent with most approaches taken in the literature. As an example of a simple route, i.e. a linear chain of directed Route Segments, consider the commuter train line S6 in Munich from Erding to Tutzing. Routes may also be cyclic, as for example part of the underground in London (cf. Fig. 1). Although we will sometimes refer to entities in the physical world (such as the train lines or the underground in this example), our interest lies in the knowledge that an agent has to possess to successfully navigate in space. Concepts such as a Place thus have to be interpreted in terms of an agent's knowledge, not in terms of particular physical locations, even though they often go hand in hand.

a   b

Fig. 1. Simple Route (a) and Cyclic Route (b)

A **Path along a Route** is distinct from a Route. It embodies the dynamic usage of a Route or a contiguous part of it. An example would be a ride on the train S6 from Munich Main Station to Tutzing. A Path may be periodic (even infinite) on a cyclic Route (for a detailed discussion of formalisms to denote paths and spatial trajectories see Eschenbach et al., 2000).

A **Route Segment** consists of two Places, a Source and a Target, which are connected by a Course (Places are discussed further below). As is evident from much of the research on route knowledge, the Course information is not always reversible. The connection between Source and Target is thus directed. Being able to walk from A to B does not imply any knowledge about how to get back from B to A.

In between the Source and the Target, the Course contains information that allows a subject to follow the Route Segment. This may be a particular trajectory, e.g. by replaying a specific motor program, or following a gradient hill-climbing strategy, such as walking towards the bakery by the increasing intensity of the scent. An ant might follow its homing vector, trying to minimise the distance to its nest entrance while integrating its path. A robot, on the other hand, might use a simple following strategy, e.g. "wall following" (Krieg-Brückner, 1998; Krieg-Brückner *et al.*, 1998). It is important to note that using the abstract notion of a Course allows the model to be adapted to very different navigational strategies.

A Route Segment contains the necessary information to make a tactical decision at the Source how to enter this particular Route Segment as a continuation in a Path. This Entry to the Route Segment is non-trivial. Depending on the scenario, the specification of the Entry may differ substantially. For a human walking through a system of passages in a town, this may be the decision or action "next right" (cf. Fig. 2a), "second right" (Fig. 2b) or "right when facing the city hall" (Fig. 2c). When steering a boat, it might entail changing one's course to 135° SE (Fig. 2d).

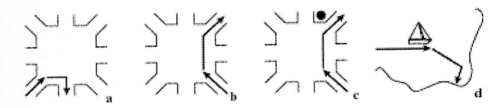

Fig. 2. Entries and Exits of Route Segments

In a similar way, the Exit of the Route Segment contains the necessary information to arrive at the target in a well-defined way. In the above examples, taking the "next path to the right" assumes that "next right" is unambiguously specified. However, this does imply that the traveller does not walk into an intersection backwards or sideways, but instead is walking forwards to avoid a conflict between a body-defined "left" and "right" and the "left" and "right" determined by the travel direction. Another example is the final Route Segment when docking a boat. One usually wants to end up parallel to the dock to avoid damage.

In some scenarios, some of the attributes may be superfluous: in a labyrinth, for example, Course and Exit may be empty, only the tactical decision when entering a Route Segment is necessary. Similarly, some information may be intentionally redundant. In some cases, either Entry or Exit, or both, may be used to model a well-defined position or bearing: when navigating in space, a (new) direction may be part of the Exit (e.g. the harbour entrance) or the Entry to a Route Segment (e.g. the compass bearing for a new course); in this model we prefer to associate the new direction to the Entry when it models a tactical decision related to the next rather than the previous segment.

Kinds of Route Segments. As is evident from the range of possibilities above, the particular information associated with Entry, Course or Exit depends on the context and grain of detail, i.e. the particular scenario one is trying to model. Let us call this the Kind of the Route Segment. Examples for common Kinds of Routes would be CommuterTrainLine, ShipRoute, FootPassage, CityRoad, Highway or Labyrinth, distinguished by different human task-related behaviour or navigational tactics involved (Krieg-Brückner, 1998). For other animals, common Kinds of Routes might entail UndergroundPassage, TreeBranches, or MigrationRoute. Route Kinds have to be consistent with Place Kinds, of course (compare also *Transfers* below); in general, the information associated with each component of a Route Segment has to be consistent with its Kind.

Places. Route Segments are linked by Places, e.g. concatenated to a simple Route: the Target of the preceding Route Segment has to be the same as the Source of the next. When constructing Route Segments incrementally, this is quite straightforward as we will continue from the Place we arrived at. For more complex Route Graphs, this is not so obvious, see *Place Integration* below.

Of course, the notion of a Place has different implications depending on the scenario at hand. In rats, for example, the concept of place representations has been used to describe the response pattern of single neurons which fire whenever the rat finds itself in the same position in its environment (O'Keefe & Nadel, 1978). However, this location is probably not absolutely defined, but instead in relation to environmental cues (such as distal landmarks or the geometry of the surroundings, etc.). If the rat knows its position with regard to these cues, however, the neuron's response pattern can also be driven by egomotion, with no further environmental cues. This implies that a place is defined as a physical location with respect to an allocentric reference system, which in turn is updated by environmental features. Similarly, navigational charts for boating often refer to the relative position of salient landmarks to define the place of bearing changes.

In contrast, bees and ants do not only respond to the relative position of landmarks, but also to the absolute orientation of the configuration (Collet & Baron, 1994). Even more so, a hiker using a global positioning device will also rely on the global coordinates to determine his or her location. Two Places can therefore be identical in one situation and different in another, depending on the definition of a Place.

In another scenario, the dancing instructions "two steps forward, side-step, turn" might be interpreted as a Route across the dance floor, where each Place is defined solely in terms of the preceding Route Segment. In this example, there is no reference to the environment at all.

Reference System and Position. Quite importantly, each Place comes with its own local Reference System (RefSystem). Uniform global reference systems will be introduced in Section 3.4. The term reference system is meant to be very general. For example, in case of the desert ant it may refer to absolute compass bearings, derived from the direction of polarised light. It may also be a constellation of *landmarks* that can be seen in the distance (e.g. church squires, towers, celestial bodies such as the sun), or a particular collection of *routemarks* which characterise the Place in a passage scenario (e.g. the city hall, the ice-cream parlour)[2]. Probably the most basic RefSystem for a Place is the direction of travel in which one approaches that Place – which is often implied when giving directions.

Relative to its RefSystem, a Place is characterised by a RefPosition, i.e. a well-defined reference *position* in it and, in general, a *bearing* (cf. the crosses plus arrowheads in Fig. 3). As examples consider "in front of the city hall, facing the main entrance", "on Königsplatz, leaving Luisenstraße, coming from the direction of the main station", or "in the centre of the market place, bearing North". Other ways to indicate a RefPosition include the snapshot model (cf. Section 2.2), whereby a RefPosition is defined as the position where the image stored for that position matches the current percept, or simply by referring to a salient point.

The local RefSystem is not universally defined for each Place. Depending on the kind of RefSystem, Places may have different Kinds, consistent with the Kinds of Route Segments above. Thus the *same physical location* might have multiple RefSystems associated with it for different Routes; in our model, we distinguish these as *different* Places unless they are integrated to the same RefSystem (see Place Integration). For example, when using the direction of entry as a local RefSystem, a Place is automatically accompanied by a different RefSystem whenever it is approached from a different direction. This means, of course, that the interpretation of instructions such as "next left" depends on the preceding Route Segment.

In Figure 3, a few examples of Places and their corresponding RefPositions and RefSystems are depicted. As the two Figures 3b ("second right") and 3c ("right when facing the city hall") show, the same physical location might have different RefPositions and RefSystems even if the trajectory is very similar.

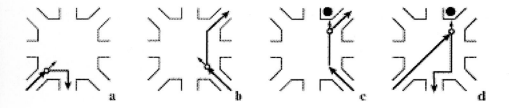

Fig. 3. Places with Individual Reference Systems

3.2   Route Graphs

Union of Routes into Route Graphs. The representation of Places and Route Segments corresponds to the mathematical notion of a directed graph with a set of nodes (Places) and edges (pairs of Source and Target in Route Segments), both enriched with their particular attributes (there may be several edges between a given pair of nodes, in both directions). The union of different Routes into Route Graphs therefore corresponds directly to the union of the corresponding sets of nodes and sets of edges, respectively. This union of different Routes into a Route Graph is often non-trivial. In this section we will deal with the most prevalent problems that might occur when trying to integrate different Routes into a coherent whole.

In Fig. 4, two previously separate Routes are joined into one Route Graph, sharing a common sequence of Places and Route Segments (i.e. a sub-Route). As an example consider the commuter train lines S6 from Erding to Tutzing and S8 from Munich Airport to Pasing; these share the same track and stations between Leuchtenbergring and Pasing[3]. Of course, it is essential for the union of these two Routes that the identity of the Places and Route Segments of the two sub-Routes is noticed. In the case of two subway-stations with the same names, this might be fairly obvious. However, in the general case, the recognition of one Place as a known Place which was visited before might be much more demanding.

Fig. 4. Union of two separate Routes (a) into one Route Graph (b)

Place Integration. In the scenario of Fig. 3a and 3b, for example, two positions for two Routes are marked. In the current model, they thus signify two separate Places and are therefore two unrelated parts of two different Routes. The two Places might share a common RefSystem (e.g. a collection of route marks) but each has its own RefPosition. The way the union of Routes into a Route Graph was introduced above, these two Routes would be treated as separate, with no relationship between them. What is needed is a way to integrate Places which only differ in their respective RefSystem or RefPosition – and this is indeed a non-trivial activity.

---

[3] It is a choice of modelling, whether the routes S6 and S8 are kept separate (and drawn side-by-side in different colours, as e.g. in the diagram of the Munich commuter train system), or whether the same track is regarded as an identical sub-route.

To integrate two Places, one thus has to come up with a common RefSystem and to agree on a common RefPosition. Choosing the same RefPosition "in front of the city hall" in Figures 3c and 3d, we are able to integrate the two Places.

Depending on the scenario, Place integration may be easy or require complicated updating of Route Segment attributes to conform with the new Place. Exits and Entries of the various Route Segments associated with a Place all refer to the same RefPosition in its RefSystem. For each Place that is affected by an integration, each Exit of every Route Segment leading towards it and each Entry of every Route Segment leading away from it has to be re-computed. In the example above, the Exit of the left-hand Route Segment in Fig. 3a and the Entry to the segment leading downwards have to be re-computed (as is depicted in Fig. 3d) relative to the new RefPosition.

Choice of Route Continuations in a Graph. In general, a common RefPosition and RefSystem is necessary to achieve the increased functionality of a Route Graph: after Place integration, we are able to continue with *either* of the two Route continuations.

Our own experience reveals that we often have a hard time with Place integration, having to "look around" to recognise a Place to be the same as one that was previously reached by another Route. This is especially obvious when trying to retrace ones Path through an unfamiliar environment, such as a forest, with new passages never seen before opening up when walking back. One reason for this problem might lie in the limited field of vision for humans, which dramatically changes the visual input depending on the walking direction. Other animals, such as the desert ant, have a 360° visual field, which makes them less vulnerable to this problem.

A second problem for Place integration are false matches (undesired aliases) for a Place. In modern buildings, for example, one intersection might appear to be the same as the intersection encountered just minutes ago, although they are in different parts of the building. Whereas humans can be confused in such a situation, the desert ant can usually rely on its well-developed path integration system, not to be fooled by the similar appearance.

3.3   Layers and Transfers

Modelling Aspects by Layers. As we have mentioned above, it is often necessary to distinguish between different Kinds of Routes when modelling different scenarios. One possible way to conceive of these different Kinds of Routes can be found in separate Layers. This approach is used, for example, in constructing cartographic maps of an area, where the information for water and power lines, for the sewer system, for roads, train tracks, and other significant entities are kept separate on different maps. Each layer represents a different aspect of the environment, in analogy to aspect maps (Berendt *et al.* 1998). In our model, each Layer represents a Route Graph of a particular Kind.

Munich main station, for example, would then correspond to different Places on different Layers. When arriving by train, the corresponding RefPosition might be "at the head of the tracks, facing the timetable". If travelling via commuter train, it might be "underground, in the centre of the platform in the direction of Pasing". If travelling by foot, the entire train station might correspond to a Route Graph, instead of a single Place. To achieve appropriate links in each Layer, these Places have to be distinct.

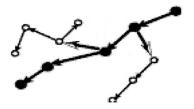

Fig. 5. Transfer Routes (half-filled arrows)

Abstraction Hierarchies and Transfers between Layers. Obviously, modern day travel requires the use of a combination of different Kinds of Route Graphs at different levels of detail. When travelling from Hamburg to Munich, for example, one will probably have to walk towards the subway, ride the subway to the train station, take a fast train from Hamburg to Munich, get off the train in Munich, walk downstairs to the commuter train, take it for a few stops, maybe changing trains, and finally walk towards one's final destination. Understandably, the level of detail and the kind of Route Graph needed for walking towards the commuter train is different from the one dealing with the train ride. So far, the model cannot deal with transitions between different kinds of such Route Graphs since we tacitly assumed that Source and Target nodes of a Route Segment are of the same Kind; such Route Graphs shall be called *homogeneous*. To allow transitions in heterogeneous Route Graphs, need a special type of Route Segment, whose Source and Target nodes can be of different Kinds. We will call these special Route Segments "Transfers". To allow different complexities of Transfers, they can also consist of a whole Route or Route Graph.

In Fig. 5, such Transfers are depicted by half-filled arrows. Transfers are also directed (e.g., highway ramps). They may be one-to-one, one-to-many (e.g., Exit Transfers) or many-to-one (e.g. Entry Transfers). Transfers do not only occur when switching the mode of transportation or ego-motion, though. In some cases, place integration might not be possible even within the same Kind of Route Graph or Layer. In these cases, Transfers between Layers then also serve as refinements or abstraction projections. One such example is the set of entry and exit ramps of a highway intersection. In this case, the exit ramp taken in one direction is physically different from the entry ramp on the other side of the intersection. However, we still wish to think of the highway crossing as the same Place at a higher level of abstraction. In the current model, this would be modelled by a second, more abstract Layer. Both, the entry and the exit ramp would be connected by a Transfer to a single Place at this higher level of abstraction.

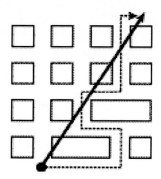

Fig. 6. Co-ordination Between Navigation in Passages and Navigation in Space

The same technique may also be used to model the relationship between *planned route – actual route,* or *navigation in passages* in co-ordination with *vector navigation.* The straight arrow in Figure 6, for example, represents the desired direction in which a person might want to travel, while the dashed arrow denotes a path within a city block environment (where the direct route is blocked). While the actual travel is performed within the physical constraints that can be represented as a Route Graph in the street layout, the mean direction of travel is determined by the dynamically changing navigation vector in a different Layer. In this way, different navigational strategies, which do not necessarily have to be based on route knowledge, can be integrated within the model's framework (e.g. by continuously attempting to do place integration dynamically, or by minimising the length of a co-ordination Transfer between the layers, in a common RefSystem). Of course, the example here is similar to the navigational strategy employed by the desert ant returning home (cf. Section 2.2).

3.4   From Route Graphs to Overviews

In Section 2.1, we already touched upon the distinction between route and survey knowledge in cognitive psychology. We will finish this approach of formalising route knowledge with a few remarks towards the development of a particular kind of survey knowledge, which we will term *overview.* Most, but not all of the examples used throughout this chapter, have dealt with the identification of a certain Place by relying on cues perceivable from one's actual location. This situation has been termed "field perspective" in contrast to so-called "observer perspective", which roughly corresponds to a view from above (Nigro & Neisser, 1983; Schweizer *et al*, 1998). Interestingly, humans can readily switch between both perspectives. Saade and Werner (in press), for example, have found that spatial knowledge acquired in field perspective can be used in situations requiring an observer perspective, and vice versa. Humans are thus able to construct an overview from information in a route graph. An overview may be thought of as a bird's view from the top straight down, or a view from an elevated position (mountain, tower). An overview may also denote an artificially constructed view, as an approximation of a map. In our model, an overview can be modelled by an **Overview Graph**.

Uniform Reference System. The essential property of an **Overview Graph**, or Overview for brevity, is a uniform **RefSystem** from the observers **RefPosition** as far as the view allows. The **RefSystem** is therefore not completely "global" but restricted to the sub-graph of the original graph from which the **Overview** is constructed. Consequently, the **Overview Layer** requires a re-computation of the local **RefSystems** of all **Places** included in the **Overview**, based on the uniform **RefSystem** of the observer's location, e.g. viewpoint. The **RefPositions** for each **Place** are not necessarily affected by this transformation (whereas the bearing often is).

The variety of possible **RefSystems** is again large. A special case can be seen in the introduction of a Euclidean space, such as $\Re^2$ or $\Re^3$, which allows an approximation to common cartographic maps.

In the uniform **RefSystem**, the construction of new relations between **Places** becomes easily possible: new distances or angles can be denoted in pseudo-segments. Similarly, deviations and shortcuts can now be planned. Note that the relation between the original (sub-)graph and the **Overview Graph** can be maintained in different **Layers**, with appropriate **Transfers** for co-ordination.

4   Closing Remarks

Our main goal in this paper was to present a common, simple formalism to express the key notions of route based navigation as they occur in different disciplines. We hope that this endeavour will help researchers focussing on different research questions pertaining to spatial navigation to communicate their findings and theories more efficiently. Of course, we were not able to present thorough overviews of the navigational literature in all different areas. Instead, we tried to present a number of exemplary research questions and results from human, animal, and robot navigation to emphasise the close resemblance of navigational problems encountered in each field.

We are well aware that the general modelling framework presented above leaves much freedom to model the same scenario in different ways. There are often trade-offs, e.g. between complexity of generating a route graph vs. using it for navigation, between modelling by place integration or explicit transfers, or between several levels of granularity. It is not immediately obvious, for example, whether several trains on the same track (e.g. a fast train and a slow train between Hamburg and Munich) correspond to the same or different routes; this is rather a matter of abstraction. Similarly, a path along the route from Hamburg to Munich with or without changing trains of the same kind might make a difference; is it the same or two different routes?

Obviously, the model does not provide answers to the particular navigational strategy humans or other organisms use. However, the explicit definition of the basic elements and the preconditions for integrating local knowledge put forward in the model might help to specify the questions to ask. From a psychological point of view, for example, many questions remain open. To what extent are separate routes learned before they are integrated using place integration? What are the mechanisms that humans use to identify a place? How and when are the adjustments made to the local reference systems so that different routes can be combined to a route graph? What kind of information is used to specify the course of a route segment, after a particular route segment has been entered? How is survey knowledge related to overviews constructed from sub-graphs with a common reference system? These questions and more, we feel, will benefit from a common language of route based navigation – thus enabling an interdisciplinary research effort.

Acknowledgements

This paper is the result of an interdisciplinary dialog within the context of the priority program on "Spatial Cognition" funded by the German Science Foundation. We gratefully acknowledge many fruitful discussions with our colleagues. We especially thank Thomas Röfer for his comments on the section on robot navigation, which is in part based on his thesis work, and Carola Eschenbach, Ute Schmidt, and an anonymous reviewer for their detailed comments on an earlier version of this paper.

References

Able, K. P. (1996). Large-scale navigation. *The Journal of Exp. Biology, 199*, 1-2.
Allen, G.L. (1999). Spatial abilities, cognitive maps, and wayfinding: Bases for individual differences in spatial cognition and behavior. In R. Golledge (ed.), *Wayfinding behavior* (46-80). Baltimore: Johns Hopkins.
Berendt, B., Barkowsky, T., Freksa, C., & Kelter, S. (1998). Spatial representation with aspect maps. In C. Freksa, C. Habel, & K. F. Wender (Eds.), *Spatial cognition - An interdisciplinary approach to representing and processing spatial knowledge.* Berlin: Springer.
Cartwright, B.A. & Collet, T.S. (1983). Landmark Learning in Bees. In *Journal of Comparative Physiology, A 151*, 521-543.
Collet, T.S. & Baron, J. (1994). Biological compasses and the coordinate frame of landmark memories in honeybees. *Nature, 368*, 137-140.
Duckett, T. & Nehmzow, U. (1997). Experiments in evidence based localisation for a mobile robot. In *Spatial Reasoning in Mobile Robots and Animals* (25-34). Manchester University. AISB-97 Workshop.
Etienne, A.S., Maurer, R., Georgakopoulos, J. & Griffin, A. (1999). Dead reckoning (path integration), landmarks, and representation of space in a comparative perspective. In R. Golledge (ed.), *Wayfinding behavior* (197-228). Baltimore: Johns Hopkins.
Eschenbach, C., Tschander, L., Habel, C., & Kulik, L. (2000). The specifications of paths. *This issue.*
Franz, M. O., Schölkopf, B., Georg, P., Mallot, H. A., and Bülthoff, H. H. (1997). Learning view graphs for robot navigation. In W. L. Johnson (ed.), *Proc. 1st Int. Conf. on Autonomous Agents* (138-147). New York. ACM Press.
Fu, D.D., Hammond, K.J., & Swain, M.J. (1996). Navigation for everyday life. *Technical Report 96-03. Dept. of Computer Science, University of Chicago.*
Gillner, S. & Mallot, H.A. (1998) Navigation and acquisition of spatial knowledge in a virtual maze. *Journal of Cognitive Neuroscience, 10*, 445-463.

Golledge, R.G. (1999). Human wayfinding and cognitive maps. In R. Golledge (Ed.), *Wayfinding behavior* (5-45). Baltimore: Johns Hopkins.

Gutmann, J.-S. & Nebel, B. (1997). Navigation mobiler Roboter mit Laserscans. In P. Levi, T. Bräunl, and N. Oswald (eds.), *Autonome Mobile Systeme* (36-47). Informatik aktuell, Berlin, Heidelberg New York. Springer.

Habel, C. (1988). Prozedurale Aspekte der Wegplanung und Wegbeschreibung. In H. Schnelle & G. Rickheit (eds.), Sprache in Mensch und Computer (107-133). Opladen: Westdeutscher Verlag.

Herrmann, Th., Buhl, H.M. & Schweizer, K. (1995). Zur blickpunktbezogenen Wissensrepräsentation: der Richtungseffekt. *Zeitschrift für Psychologie, 203*, 1-23.

Judd, S.P.D., Dale, K., & Collett, T.S. (1999). On the fine-structure of view-based navigation in insects. In R. Golledge (ed.), *Wayfinding behavior* (229-258). Baltimore: Johns Hopkins.

Krieg-Brückner, B. (1998). A Taxonomy of Spatial Knowledge for Navigation. In: Schmid, U., Wysotzki, F. (Eds.). *Qualitative and Quantitative Approaches to Spatial Inference and the Analysis of Movements.* Technical Report, 98-2, Technische Universität Berlin, Computer Science Department.

Krieg-Brückner, B., Röfer, T., Carmesin, H.-O., Müller, R. (1998). A Taxonomy of Spatial Knowledge for Navigation and its Application to the Bremen Autonomous Wheelchair. In Freksa, Ch., Habel, Ch., Wender, K. F. (Eds.), *Spatial Cognition. Lecture Notes in Artificial Intelligence 1404* (373-397). Springer.

Kuipers, B. (1998). A hierarchy of qualitative representations for space. In Freksa, Ch., Habel, Ch., Wender, K. F. (Eds.), *Spatial Cognition. Lecture Notes in Artificial Intelligence 1404.* Berlin: Springer.

Kuipers, B. J. & Byun, Y.-T. (1991). A robot exploration and mapping strategy based on a semantic hierarchy of spatial representations. *Journal of Robotics and Autonomous Systems, 8*, 47-63.

Loomis, J.M, Klatzky, R.L., Golledge, R.G., & Philbeck, J.W. (1999). Human navigation by path integration. In R. Golledge (ed.), *Wayfinding behavior* (125-151). Baltimore: Johns Hopkins.

Lynch, K. (1960). *The Image of the City.* Cambridge: MIT-Press.

McDonald, T.P. & Pellegrino, J.W. (1993). Psychological perspectives on spatial cognition. In T. Gärling & R.G. Golledge (Eds.), *Behavior and Environment: Psychological and geographical approaches*, p. 47-82. Amsterdam: Elsevier.

Möller, R., Lambrinos, D., Pfeifer, R., Wehner, R., & Labhart, T. (1998). Modeling Ant Navigation with an Autonomous Agent. In: *From Animals to Animats, Proc. Fifth International Conference of The Society for Adaptive Behavior* (185-194). MIT Press.

Montello, D.R. (1998). A new framework for understanding the acquisition of spatial knowledge in large-scale environments. In M.Egenhofer & R.G.Golledge (Eds.), *Spatial and temporal reasoning in Geographic Information Systems* (143-154). Oxford University Press.

Musto, A., Stein, K., Eisenkolb, A., Röfer, T. (1999). Qualitative and Quantitative Representations of Locomotion and their Application in Robot Navigation. In: *Proc. of the 16th International Joint Conference on Artificial Intelligence* (1067-1073). Morgan Kaufman Publishers, Inc. San Francisco, CA.

Nigro, G. & Neisser, U. (1983). Point of view in personal memories. *Cognitive Psychology, 15*, 467-482.

O'Keefe, J. & Nadel, L. (1978). *The hippocampus as a cognitive map.* Oxford University Press.

Poucet, B. (1993). Spatial cognitive maps in animals: New hypotheses on their structure and neural mechanisms. *Psychological Review, 100*, 163-182.

Röfer, T. (1998*). Panoramic Image Processing and Route Navigation.* PhD thesis. BISS Monographs 7. Shaker-Verlag.

Röfer, T. (1999). Route Navigation Using Motion Analysis. In: Freksa, C., Mark, D. M. (Eds.), *Spatial Information Theory, Proc. COSIT''99. Lecture Notes in Computer Science, 1661*, 21-36. Berlin: Springer.

Rothkegel, R., Wender, K.F., & Schumacher, S. (1998). Judging spatial relations from memory. Freksa, Ch., Habel, Ch., Wender, K. F. (Eds.): Spatial Cognition. *Lecture Notes in Artificial Intelligence 1404.* Berlin: Springer. (79-106).

Saade, C. & Werner, S. (in press). Flexibilität mentaler Repräsentationen räumlicher Information in Abhängigkeit von der Erwerbsperspektive. *Zeitschrift für Exp. Psychologie*

Sadalla, E.K., Staplin, L.J., & Burroughs, W.J. (1979). Retrieval processes in distance cognition. *Environment and Behavior, 12,* 167-182.

Schölkopf, B. and Mallot, H. A. (1995). View-based cognitive mapping and planning. In *Adaptive Behavior* 3 (311-348).

Schweizer, K., Herrmann, T., Janzen, G., & Katz, S. (1998). The route direction effect and its constraints. In C. Freksa, C. Habel & K.F. Wender (eds.), Spatial Cognition. *Lecture Notes in Artificial Intelligence 1404*, 19-38. Berlin: Springer.

Shemyakin, F.N. (1962). Orientation in space. In B.G. Anan'yev et al. (Hrsg.),*Psychological Science in the U.S.S.R:* (Bd. 1, S. 186-255). Washington, D.C.: U.S.Department of commerce, Office of Technical Services.

Siegel, A.W. & White, S.H. (1975). The development of spatial representations of large-scale environments. In H.W. Reese (Ed.), *Advances in Child Development and Behavior* (vol.10, S. 9-55). New York: Academic.

Thorndyke, P.W. & Hayes-Roth, B. (1982). Differences in spatial knowledge ayquired from maps and navigation. *Cognitive Psychology, 14,* 560-589.

Trullier, O., Wiener, S. I., Bertholz, A., and Meyer, J.-A. (1997). Biogically based artificial navigation systems: Review and prospects. *Progress in Neurobiology, 51*, 483-544.

Tversky, B. & Lee, P.U. (1998). How space structures language. In C. Freksa, C. Habel & K. F. Wender (eds.), *Spatial Cognition* (157-175). Berlin: Springer.

v. Frisch, K. (1967). *The dance language and orientation of bees.* Oxford University Press, London.

Wehner, R. & Menzel, R. (1990). Do insects have cognitive maps? *Ann. Rev. Neuroscience. 13*, 403-413.

Wehner, R., Michel, B., & Antonsen, P. (1996). Visual navigation in insects: Coupling of egocentric and geocentric information. *The Journal of Experimental Biology, 199*, 129-140.

Werner, S., Krieg-Brückner, B., Mallot, H.A., Schweizer, K., & Freksa, C. (1997). Spatial cognition: The role of landmark, route, and survey knowledge in human and robot navigation. In M. Jarke, K. Pasedach, & K. Pohl (eds.), *Informatik '97* (41-50). Berlin: Springer.

Wiltschko, R. & Wiltschko, W. (1999). Compass orientation as a basic element in avian orientation and navigation. In R. Golledge (ed.), *Wayfinding behavior* (259-293). Baltimore: Johns Hopkins.

Using Realistic Virtual Environments in the Study of Spatial Encoding

Chris Christou[1] and Heinrich H.Bülthoff [2]

[1] Unilever Research, Port Sunlight Laboratory, Wirrall, CH63 3JW
[2] Max-Planck Institute for Biological Cybernetics, 72076 Tübingen, Germany
heinrich.buelthoff@tuebingen.mpg.de

Abstract. Computer generated virtual environments have reached a level of sophistication and ease of production that they are readily available for use in the average psychology laboratory. The potential benefits include cue control, incorporation of interactivity and novelty of environments used. The draw-backs include limitations in realism and lack of fidelity. In this chapter we describe our use of virtual environments to study how 3D space is encoded in humans with special emphasis on realism and interactivity. We describe the computational methods used to implement this realism and give examples from studies concerning spatial memory for object form, spatial layout and scene recognition.

1   Introduction

The study of human spatial cognition is the study of learning, memory and problem solving involving spatially extended materials or environments. Empirical studies in spatial cognition proceed, as within many other domains of psychological science, according to some stimulus-response paradigm in which a mental process is probed by making inferences from the subject's response to a given carefully controlled stimulus. What constitutes a stimulus depends on the experiment and the task in hand and in the past has ranged from the very abstract to the very elaborate. Elaboration in this case has meant carrying out experiments in the real world where sensory cues are difficult to control. We consider in this paper a possible extension to the tools available to the experimental cognitive psychologist afforded by advances in computer graphics and computer simulation. We argue that computer modelled virtual environments (or virtual environment simulations, VES) may be used to create realistic scenarios which simulate the visual cues available in the real world but in which control of cues is always possible. Furthermore, spatial cognition tasks may be designed which more closely reflect natural tasks. To demonstrate this here we describe our own efforts in the construction and use of a three-dimensional computer generated environment called *The Café Lichtenstein* with which we have performed a number of experiments. These experiments have concentrated mainly on investigating the role of spatial reference frames in human memory for object structure, layout of multiple objects and large-scale scenes. We begin with an introduction to the general terminology and

Ch. Freksa et al. (Eds.): Spatial Cognition II, LNAI 1849, pp. 317–332, 2000.

concepts used in such simulations, we consider some of the benefits afforded to cognitive science through the use of VES and then demonstrate their use in several spatial cognition experiments.

2   Basics of Virtual Reality and Virtual Environment Simulation

2.1   Definitions

What do we mean by a virtual reality or a virtual environment? One definition is that it is a state of affairs, a depiction of objects or of space, which has no physical basis. A picture or a painting depicts a virtual space, so does a mirror but we cannot touch the objects depicted there because only the sense of sight is being stimulated. How do we normally determine that we are observing a real object and are not experiencing an illusion or a hallucination? It is, of course, because we can touch and interact with real objects. If we explore this question more closely however, we realize that the basis of our trust in the existence of what we experience is only through sensory feedback resulting from interaction. That is, through being able to see an object, touch it, smell it, perhaps to break it. Increasingly computer-based technology is becoming more sophisticated at simulating all of those perceived events that constitute what we call reality. Theoretically, the only difference may be that the real world consists of physical constituents such as atoms and electrons, from which sensory qualities derive, whereas virtual worlds consist of *simulations* of the originating physical processes behind these sensory qualities. Our aim here is to demonstrate how sophisticated such simulations can be and provide examples of their use in studying spatial cognition.

The terms Virtual Environment Simulation (VES) and Virtual Reality (VR) are used here to relate to the same thing, although many people now prefer the former term having realized that definitions of what "reality" actually consists of are somewhat problematic. The term Virtual Reality has historical precedence and perhaps a broader meaning that has also included the possibility of interaction within the simulated space. In general, a definition couched in terms of interactivity within a simulated three-dimensional (3-D) computer generated space captures the essence of what we call Virtual Environment Simulation.

The ontology of a VES includes its actors or players, an environment or geometry in which the actors behave, and a set of rules of behavioural dynamics attached to the actors (see [1]). Such behaviour includes changing the spatial location and orientation of the actor and the manipulation of objects, or even other actors. If we simplify the scenario somewhat and consider an actor as consisting only of a simulated eye or camera that projects the environment onto an imaging device, then the possible actions that can be performed by the actor are restricted to changing its viewpoint. In essence a VES requires three differentiable components: the motion input or interactive control devices, the simulated environment itself, and finally, the rendering of the environment. Accurate simulation of the processes involved in each of these determines how effective the VES will be. Each of these components is considered in a more detail below.

2.2   Motion Input

One of the most important advantages of a VES is that it provides the operator the ability to interact or perform particular behaviour within the virtual space. The behaviour of the actors in a VES is controlled by the operators (the human user or subject). If this behaviour consists solely of the movement of a camera through the environment, the operator can control the motion of the camera by using a variety of motion input devices and motion trackers. Motion input devices are hand operated and include the joy-stick, button box, SpaceBall (Spacetec IMC Corp., USA) or 3-D SpaceMouse (Logitech, USA). The latter are essentially extensions of the common computer mouse but map not only movement across a 2-D surface but through a 3-D space. Usually, these devices provide the ability not only to translate in 3-D but also to rotate about all three axes providing 6 degrees of freedom of movement. They are used predominantly for desktop simulations in which the environment is viewed by the operator sitting in front of a screen. Such screens may range from small cathode ray tube based monitors or large-scale 180° projection screens.

Motion trackers, the other form of motion control devices, are passive recorders of movement that detect the motion of sensors placed on different parts of the body and in the physical space of the observer. Such devices can be implemented by detection of small changes in the electromagnetic field generated by the sensors with respect to a transmitter placed in a fixed position in real space. Therefore, motion trackers let the operator move around more freely but for immersive simulations require some form of head mounted display (HMD) on which the environment is displayed according to the movements of the operators' body or head.

2.3   The Simulated Environment

The visual properties of objects convey their 3-D nature to observers. For instance variations in shading and texture contain cues to three-dimensional extent or expansion in depth. Similarly, navigable or large-scale space that one may perform in is defined by the properties of the objects within it. Our natural environments are full of detail such as textures on roads or patterns on wallpaper or a carpet, but also they contain numerous objects whose relative position helps us to judge relative distance and depth. One common influence of this for example is the "moon illusion". The moon seen close to the horizon appears larger than the moon seen directly above although the variation of the moons' distance from the earth is relatively small. This effect has been attributed to size-distance scaling [2]; the retinal extent of an object is scaled according to distance in order to calculate its actual size. If only retinal extent is available then a small object at a close distance may appear relatively large in actual extent. Similarly, the moon at the horizon is scaled according to the landscape rich in depth cues (making it appear further away) whereas when viewed at its zenith few depth-cues are available (making it appear closer). Since the retinal image of the moon in both cases is the same, a different size-distance scaling is applied making the horizon moon appear larger than the zenith moon. What we learn from this is that the simulation of space is relative to the objects that populate it. Setting up the condition for a spatial cognition experiment in virtual space requires an adequate level of detail that conveys the impression of extent-in-depth, even though depth perception may not be the direct object of enquiry.

2.4   Realism

Realism in simulations relates to a direct comparison between the simulation and the natural world; a question of fidelity. There are varying degrees of realism in computer graphics. Much of the current effort in the field of computer graphics research is directed towards achieving photo-realism; that is, simulating in a picture of objects the same visual impression that might be observed in a photograph of the objects in the real world. This can be achieved by the depiction of the visual characteristics of real scenes and the effects of the interaction between objects and light before this light reaches our eyes. Such detail includes the modulation of light intensity according to the manner in which light is reflected at the surface of an object. Different materials reflect light differently - accounting for, for instance, the highlights seen on shiny metals as opposed to diffusely reflecting surfaces such as concrete and fabrics. Such cues are regularly used by painters who want to portray materials of objects accurately. One means of achieving such realism in computer graphics has been to model the physical principles of light-surface interaction. One example of this is the modelling of diffuse interreflection of light using the radiosity method [3]. Light reaches our eyes after a complex process of reflection between ordinary surfaces in an environment. This interreflection means, for example, that surfaces not directly illuminated by a light source are still visible owing to reflected light from nearby surfaces that are directly illuminated. Proper simulation of light interreflection improves the overall realism of any complex [4]. Currently, many software programs exist which can perform such simulations. The necessary calculations used in modelling light interreflection are computationally expensive and cannot be performed in real time. However, there are methods of pre-calculation and incorporation within walk-through simulations (see below).

2.5   Real Time Rendering

Real time graphical rendering of a depicted scene is essential for interactive VES. For any particular motion input by the operator, the actor within the scene (for instance the effect on the camera) must be reflected quickly enough to give the impression of direct control. In the real world when we turn our head the field of view changes immediately because light travels so quickly. In a VES every movement of a camera requires the scene to be rendered again on the visual display. Rendering involves calculating what is to be observed in the scene from a given simulated vantage point and projecting the visible surfaces onto the pixels (picture elements) which constitute the display device. Surfaces in VES are usually simulated by large numbers of 2-D polygonal elements and those elements, which are visible to the camera, must be projected and drawn in the display. This process requires thousands, if not millions, of calculations and these calculations increase with increasing requirements in realism. In order to maintain real-time feedback according to operator movements we require a video update rate that is fast enough for the changes in the scene to appear smooth and not to lag behind. Such a time lag between action and visual feedback may result in simulation sickness [5] and decreases the sense interactivity or immersion in the scene. In reality, the interactivity of the visual display is a function of the complexity of the scene (i.e. the number of primitives to be portrayed) and the desired level of visual realism. Possible solutions to such problems, beyond eliminating the time lag itself, include predicting future actor states from previous ones (e.g., [6]).

3   Pros and Cons of Virtual Environments Simulations

The use of computer generated stimuli for visual psychophysics is now widespread and many experiments investigating integration of visual cues (shading, texture, stereo) have been made possible by the use of computer-generated graphics [7]. There are a number of ways in which using a VES may be beneficial. Firstly, one may consider any virtual scene as simply an abstract form of stimulus like, for instance, a line-drawing used in object recognition which is used to eliminate surface cues in order to concentrate on the saliency of outline contour in defining shape. Using a VES gives one the opportunity to eliminate or control for unwanted cues while preserving the three-dimensional nature of the stimulus. This benefit is demonstrated in the study of scene recognition, where one may want to keep illumination conditions constant throughout an experiment. This of course is very difficult to achieve in real environments where the weather conditions and time of day result in continually varying levels of illumination.

As well as simulating the naturalistic contexts under which cognitive tasks are performed, VES also allow us to look at the dynamics and behavioural aspects of learning and encoding. The benefits afforded to the active observer in computation vision are discussed by Aloimonos [8] who points out that a perceptual system that operates only on static stimuli is less flexible than one that can actively change the "geometric parameters of its sensory apparatus". Such systems can also encode what kinds information can in principle aid a given problem-solving task even though this information is not explicitly contained in the current stimulus. Thus, the system is able to change its sensory parameters in a way that maximizes the sensory information available to it or make available new information. In support of similar principles in human perception-action coupling, it has been argued that human perception operates by hypothesis testing where a particular state of affairs in the world is verified using predictive behaviour [9],[10]. For instance, many visual inversion illusions (figure-ground or convexity-concavity ambiguities) arise in static stimuli and can sometimes be eliminated by, for instance, small head movements that can provide additional depth cues (e.g., [11]). Given the importance of interactivity for perception, this important implications for the ability to encode and remember spatial relations [12],[27].

In terms of performance (or response) measurements, using VES allows the stimulus creation process to be automated and be entirely personalized for each observer. This may be achieved by monitoring, and in some cases storing, views of what the observer is seeing at any given time. As an extension of this, it is in principle possible to monitor the observers simulated or actual body movements continually as they perform some action or during problem solving. In this manner, continuous parametric data may be obtained and natural problem solving strategies observed, which can be more informative than verbal responses in forced-choice paradigms.

Another benefit of VES that we consider valuable is the ability to create any, and as many, novel environments as necessary for a particular application. Creating a virtual environment is comparatively easy compared to real, purpose built, constructions. It is also very easy to generate environments that are essentially similar in all respects except those that serve as experimental variables. Furthermore, it is now possible to create artificial environments of astounding realism using software that accurately simulates illumination effects and surface texture (see below).

There are however a number of barriers to using VES for spatial cognition studies. The most common is that the technology is new and many researchers are sceptical regarding the validity of results obtained under such artificial conditions. One remedy for this appears to be *bridging experiments,* in which the performance of subjects in particular tasks is compared in both VES and in equivalent physical setups [13],[14]. As long as such comparisons are fair, they may serve to reveal differences in strategy or failures of measurement in one or other of these situations. Another problem relates to the precision obtainable in motion tracking or input devices that have mainly been developed for consumer markets rather than carefully controlled scientific experiments. The quality of the scenes used is also at issue. The desire to maintain realistic update rates and reduce time lag may result in the use of degraded or unrealistic scenes which may produce results that do not reflect performance in the real world. In the next sections of this paper we hope to demonstrate that these are not unassailable problems. We describe experiments in which the use of VES has served considerable benefit while preserving realism and interactivity.

4   An Example Environment – The Virtual Café

In the rest of this paper we shall be concerned with a virtual environment which we call the Café Lichtenstein. The café is based on a *Wirtshaus* (English = Inn) in the mediaeval town of Tübingen, Germany and was modelled as part of an attempt to study spatial encoding of large-scale environments after interactive learning. We based the café on a real house for two main reasons. Firstly, an indoor environment was required in order to limit the context of the training situation and it was also necessary that this indoor environment, whilst being novel to our observers, should also be believable in its geometry and extent. Secondly, the actual house is extremely irregular in its original construction. This was particularly useful for providing an environment that was still sufficiently rich in structural detail even when not furnished (a box room would make a poor context for study spatial encoding).

Fig. 1. An architectural drawing used to create the model

The model was constructed from original architectural drawings obtained from the city offices (see Figure 1). The modelling was performed on an Intergraph TDZ 2000 graphics workstation running Windows NT using the 3D Studio Max (AutoDesk Inc., USA) modelling software. The model is modular in nature, which allowed us to separate and use particular components of the entire structure in isolation.

A example of the level of realism attainable from this modelling is shown in Figure 2 which depicts a single rendered frame taken from the back of the downstairs café area. The illumination of this scene was simulated using the Lightscape Visualization System (Discreet Logic, USA) to model the light distributed from the several simulated light sources. The properties of these lights (including photometric properties and directionality of emitted light) were carefully modelled. The Lightscape program implements a physically based model of the diffuse interreflection of light. This café area is made up of approximately 32 thousand polygonal elements.

Fig. 2. The ground floor of "The Café"

5   Spatial Encoding and View-Dependent Recognition

5.1   Overview

Our ability to recall the structural makeup of objects and the layout of scenes shows that we posses some form of spatial representation. There are competing theories as to the form taken by spatial representations of space. One theory considers that the observer moving through the environment extracts geometrical relations between objects and these are used to construct a single explicit structural description that is used for the purposes of recognition and remembering. This view was popularised by David Marr [15],[16] and became especially popular in computational models of vision and object recognition because of its modular nature and its grounding in neurophysiology. It proposes for instance that the initial neural processes that make vision possible include the extraction of edges and segments and the derivation of depth information from the retinal stimulus. This eventually allows structural mental models to be constructed. However, other researchers have argued that human representation of space does not need this elaborate extraction of detail. It can be facilitated by operations on just a few stored retinal images (e.g., [17]). In this sense the mental representation encodes spatial relations only implicitly and the real work involved during recognition occurs only when necessary. Evidence for this theory was provided from the mental rotation experiments devised by [18]. In these experiments, it was reported that the time to recognize depth-rotated versions of familiar (previously seen) 3-D geometric forms was a function of the amount of mis-orientation, in depth, away from the familiar view. It was therefore proposed that recognition involves a process of mentally rotating the visual stimulus until it matched a stored representation in memory. In this way recognition is said to be view-dependent.

We believe that much of the controversy in this field has been generated by arguments concerning the appropriateness of stimuli used and the conditions under which experiments have been carried out. Biederman and Gerhardstein [19], for instance, argue that view-dependent performance is usually apparent for "artificial" stimuli that do not contain salient components, as found in the familiar objects we encounter in daily life (but see [20]). However, one problem with carrying out experiments on familiar objects is that the initial process of getting to know the object is neglected. This learning process may be very informative about the development of spatial encoding of objects and scenes beginning with initial perceptual glimpses to eventual overall familiarity. The best way to do this we think is to use VES to facilitate the natural process of learning but within a carefully controlled spatial context, perhaps involving "artificial" or contrived 3-D geometric forms. In the next sections of this paper, we describe experiments that address the issue of view dependency in the recognition of novel stimuli by exploiting the use of VES for providing realistic contexts and natural learning afforded by interaction.

5.2   Object Recognition within Context

Previous studies using novel 3-D geometric forms have provided the most convincing evidence that representation is view dependent [17],[21],[22]. Although an increasing number of object recognition studies have utilized virtual, or computer generated objects such as these (e.g., [22]) no context was provided in such experiments and no interactive control of learning was facilitated. We hypothesized that much of the observed inability to recognize novel views found by the studies might have been the result of limited information regarding context and observer viewing direction. In the real world we are always aware of where we are and from what direction we look at objects of interest. Thus, spatial reference frame knowledge may provide important information in object recognition. We therefore carried out some of our own experiments using similar geometric forms used in previous studies but within a rich spatial context [23]. Our experiments were novel in two respects. First, in the interactive manner in which observers learned the objects; by manipulating their movements around the objects using a SpaceMouse and in the spatial context in which the objects were learned.

Fig. 3. Shows one instance of the "paper-clip" objects placed on the pedestal of the virtual living room and depicted from viewing directions differing by 90º rotations. Observers familiarized themselves with the objects by interactively moving around them within a limited range of directions. The two rows of pictures depict example stimuli from the two experimental conditions

The room we used for this study was also based on The Café Lichtenstein model, albeit somewhat loosely as it included much modern furniture, but utilized high quality illumination and realistic decoration. The Lightscape software was used to illuminate the room model. The illumination was provided by several simulated extended lightsources positioned around the room. The shading calculations were performed prior to the use of the model in the experiments. Once the radiosity calculations (based on the specified light sources) were carried out, appropriate intensity values could be attributed to each of the 17,000 polygons that made up the scene. The experimental run-time software, written using the Sillicon Graphics IRIS Performer library, could then load these polygons and render them from any viewing position with the appropriate shading calculated previously using the radiosity method. The wire-like geometric forms shown on the pedestal in Figure 3 were loaded individually as and when required and were merged with the rest of the 3-D scene prior to being rendered.

Subjects were trained to recognize four of the wire-like paperclip objects from one particular direction within the simulated room. Subjects could see each object individually by pressing buttons on a keyboard and, within certain bounds, could smoothly rotate around each object using the SpaceMouse. This facilitated what we believed was a natural means of learning to discriminate between four rather unnatural looking objects. After being trained to recognize each object from the "familiar view", we began the main data collection part of the experiment. Now subjects were shown random views of the objects from all possible directions, including minor perturbations in viewing elevation and rotations up-to 360° and they had to identify each object. In all cases the subjects view was changed but the objects' orientation remained fixed within the room.

The results from this experiment showed that the proportion of errors varied as a function of the degree of orientation shift away from the familiar viewing direction, just as in previous experiments. In addition, the front (familiar) and rear views were always recognized with fewer errors than the side (oblique or 90°) views, also as in previous experiments. Although the spatial context did not eliminate view-dependency, when we performed the same novel direction test without the contextual background, performance was consistently worse for all rotations. That is, fewer errors were made when the original room scene was visible in the background to each object. This may have revealed, indirectly, the extent of the shift in the observer's view of each object. Further tests looked at whether this was indeed the case or whether the benefit of having the room present was some artifact of the experimental method. In these tests, the orientation shift in the observers' view sometimes coincided with a random rotation of the room with respect to the object. In all cases, the room was present in the background but in half the tests it did not correspond to the change in the observers view. In this case the number of errors was again consistently higher, which indicates that subjects could have been using the background detail to reduce uncertainty in deciding whether the current view of each object was consistent with a rotation in their viewpoint.

5.3   Recognition of Spatial Layout Using Multiple Objects

The observed view-dependent recognition performance described above has also been found in more complex stimuli involving the 3-D spatial layout of several objects. The general paradigm used previously requires subjects to view the spatial layout of a collection of objects from a given viewpoint and then imagine or point to individual objects from a new (unfamiliar) vantage point (e.g., [24]). Results have shown that people's ability in such displacement tasks is, as in object recognition, dependent on any shift in their viewpoint away from the familiar or training direction. More recent experiments have shown that facilitating actual movement to these new viewpoints (by allowing subjects to walk to the new viewpoint) eliminates view dependency [25]. This suggests that non-visual, perhaps proprioceptive or vestibular, information may be used when making decisions about spatial orientation and spatial layout.

We have attempted to simulate spatial layout tasks using a virtual environment scenario [26]. The environment (see Figure 4) consisted of another room of the Café consisting of numerous items of furniture and illuminated, once again, using the Lightscape radiosity software. A circular table in the centre of the room was used as our platform about which a camera could be rotated allowing our subjects to observe collections of 5 familiar objects at a time. These objects were 3-D models of objects such as bottles, torches etc. A curtain could be lowered from the ceiling to obscure the objects when required in the experiment. The movement of this curtain and the changes in viewpoint were smooth animations performed by updating the view from the camera 30 times a second. We used a database of 32 three-dimensional objects that could be loaded into the simulation as and when required. The simulation was viewed through an 80cm long viewing tube. The choice of familiar viewpoint from which each set of objects was observed was completely randomized.

Fig. 4. View of the room chosen for the spatial layout study showing the round table on which the objects were placed

The basic sequence of events (Figure 5) consisted of showing the observers a single view of five objects after which the curtain was lowered followed by a rotation of the observer's viewpoint around the table or a rotation of the table (and objects) while the observer remained still. The curtain was then drawn up to reveal the objects again and subjects had to say whether one of the objects had moved in the interim period. Our initial results show that subjects who are visibly rotated to a new viewpoint can detect the movement of one of the test objects better than subjects who perform the same task from the original viewpoint but with an equivalent rotation of the objects themselves. These results are similar to those of [26] in that subjects who perform these judgements after their own movement to a new position is better than for subjects who perform the task from the original fixed position. Unlike the results of [26] however this result was obtained because of *simulated* movement rather than *actual* movement of the observer. Although the reason for this facilitation is unclear, it suggests that the crucial information might be the perception of movement itself.

Fig. 5. Views showing the three main stages of each trial in our spatial layout experiment

5.4   View Dependency in Scene Recognition

Another VES-based experiment sought to determine if restrictions on viewing ability during training reflect directly on scene recognition performance when the object of learning is a large-scale or navigable environment rather than a single object or a collection of objects. In order to study view dependency in scene recognition it was necessary to use a novel environment that was highly controlled in terms of its content. We wanted to restrict encoding and recognition to rely solely on structural information rather than, for instance, surface properties such as colours or textures. For this purpose we developed the Attic of the Café Lichtenstein (see Figure 6). Because we typically accustom ourselves with an environment by performing movements within it we allowed our observers to perform simulated movement across a restricted "walk-way".

The acquisition stage of this experiment involved subjects controlling a "virtual eye" or viewing camera that had a fixed distance from the ground and that could be moved along the "walk-way" between two flights of stairs. Observers' task during training was to explore the environment to find and acknowledge spatially localized two-digit coded markers. This ensured that all observers were at least aware of the same parts of the scene on which they were to be tested. However, because the entire environment was searched for these markers they could also form a complete representation of its spatial layout. The question we wanted to address was whether after such extensive exploration their recognition performance still reflected the viewing restrictions that we imposed on them.

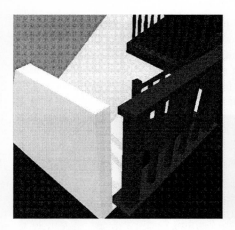 

Fig. 6. Familiar (left) and novel (right) direction views of the virtual attic. Images such as these served as stimuli in the recognition stage of our experiment

The main restriction was that they could not fully rotate their view and therefore look back at the environment from a new perspective. This restriction meant that we could later test their recognition ability for the environment not only from the familiar (forwards) direction observed during learning but also from novel (looking back) directions too. Figure 6 (left) shows a typical stimulus from the familiar direction and Figure 6 (right) a novel direction image of the same location.

We tested subjects ability to remember the training environment by using a forced choice same/different task in which they had to judge whether a given randomly presented image was taken from the "target" attic or from a very similar, yet structurally different, "distractor" attic. Using computer-modelling means that the properties of such "distractor" environments can be carefully controlled. We also tested their ability to generalize recognition of familiar locations seen from a new perspective by including novel views in the test, which were individually generated according to what subjects saw during learning. Our results showed that, as expected, familiar views were most easily recognized although novel perspective views were also recognized better than chance [27]. However, subjects were also able to identify the correct floor-plan of the learning environment. These results demonstrate both a view-specific dominance in picture recognition but also an ability to use this view-specific representation to perform tasks requiring global knowledge such as identifying topography and the identification of novel views.

6   Summary

We hope to have given a brief glimpse of the possibilities afforded to the cognitive psychologist by the relatively new tool of Virtual Environment Simulation. The potential benefits of VES based research relate to cue control, flexibility, sense of realism and ease of data collection. In many cases using a VES not only allows better control over experimental variables but could also extend research into areas and issues, which would otherwise be impossible to study. As an example, it is possible to alter movement dynamics and sensory feedback in such a way that the usual bodily responses to sensory stimuli need to be re-calibrated providing an opportunity of studying the connection between different sensory information and proprioception. However, many issues must still be addressed including the transfer from real-world performance to virtual-world performance in customary behavioural tasks, the effects of discordant sensory information and, in desktop-based simulations, the lack of motor feedback.

Given the benefits of VES, the chief concern in many laboratories is the cost effectiveness of the investment. However, most of the tools needed for implementing VR (such as input devices and graphical modellers) are all commercially available. There is a modest investment in terms of programming effort but the major costs are incurred by the computers needed to run such computationally intensive applications. Clearly when it comes to computer power "bigger is better" in the sense that the more computing power you have the more realism you can introduce. However this depends on individual experimental requirements. The experiments that we have described here were implemented and run on desktop computers that are now common in many psychology laboratories. In our case realism in terms of visual richness of detail and interactivity were both very important but we have been able to meet these requirements with an investment well within the budget of a modest laboratory.

References

1. Ellis, S.R. Prologue In: Ellis, S.R., Kaiser M.K. & Grunwald, A.J.: *Pictorial Communication in Virtual and Real Environments*. Taylor & Francis, London (1991)
2. Kaufman, L. & Rock, I.: The moon illusion. Science. 136 (1962) 953-961
3. Goral, C.M., Torrance, K.E. & Greenberg, D.P.: Modelling the interaction of light between diffuse surfaces. ACM Computer Graphics – SIGGRAPH. 18(3) (1984) 213-222
4. Christou, C.G. & Parker, A.J.: Visual realism and virtual reality: a psychological perspective. In: Carr K and England R (eds.): Simulated and Virtual Realities. Taylor and Francis, London (1995) 53-80
5. Oman, C.M.: Motion sickness: A synthesis and evaluation of the sensory conflict theory. Canadian Journal of Physiology and Pharmacology. 68 (1990) 264-303
6. Friedman, M., Starner, T. & Pentland, A.: Synchronization in virtual realities. Presence. 1(1) (1992) 139-143
7. Bülthoff, H.H. Shape from X: Psychophysics and Computation. In: Landy, M. and Movshon, A. (eds): Computational Models of Visual Processing. M.I.T. Press. (1991) 305-330
8. Aloimonos, Y.: Active Perception. Lawrence Erlbaum Associates, London (1993)
9. Gregory, R.I.: Perceptions as hypothesese. Philosophical Transactions of the Royal Society. B290 (1980) 181-197

10. Mourant, R.R. & Grimson, C.G.: Predictive head-movements during automobile mirror-sampling. Perceptual and Motor-Skills. 18(1) (1977) 21-25
11. Lindauer, M.S.: Expectation and satiation accounts of ambiguous figure-ground perception. Bulletin of the Psychonomic Society. 27(3) (1989) 227-230
12. Peruch, P. & Gaunet, F.: Virtual environments as a promising tool for investigating human spatial cognition. Current Psychology of Cognition. 17 (4-5) (1998) 881-889
13. Nemire, K., Jacoby, R.H. & Ellis, S.R.: Simulation fidelity of a virtual environment display. Human Factors. 36(1) (1994) 79-93
14. Profitt D.R., Bhalla, M., Grossweiler, R. & Midgett, J.: Perceiving geographical slant. Psychonomic Bulletin & Review. 2 (1995) 409-428
15. Marr, D.: Vision. Freeman, San Francisco (1982)
16. Marr, D. & Nishihara, H.K.: Representation and recognition of the spatial organization of three-dimensional shapes. Proceedings of the Royal Society of London. B200 (1978) 269-294
17. Tarr, M.J.: Rotating objects to recognize them: A case study of the role of viewpoint dependency in the recognition of three-dimensional objects. Psychonomic Bulletin and Review. 2(1) (1995) 55-82
18. Shepard, R.N. & Metzler, J.: Mental rotation of three-dimensional objects. Science. 171 (1971) 701-703
19. Biederman, I. & Gerhardstein, P.C.: Recognizing depth rotated objects: Evidence and conditions for three-dimensional view-point invariance. Journal of Experimental Psychology: Human Perception and Performance. 19 (1993) 1162-1182
20. Tarr, M.J. & Bülthoff, H.H.: Is human object recognition better described by geon structural descriptions of by multiple views? Comments on Biederman and Gerhardstein (1993), Journal of Experimental Psychology: Human Perception and Performance. 21(6) (1995) 1494-1505
21. Rock, I. & DiVita, J.: A case of viewer-centered object perception. Cognitive Psychology 19 (1987) 280-293
22. Bülthoff, H.H. & Edelman, S.: Psychophysical support for a two-dimensional view interpolation theory of object recognition. Proceedings of the National Academy of Sciences USA. 89 (1992) 60-64
23. Christou, C.G., Tjan, B.S. & Bülthoff, H.H.: Viewpoint information provided by a familiar environment facilitates object identification. Max-Planck Institute for Biological Cybernetics, Technical Report No. 68, Tübingen Germany (1999)
24. Diwadkar, V.A. & McNamara, T.P.: View dependence in scene recognition. Psychological Science. 8(4) (1997) 302-307
25. Simons, D.J. & Wang, R.F.: Perceiving real-world viewpoint changes. Psychological Science 9 (1998) 315-320
26. Christou, C.G. & Bülthoff, H.H.: Perception of spatial layout in a virtual world. Max-Planck Institute for Biological Cybernetics, Technical Report No. 75, Tübingen, Germany (1999)
27. Christou, C.G. & Bülthoff, H.H.: View-dependency in scene recognition after active learning. Memory & Cognition. 27 (1999) 996-1007

Navigating Overlapping Virtual Worlds: Arriving in One Place and Finding that You're Somewhere Else

Roy A. Ruddle

School of Psychology, Cardiff University, Cardiff, U.K. CF10 3YG
ruddle@cardiff.ac.uk

Abstract. When a person moves through an overlapping environment they can travel in a closed, Euclidean loop but still end up in a different place to where they started. Although such environments are unusual, they do confer potential advantages for navigation. Three independent attributes of spatial overlap, as applied to 3-D virtual environments (VEs), are described, together with their likely effects on navigation. An experiment that investigated one type of overlap (loop connectivity) is described. Participants learned spatial knowledge more slowly in an overlapping VE than in a conventional VE, but the differences were small and, after initial navigation, not significant. Therefore, there seems to be no cognitive barrier to the useful implementation of overlapping VEs within a wide variety of applications.

1 Introduction

A central characteristic of conventional virtual environments (VEs) is that they are constructed using real-world laws of spatial structure. These environments are either models of places that exist in the real world (e.g., Witmer, Bailey, Knerr & Parsons, 1996; Tate, Sibert, & King, 1997), or environments that could be constructed in the real world (e.g., Tlauka & Wilson, 1994).

Two novel classes of VE sometimes violate these laws. The theoretical and practical effects on navigation of the first of these (hyperlinks in VEs) are dealt with in detail by Ruddle, Howes, Payne, and Jones (1999). The second class is spatial overlap, where two or more different places occupy the same position in Euclidean space. Hyperlinks are already used to connect together VEs that are accessed via the World-Wide Web. At present spatial overlap only appears in some computer games and requires alterations to conventional 3D computer graphics algorithms. Both hyperlinks and overlap may be used to reduce navigation time in VEs, and increase design flexibility, but will only become useful if these advantages are not outweighed by the extra, cognitive difficulties that are likely to be caused to a person while navigating. Central to these difficulties in an overlapping VE is the fact that the person can leave one place, travel in a closed loop in Euclidean space, and still arrive somewhere different to where they started out.

In this paper we present three independent attributes of spatial overlap, and describe an experiment that investigated participants spatial knowledge when they repeatedly navigated conventional and spatially overlapping VEs. First, however, we set the scene by summarizing studies that have investigated navigation in conventional VEs.

Ch. Freksa et al. (Eds.): Spatial Cognition II, LNAI 1849, pp. 333-347, 2000.

2   Navigation in Conventional VEs

There is considerable potential for the use of large-scale VEs in training applications such as training, tourism and design (large-scale environments are those in which a person cannot resolve all the detail necessary for efficient navigation from a single human's-eye viewpoint, see Weatherford, 1985). Usually, VEs for those types of application are created using real-world spatial structures.

Experimental studies have shown that navigation of conventional VEs relies on and is associated with cognitive processes similar to those employed in real-world settings. For example, when people navigate VEs they form similar spatial representations to those formed in real-world environments (Hancock, Hendrix & Arthur, 1997; May, Péruch & Savoyant, 1995; Tlauka & Wilson, 1996) and to some extent can transfer knowledge learned in a VE to an equivalent, real-world environment (Bliss, Tidwell & Guest, 1997; Witmer et al., 1996). Although people seem to be more disoriented when they initially navigate a VE than when they navigate in the real world, they ultimately tend to develop efficient route-finding ability (they learn to travel by the shortest route) and accurate survey (map-type) knowledge (Ruddle, Payne & Jones, 1997). In other words, the problem of navigation in VEs seems not to be *whether* people are always disoriented, but *how long* people take to develop accurate spatial knowledge.

The rate at which people learn spatial knowledge is affected by the type of interface they use. With desk-top VEs (typically a mouse and keyboard interface) people have to use abstract interfaces to control their movements but when people use a helmet-mounted display (HMD) changes in view direction are made using natural, physical movements. In a study by Ruddle, Payne and Jones (1999b), participants navigated VEs significantly more quickly when they used a helmet-mounted display (HMD) than when they used a desk-top display. The speed increase arose because the HMD allowed natural body movements to be exploited whereas, with the desk-top display, the equivalent of a glance over the shoulder became more like an implicit instruction to "rotate until you are facing the intended direction and then rotate back". In the study by Chance, Gaunet, Beall, and Loomis (1998) participants who physically walked around a VE while wearing an HMD (physical translational and rotational movements) made significantly more accurate estimates of direction than participants who used an abstract interface. The use of walking interface can also increase the accuracy with which people transfer spatial knowledge that is learned in a VE to the real world (Grant & Magee, 1998).

None of the techniques described above, or others such as the introduction of supplementary landmarks (Darken & Sibert, 1996; Ruddle et al., 1997; Tlauka & Wilson, 1994), have produced a large increase in people's rate of spatial learning. Spatial overlap introduces a fourth dimension to an environment and, hence, greater navigational complexity. Attributes of spatial overlap are described in the following sections, together with an experiment that investigated some of the effects that overlap had on participants' navigation.

3   Spatial Overlap

Spatial overlap exists where two or more points in a VE occupy the same position when they are mapped to 3-D Euclidean space. One well-known example is the *Tardis* in the science fiction television program *Doctor Who*, which is much larger on the inside than it appears to be from the outside. Each point in an overlapping VE must be defined in four dimensions. Three of the dimensions define a point's *actual* position in Euclidean space (the XYZ coordinates) and the fourth is a visual zone that the VE software uses to render the correct scene to the display as a person moves through the environment. The complexity of an overlapping VE can be quantified using three attributes of overlap, which are presented in Table 1 and discussed in the following section.

Table 1. A taxonomy of attributes of overlapping space that may affect navigation in VEs

Attribute	Ease or speed of navigation		
	Less		More
Overlap density	multiple-overlap	--	1 (no overlap)
Connectivity	loop	multi-story building	street
Position	1 actual and many apparent positions	--	Only 1 position (the actual position)

3.1   Attributes of Spatial Overlap

3.1.1 Overlap Density. The amount of overlap that is present in an environment may be measured by calculating the overlap density of all points in a Euclidean map of the VE. Non- overlapping points have a density of 1. However, as will be explained below, the attributes of connectivity and position are also required to define the overall complexity of overlapping space.

3.1.2 Connectivity. The connectivity between overlapping sections of an environment can be defined using the terminology of graph theory. To do this, we divide a VE into zones of conventional (non-overlapping) space. Then, we classify the connectivity of the zones. The effects of different classes of connectivity are best illustrated by using examples.

The simplest form of connectivity is *street overlap*, which is characterised by a single thoroughfare (the street) that is bordered by a row of buildings, arranged side by side. The buildings overlap with each other but none of them overlaps the street. Therefore, to travel between overlapping zones of the VE a person must go via some non-overlapping space (the street). This type of structure has the advantage of compressing the distance that a person must travel, while maintaining a simple and intelligible structure.

In more complex forms of overlap a person can travel directly between overlapping zones of space (see Figure 1). *Multi-story building overlap* is, as the name suggests, analogous to a multi-story building, except that it is constructed on a single level. The different "floors" overlap with each other in Euclidean space, and a person travels between adjacent floors by moving through a connection which preserves visual continuity (it may have the appearance of a normal door, or it could be distinctive). An advantage of that type of arrangement, compared with a non-overlapping version of the same environment, is that the entire VE can be navigated by movement in two, rather than three, dimensions, thereby simplifying the user (movement) interface. More complex still is *loop overlap*, in which a person can travel though a sequence of overlapping zones and return to their start point without retracing their steps. The difference between this example and loop structures that occur in the real world is that, in the former, the loop is formed by regions of space that overlap with each other. A VE with multi-storey building overlap can be converted to loop overlap by adding a connection that allows the person to travel directly between the "top" and "bottom" floors.

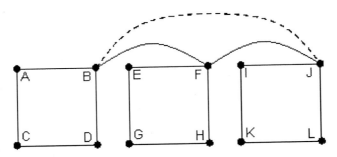

Fig. 1. An example topology of multi-story building overlap. The 'building' has three zones ('stories') of 4 nodes. The zones all occupy the same Euclidean space (overlap density = 3) and are connected by the links BF and FJ. If the link BJ is added (shown dashed) then the structure changes to loop overlap

3.1.3 Position. At all times, VE software has to know the position of a person in an environment (the viewpoint) to render the correct scene to the display. This position is known as the person's actual position. When continuous movement occurs the change in a person's actual position from one graphics frame to the next is the same as the change that could be inferred by the person from the (small) changes that occur in the two scenes that are displayed. More complex perceptual spaces can be created if overlap is combined with visually continuous hyperlinks, because a person's actual and apparent changes of position are no longer equal. This is also best explained using an example.

The hyperlink that connects Rooms A and B in Figure 2 could be designed to look and act like a door. A person who stands in Room A and looks through the door would see an image that looks like Room B (actually, it *is* Room B), even though the room that is actually adjacent is Room C. Whenever a participant crosses the threshold of the hyperlink that connects Room A to Room B (the threshold is at point P_a), the VE software makes a sudden change to the participant's actual position in the

VE and moves them to point P_b, which is the equivalent position in Room B. Despite the large change in the person's Euclidean position, the change in the scene that is displayed is small (i.e., visual continuity) and is similar in magnitude to the change that takes place when the person moves between adjacent rooms via a conventional door. By contrast, when a participant traverses a "normal" (i.e., visually discontinuous) hyperlink, the scene changes completely.

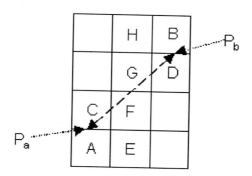

Fig. 2. An environment that contains a connection that allows a person to move between rooms that are spatially non-adjacent, but which preserves apparent continuity of movement

In the above example, the person can also reach Room B from Room A by traveling through Rooms E, F, G and H. This means that Room B appears to exist in two different positions, relative to Room A, and only one of those is the actual position of Room B that is stored in the VE's database. In other words, Rooms B and C, and Rooms A and D, *appear* to overlap, but don't *actually* overlap. Although the combination of apparent spatial overlap and visually continuous hyperlinks seems unusual, they form a feature that could be useful because data from experimental studies indicate that people learn spatial knowledge more quickly if visual continuity is preserved (see Ruddle, Howes et al., 1999). Videos that illustrate loop-type actual overlap, and apparent overlap ("visually continuous non-planar hyperlinks") are stored at http://www.cf.ac.uk/uwc/psych/ruddle/C-HIVE/DVE/).

By now the reader will probably appreciate how difficult it can be to discriminate between different types of overlap, and discontinuities introduced by hyperlinks. The equations presented in the next section help to address this problem.

3.2   Equations of Movement

When a person travels around a VE they move in discrete steps, although in conventional VEs the size of the steps is very small. Therefore, a general equation of movement that applies to VEs is:

$$S = Q - P \ .$$

$$(1)$$

(S = vector sum of all the steps a person takes from point P to point Q)

In a VE, there are two ways of measuring the steps that are used to calculate S. The first of these is to consider the actual movements that a person makes. That is, the changes in the person's Euclidean position that are calculated by the VE software and used to specify the position of the person's viewpoint as they move from P to Q. This leads to a new version of the equation, in which the subscript a indicates that the calculations are performed using actual positions:

$$S_a = Q_a - P_a \ . \tag{2}$$

When a person makes a step in a VE, the scene that is shown on the VE display changes slightly and, in theory, the person could use information from that change to determine how far and in which direction they have moved (we do not claim that people can do this accurately, simply that there is sufficient visual information to do so). The second method of measuring the steps is to calculate S in terms of the changes in the visual scene as the person moves from P to Q, and this is denoted using the subscript v:

$$S_v = Q_v - P_v \ . \tag{3}$$

Three statements about these equations (see Table 2) can be made that refer to conventional VEs, and the statements also apply to the real world. First, S_a always equals S_v. Second, if, for any sequence of steps between two points, $S_a = 0$, then the two points are actually the same point ($P == Q$). Third, the converse is also true (if $S_a \neq 0$, then $P \neq Q$).

Table 2. Equations of movement that distinguish conventional VEs, actual and apparent spatial overlap, and hyperlinks

Type of environment	Actual movement (S_a)	Visually perceived movement (S_v)	Actual vs. visual movement
Conventional VE	If $S_a = 0$, $P == Q$	If $S_v = 0$, $P == Q$	$S_a = S_v$
Actual overlap	$S_a = 0$, but $P \neq Q$	--	--
Normal hyperlink (visually discontinuous)	--	S_v is indeterminate	--
Apparent overlap	--	--	$S_a \neq S_v$

If a VE contains actual overlap and supports only continuous movement then $S_a = S_v$, as it does for a conventional VE. However, there will be some pairs of points for which $S_a = 0$, but $P \neq Q$. The Euclidean coordinates of P and Q will be the same, but their visual zone (the fourth dimension) will be different. Normally a hyperlink contains no visual information that a person can use to calculate their change in position because the scene that is displayed at the two ends is totally different. This means that S_v is indeterminate but, by contrast, S_a is always determinate. It has to be if the VE software is to render the correct scene when the person moves. When different places in an environment apparently overlap, but none of them actually overlap, then the actual movement equation of a conventional VE still applies. However, the changes in position that a person perceives from changes in the visual scene (S_v) are different from the person's actual changes of position (S_a).

Referring back to the title of this article, when actual overlap occurs a person can make Euclidean movements to one place and find that they've arrived somewhere else. With visually discontinuous hyperlinks the person would have no idea of the position of the place they have arrived in, and with apparent spatial overlap the person simultaneously appears to be in two different Euclidean positions!

The following experiment investigated the effect of actual, loop-type overlap on participants ability to search for objects in virtual buildings. A conventional (non-overlapping) virtual building was used as a control condition.

4   Experiment 1

Two versions of three different virtual buildings (Buildings A, B1 and B2; see Figure 3) were created. The conventional version of each building contained a chessboard-type arrangement of rooms and the other version was an overlapping chessboard. The connectivity of the rooms was the same in both versions. In the overlapping versions the overlap density was either 2.0 (part of Building A) or 4.0 (the remainder of Building A, and the whole of Buildings B1 and B2).

4.1   Method

4.1.1 Participants. Twenty-four participants (10 men and 14 women) took part in the experiment. All the participants were either students or graduates, volunteered for the experiment, were paid an honorarium for their participation and had not taken part in any other VE navigation experiments at Cardiff. Their ages ranged from 16 to 25 years ($M = 19.8$). Each participant first navigated Building A. Then they either navigated Building B1, followed by Building B2, or they navigated Building B2 and then Building B1. Each participant was randomly assigned to one of eight groups, which were used to counterbalance the version of Building A that they navigated, and the order of navigation and version used in Buildings B1 and B2. One participant asked to withdraw after completing their searches in Building A and was replaced in the experiment.

4.1.2 Materials. The experiment was performed on a Silicon Graphics Crimson Reality Engine, running a C++ *Performer* application that we designed and programmed. A 21-in. (53 cm) monitor was used as a display and the application update rate was 20 Hz.

Pilot studies were used to help determine an appropriate number of rooms for the VEs, and hence one component of navigational complexity. Barriers to movement are another component of complexity, and these were created by omitting the connections between some adjacent rooms. To travel between each object pair used in the fourth stage of the test procedure (see below) a participant had to enter a minimum of three rooms (excluding the starting room). The connections that were omitted were chosen so that each pair was connected by a unique, shortest route.

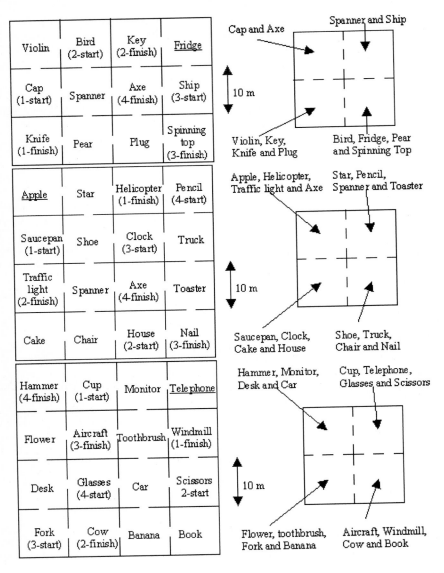

Fig. 3. Plan view of normal and overlapping versions (left and right, respectively) of Building A (top), Building B1 (middle) and Building B2 (bottom). For each building the connectivity of the two versions was identical but the doors (the gaps in the walls) were in different places. In the normal version each wall contained either zero or one door. In the overlapping version some walls contained two doors (see Figure 4). The any-order searches started in the room that contains the underlined object name. The object start/finish pairs are shown in parentheses

All the connections were doors. Each door opened automatically, by an amount that depended on a participant's distance from the door. The door was closed if the participant was more than 2 m from the door and fully open if the participant stood in the doorway. Each room of each building contained a scaled, 3-D model of a different everyday object and these were placed on top of gray box that was 1 m in height (see Figure 4).

Find in ANY order: Axe Clock Helicopter House Nail Pencil Saucepan Traffic_light

Fig. 4. An interior view of the overlapping version of Building B1. The participant is in the room that contains the shoe. From left to right, the doors lead to the clock, saucepan, spanner and star

To define what was seen on the monitor, the application had to specify the field of view to be used and the height above the buildings' floor at which viewing took place (effectively a participant's virtual "eye" height). The FOV was set to be 90°. This allowed a participant to see two doors simultaneously, but was substantially different from the physical FOV of the monitor (approximately 45°). Each participant's virtual eye height was set equal to their actual eye height. The center of the screen was marked using a small green square.

The interface used the mouse and five keys on the keyboard, and allowed a participant to look around while traveling in a straight line. If the participant held down the left or right mouse buttons, then the azimuth of their view direction changed at a rate of 60 degrees/s. By moving the mouse to change the offset of the cursor from the center of the screen, the participant could vary the azimuth and altitude of their direction of view by $\pm 90°$. Four of the keys allowed the participant to slow down, stop, speed up, and move at the maximum allowed speed (4.8 km/h). The fifth changed the participant's direction of movement to the current view direction. Participants were prevented from traveling through walls by a collision detection algorithm.

4.1.3 Procedures. Participants were run individually. They were told that the experiment was being performed to assess people's spatial knowledge in VEs, all the VEs had one story, some of the VEs might contain discontinuities, and that their primary goal in the tests was to enter as few rooms as possible. Rooms were counted each time they were entered. In the experiment, a participant first practiced the interface controls, using a conventional VE that contained five rooms, and then navigated the three test VEs. One role of the first test VE (Building A) was to familiarize participants with the procedure used in the test and to help control for the general learning effects that have been found in other studies of VE navigation (Ruddle, Payne, & Jones, 1998; Stanton, Wilson, & Foreman, 1996). The elapsed time for each participant's practice and tests was approximately 3 hr.

The test procedure was the same for all three test buildings and was divided into four stages. A participant completed all four stages in one building before resting for 3 min and then starting the test in the next building. For Stage 1, a participant was taken into a one-room virtual building that contained a number of objects, which were designated as targets (six in Building A, and eight each in Buildings B1 and B2). When the participant had familiarized themselves with the appearance of the objects they started the next stage.

Stages 2 and 3 were identical. All participants started in the same room and searched for all the target objects in *any order*. These searches were designed to investigate how efficiently a participant could search the whole of a VE (the first search was *uninformed*, because the participant had no prior knowledge of the VE's layout, and the second was *informed*; this terminology is consistent with Ruddle, Payne, & Jones, 1999a). At all times a message on the monitor showed the name of the object(s) that were still to be found. When an object was found and selected its name was deleted from the screen.

In Stage 4 the participant had to repeatedly find specific target objects in a test that was designed to investigate whether the participant could learn a specific route (a *repeated-route*) through the VE and take *short-cuts*. The objects were divided into pairs, with one of each pair designated as the start and the other as the finish. For each group of participants one pair was used to test the learning of the repeated route and the other pairs were used to test the short-cuts. Different groups used a different pair for the repeated route. The detail of the procedure was quite complex, and is best illustrated by using Building B1 and participants in Group 1 as an example.

A participant in this group started in the center of the room that contained the saucepan. A message on the screen said "Revisit target saucepan (it is beneath you)" and the participant used the mouse to select the saucepan. This caused the message to change to "Revisit target helicopter". The participant then searched the VE for the helicopter and selected it. This caused the screen to go blank for 5 s, and when the VE was displayed again the participant was in the room that contained the house and the message said "Revisit target house (it is beneath you)", and so on. In total, the participant searched for eight pairs of objects in the following order: saucepan-helicopter, house-traffic light, saucepan-helicopter, clock-nail, saucepan-helicopter, pencil-axe, saucepan-helicopter, house-traffic light, saucepan-helicopter, clock-nail, saucepan-helicopter, and pencil-axe. The saucepan-helicopter pair tested the repeated-route learning and the other pairs tested the participant's ability to take short-cuts. The real difference between the repeated-route and short-cut pairs was the frequency with which, and number of times that, participants performed the searches.

4.2   Results

The any-order, repeated-route and short-cut searches were measured by recording the number of rooms that participants entered, with the data for the three short-cut pairs averaged for each search number. The distribution of the three types of search data was normalized using a square root transformation. For Building A, the data were analyzed using analyses of variance (ANOVAs) that treated the search number as a repeated measure, and the version of the building (conventional or overlapping) as an independent variable. Each participant navigated either the conventional version of Building B1 or Building B2, and the overlapping version of the other building. For the analyses reported below these data are collectively termed as referring to Building B. The data were analyzed using ANOVAs that treated the search number and the version of Building B as repeated measures, and the version of Building A that participants had used as an independent variable. There were no significant interactions, and the effect of the Building A version was also never significant.

Table 3 summarizes participants' any-order searches. Participants who navigated the conventional version of Building A entered significantly fewer rooms than participants who navigated the overlapping version, $F(1, 22) = 5.99$, $p = .02$, but there was no difference between the two searches, $F(1, 22) = 0.78$, $p = .39$. In Building B there was a main effect of search number, $F(1, 22) = 17.93$, $p < .01$, but no effect of version, $F(1, 22) = 2.66$, $p = .12$.

The percentage of rooms that participants entered while retracing their steps (outward leg plus return leg) was also analyzed. Participants retraced a lower percentage in the conventional versions of both buildings, but repeated measures ANOVAs showed that the difference only approached significance in Building B ($M = 36\%$ vs. 44%), $F(1, 22) = 3.81$, $p = .06$.

Table 3. Number of rooms entered in the any-order searches

Building and version	Search 1			Search 2	
	M	*SE*		*M*	*SE*
Building A, conventional	28.6	4.3		25.0	2.5
Building A, overlap	39.1	5.4		36.7	4.4
Building B, conventional	26.8	1.9		23.1	1.3
Building B, overlap	31.9	2.5		26.3	2.2

For the repeated-route searches in Building A, the two groups of participants entered a similar number of rooms (conventional: $M = 6.0$; overlapping: $M = 7.2$), $F(1, 22) = 0.87$, $p = .36$, and there was no effect of search number, $F(3, 22) = 2.37$, $p = .08$. In Building B, participants entered fewer rooms in the conventional condition than in the overlapping condition (see Figure 5), but the difference was not significant, $F(1, 22) = 2.47$, $p = .13$. However, there was a main effect of search number, $F(5, 22) = 6.83$, $p < .01$.

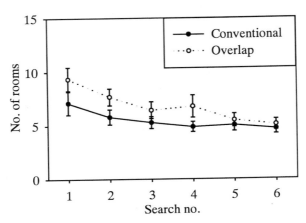

Fig. 5. Number of rooms that participants entered in their repeated-route searches in Building B. Error bars indicate standard error of the mean

For the short-cut searches in Building A, participants who navigated the conventional version entered slightly fewer rooms than participants who navigated the overlapping version ($M = 10.3$ vs. 14.2), but the difference was not significant, $F(1, 22) = 3.13$, $p = .09$. A similar pattern of results was found in Building B ($M = 8.7$ vs. 9.6), $F(1, 22) = 0.55$, $p = .46$.

4.3   Discussion

The experiment produced a very unexpected outcome. Even though a severe form of overlap (a loop structure) was used, participants entered only slightly more rooms in the overlapping version of the VEs than they did in the conventional version, and in both versions some participants had great difficulty navigating. Only the difference for participants initial searches (Building A, any-order) was statistically significant and, in general, overlap had less effect on participants' navigation than the hyperlinks used in other experiments by the author (see Ruddle, Howes et al., 1999).

Most of the difference between participants' any-order searches in the conventional and overlapping versions of Building B was caused by participants retracing their steps more often in the latter. In other words, participants reinforced their route knowledge in the overlapping environments to help compensate for the lack of meaningful survey knowledge (knowledge stored in the form of an Euclidean map).

It is interesting to note that the experimenter, who informally observed participants as they traveled through the VEs, was often unsure of which door to go through to enter a particular room in the overlapping versions of the VEs. Perhaps this was because the experimenter's knowledge of the layout was based on a memory of a plan view of the non-overlapping version. The participants were unlikely to have suffered from this problem because they only navigated one version of each VE and were never shown a plan view.

The overlapping VEs preserved visual and movement continuity but could not have been visualized clearly in Euclidean space (they would have been impossible to construct in the real world). The experiment's data demonstrates that people can adapt the type of spatial representations that they use to navigate in the real world and conventional VEs, to allow them to navigate almost as effectively in overlapping environments that are "impossible" cognitive spaces. Although this was a very surprising result, it should be borne in mind that people's spatial representations of the real world contain some systematic distortions which are non-Euclidean in nature. For example, Tversky (1981) reported that participants tended to draw sketch maps in which streets intersected at 90°, even when that was not the case, and non-parallel streets were made parallel, and Sadalla and Magel (1980) and Sadalla and Staplin (1980) showed that changes of direction and the number of path intersections influence the distance that people perceived they had traveled. Perceptual continuity is not a necessity if people are to acquire route knowledge (Allen, Siegel, & Rosinski, 1978). Neither, it seems, is Euclidean conformity.

The implication of the results is important. Even complex overlapping layouts have a minimal effect on navigation in VEs and designers should have few reservations about implementing overlap, and particularly simple forms such as street overlap, in VEs where "space" is at a premium, or there is a desire to minimize the distance that people travel. Techniques as diverse as landmarks, a compass, virtual breadcrumbs, and a virtual sun have been investigated to aid people's navigation in conventional VEs (Darken & Sibert, 1993, 1996; Ruddle et al., 1997, 1998; Tlauka & Wilson, 1994). Once successful techniques have been developed it may be possible to apply them equally effectively to overlapping environments. Overview maps could present a problem but one solution would be to display the regions of overlapping space separately, in the same way as the floors of multi-story buildings are separated on architects' plans, and link the spaces together using visual momentum (Woods, 1984; as applied by Aretz, 1991; Ruddle et al., 1999a).

Acknowledgements

This work was supported by grant GR/L 28449 from the Engineering and Physical Sciences Research Council. I am also grateful to my fellow investigators, Dr. Andrew Howes, Prof. Stephen Payne, and Prof. Dylan Jones, for their help in this research.

References

Allen, G. L, Siegel, A. W., Rosinski, R. R. The role of perceptual context in structuring spatial knowledge. Journal of Experimental Psychology: Human Learning and memory **4** (1978) 617-630

Aretz, A. J. The design of electronic map displays. Human Factors **33** (1991) 85-101

Bliss, J. P., Tidwell, P. D., Guest, M. A. The effectiveness of virtual reality for administering spatial navigation training to firefighters. Presence: Teleoperators and Virtual Environments **6** (1997) 73-86

Chance, S. S., Gaunet, F., Beall, A. C., Loomis, J. M. Locomotion mode affects the updating of objects encountered during travel: The contribution of vestibular and proprioceptive inputs to path integration. Presence: Teleoperators and Virtual Environments **7** (1998) 168-178

Darken, R. P., Sibert, J. L. A toolset for navigation in virtual environments. Proceedings of ACM User Interface Software & Technology (UIST '93). New York: ACM. (1993) 157-165

Darken, R. P., Sibert, J. L. Navigating large virtual spaces. International Journal of Human-Computer Interaction **8** (1996) 49-71

Grant, S. C., Magee, L. E. Contributions of proprioception to navigation in virtual environments. Human Factors **40** (1998) 489-497

Hancock, P. A., Hendrix, C., Arthur, E. Spatial mental representations in virtual environments. Proceedings of the Human Factors and Ergonomics Society 41st Annual Meeting. Santa Monica, CA: Human Factors Society. (1997) 1143-1147

May, M., Péruch, P., & Savoyant, A. Navigating in a virtual environment with map-acquired knowledge: Encoding and alignment effects. Ecological Psychology **7** (1995) 21-36

Ruddle, R. A., Howes, A., Payne, S. J., Jones, D. M. The effects of hyperlinks on navigation in virtual environments. Manuscript submitted for publication. (1999)

Ruddle, R. A., Payne, S. J., Jones, D. M. Navigating buildings in "desk-top" virtual environments: Experimental investigations using extended navigational experience. Journal of Experimental Psychology: Applied **3** (1997) 143-159

Ruddle, R. A., Payne, S. J., Jones, D. M. Navigating large-scale "desk-top" virtual buildings: Effects of orientation aids and familiarity. Presence: Teleoperators and Virtual Environments **7** (1998) 179-192

Ruddle, R. A., Payne, S. J., Jones, D. M. The effects of maps on navigation and search strategies in very-large-scale virtual environments. Journal of Experimental Psychology: Applied **5** (1999a) 54-75

Ruddle, R. A., Payne, S. J., Jones, D. M. Navigating Large-Scale Virtual Environments: What Differences Occur Between Helmet-Mounted and Desk-Top Displays? Presence: Teleoperators and Virtual Environments **8** (1999b) 157-168

Sadalla, E. K., Magel, S. G. The perception of traversed distance. Environment and Behavior **12** (1980) 65-79

Sadalla, E. K., Staplin, L. J. The perception of traversed distance: Intersections. Environment and Behavior **12** (1980) 167-182

Stanton, D., Wilson, P., Foreman, N. Using virtual reality environments to aid spatial awareness in disabled children. Proceedings of the 1st European Conference on Disability, Virtual Reality and Associated Technology. (1996) 93-101

Tate, D. L., Sibert, L., King, T.. Virtual environments for shipboard firefighting training. Proceedings of the Virtual Reality Annual International Symposium (VRAIS'97). Los Alamitos, CA: IEEE. (1997) 61-68

Tlauka, M., Wilson, P. N. The effects of landmarks on route-learning in a computer-simulated environment. Journal of Environmental Psychology **14** (1994) 305-313

Tlauka, M., Wilson, P. N. Orientation–free representations from navigation through a computer–simulated environment. Environment and Behavior **28** (1996) 647-664

Tversky, B. Distortions in memory for maps. Cognitive Psychology **13** (1981) 407-433

Weatherford, D. L. Representing and manipulating spatial information from different environments: Models to neighborhoods. In R. Cohen (Ed.), The development of spatial cognition. New Jersey: Erlbaum. (1985) 41-70

Witmer, B. G., Bailey, J. H., Knerr, B. W., Parsons, K. C.. Virtual spaces and real-world places: Transfer of route knowledge. International Journal of Human-Computer Studies **45** (1996) 413-428

Woods, D. D. Visual momentum: A concept to improve the cognitive coupling of person and computer. International Journal of Man-Machine Studies **21** (1984) 229–244

Influences of Context on Memory for Routes

Sabine Schumacher[1], Karl Friedrich Wender[1], and Rainer Rothkegel[2]

[1] Department of Psychology
University of Trier, 54286 Trier, Germany
{schumacher, wender}@cogpsy.uni-trier.de

[2] Max-Planck-Institute for Biological Cybernetics
Spemannstr. 38, 72076 Tuebingen, Germany
Rainer.Rothkegel@tuebingen.mpg.de

Abstract. Possible influences of contexts on memory for routes are investigated. Route knowledge was established by learning a route which was presented on a computer screen. Activation of knowledge of items along the route was tested. The main goal was to decide whether the surrounding context in the learning and the test phase has an effect on memory for routes. Beyond general context effects, we looked for a possible indirect or mediated context effect. Such a mediate context effect would occur, when memory improves also in cases where context and the to-be-remembered items are separated by a spatial distance. The results reported here provide evidence for immediate context effects. A mediate context effect is not very strongly supported.

1 Introduction

There is ample evidence in the literature that human memory is context dependent and there is in particular evidence for the influence of environmental context on memory. The environmental context effect refers to the phenomenon that individuals who learn and who are tested in the same environment perform more accurately than do individuals who are tested in a new, unfamiliar or interfering physical surrounding (Nixon & Kanak, 1981). The facilitative effect of same context on memory performance has been found with humans (Godden & Baddeley, 1975, 1980; for a review see Smith, 1988) as well as with animals (Dellu, Fauchey, Le Moal & Simon, 1997; Jobe, Mellgren, Feinberg, Littlejohn & Rigby, 1977; Zentall, 1970). Matching context information has been found to be an important factor in the successful retrieval of specific items and learning episodes (e. g. Gillund & Shiffrin, 1984). Mismatching context information at learning and at test on the other hand has been identified as an important cause for forgetting (Bjork & Richardson-Klavehn, 1989; Mensink & Raaijmakers, 1988).

In the present study the influences of context on memory for routes are investigated. Route knowledge refers to the remembered spatio-temporal relations

Ch. Freksa et al. (Eds.): Spatial Cognition II, LNAI 1849, pp. 348-362, 2000.

between objects along a path connecting two places (Siegel & White, 1975; Stern & Leiser, 1988). As Allen puts it: "...simple linear-order knowledge structures involved in navigating from a starting point to an unseen destination by means of coordinating motor behavior with a sequence of perceptual events ..." (Allen, 1987, p. 274).

The motivation for our research comes from an everyday observation: Assume someone has travelled a route that connects two places. When trying to remember this route after some time and while being at a different location, he or she will perhaps not remember every detail of the route, as for example turn-offs or particular objects. But when being on the route again and having already traveled some distance, locations, houses, turning points etc. will come to mind, which could not be remembered before. And finding the correct way is – in many cases – no problem. Most of us probably have had similar experiences (e.g. Searleman & Herrmann, 1994). Such connections between landmarks and context have also been stressed by Chown, Kaplan and Kortenkamp (1995, pp.15), "...landmarks are intimately linked to context. A good landmark in one environment may be a poor one in another environment. In addition, in a familiar environment the activation of a landmark might not even require seeing it. Conversely, seeing a familiar landmark is often enough to call to mind its setting...".

In a study by Anooshian (1996) different strategies for learning large-scale environments were explored. The results also shed some light on context effects in the domain of spatial memory. The environment used in this experiment consisted of three large laboratory rooms, which were connected by corridors. Participants learned a route that connected simulated landmarks (e.g. photographs of distinctive places in an urban environment) which were placed in the laboratory rooms. The acquired spatial information was tested two days later either in the same environment (while walking along the route again) or in a separate lab room. When tested in the same environment, participants' performance was generally better. This was the case for configurational knowledge (tested by bearing estimates), procedural knowledge (the ability to turn correctly at a particular place) and sequence knowledge (being able to name the next landmark correctly while standing in front of the preceding landmark and seeing it). Only memory performance for place knowledge did not differ when tested in the same environment (participants had to name the landmark while it was still covered) or in the lab room (participants were asked to recall the landmark names verbally). Thus, the results of Anooshian (1996) provide evidence for a general facilitating effect of context on memory of spatial information. And the context effects concerning sequence knowledge provide empirical evidence in line with the everyday observation described above.

One major question pursued in the present paper concerns a possible generalization of environmental context effects. The environmental context effect as described in the literature refers to the case, when the to-be-remembered items and the environmental context are present simultaneously. We call this the *immediate context effect*. Thus an immediate context effect would be observed where the facilitating context elements and the to-be-recalled targets had been simultaneously present during the acquisition phase. However, a facilitating effect of context might in principal also occur when the to-be-remembered items and the context elements have not been observed simultaneously but have been separated by a certain distance in space and/or time. For example, when traveling along a route, details might come to mind which still lie ahead and are not yet visible. Better memory performance in this case would also be evidence for a context effect. But in this case the facilitating context elements and the recalled objects have never been observed simultaneously. We call this a *mediate* or *generalized context effect*. Similar ideas have also been described by Chown et al. (1995). But to our knowledge there is yet no experimental evidence for a mediate context effect. By manipulating the spatial distance between cue and to-be-remembered items, the influences of the different types of context effects can perhaps be separated. Previous studies on spatial context effects for route knowledge in our lab provided evidence for an immediate context effect but we did not find evidence for a generalization (Wender, 1998).

2   Experiment 1

2.1   Method

Participants. In this experiment, 40 students participated in 30-minute sessions and were paid for their services. Most of them were psychology students at the University of Trier.

Material. Participants were presented a map-like structure on a computer screen. The map showed a country road that in the drawing had a width of 1 cm and had several curves and intersections. The road was colored in gray with white lines in the middle. A black and white picture of part of the road is presented in Figure 1.

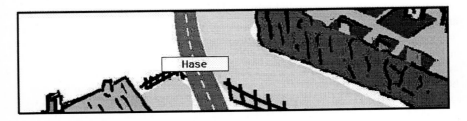

Fig. 1. A part of the road with landscape (Experiment 1)

The experiment was carried out on a Macintosh PowerPC with a 17-inch Apple color monitor. The visible part of the route was displayed in a 17.5 by 4.5 cm large window. The total route covered 29 pages of the monitor. Movement along the route was simulated by scrolling the visible part of the route continuously from top to bottom (moving background technique). Thus, the impression of riding along the road from the bottom (subjectively closer) to the top (subjectively further away) was produced. One ride along the road from start to finish lasted for approximately 2 minutes.

Twenty-nine white rectangles were posted along the street in regular distances. Each of these rectangles contained the name of an animal. These animal names had to be learned. The size of the window with respect to the spacing of the rectangles was just large enough to show one animal name at one time. The assignment of the animal names to the positions on the street was randomly determined for each participant.

To the left and to the right of the road there were additional items – the surrounding context. These additional items were colored drawings, for example of houses, trees, a play ground with a swing, a fence, a wind mill etc. These drawings constituted the possible context. They had the purpose of conveying the impression that the road was embedded in a landscape.

Two experimental conditions were introduced. In the conditions *with context*, the animal names and the landscape were visible. In the conditions *no context* only the rectangles with the animal names were shown, but neither the street nor the landscape. All other aspects of the presentation were identical.

Procedure. The experiment combined two learning phases with two subsequent test phases. The first learning phase consisted of five rides along the route. Participants initiated a ride by pressing a button on the keyboard. A ride took approximately 2 minutes of continuous movement. Thus the presentation time per stimulus was about 4 seconds. Participants were instructed to remember the route so that they would be able to describe it to a friend. They were also instructed to learn the animal names along the route.

The learning phase was followed by the test phase, a cued recall test. Each test started from a randomly determined place somewhere along the route. From there the participant had the impression of driving along the road for a short distance with two stimuli (animal names) passing through the visible window. When the third stimulus appeared on the screen the ride stopped. These three stimuli were always presented in the same order in which they had occurred in the learning phase. The task for the participant was then to recall *the following three animal names* along the road. The responses had to be typed into the computer.

By this task the distance between cue and target items was manipulated. The size of the window and the spacing of the to be learned stimuli were constructed in such a way that at one moment only one stimulus was present in the window. But as soon as one stimulus disappeared at the bottom of the window, because the map was scrolled downwards, the next stimulus entered the window from above. As a consequence, two consecutive stimuli had approximately one half of their context in common. So, if the first of these two stimuli in one particular test is the cue and if we find a context effect for the next stimulus, this would be an immediate context effect because the cue and the target share a common context. This is not true anymore for the second stimulus along the road because a cue and a second target have no context elements in common.

Four locations were predetermined from which a test ride could depart. The actual sequence of tests was randomly determined for each participant. After having finished the first test phase, a second learning phase started consisting of three further rides along the route. In the following second test phase a different set of four cues was presented.

Design. During the learning phase the surrounding context was either present or not. This constitutes the first independent variable: *learning with context vs. learning without context*. The second independent variable was whether *testing* was also conducted *with or without context*. These two factors were combined resulting in a two by two design. Independent samples of 10 participants each were assigned to each of the four conditions.

Two more factors were included. The third factor was the *time of test* (first or second test). And the last independent variable was the *position* of the remembered *item*: first or second or third item after the cue.

2.2   Results

A 2 (learning with or without context) x 2 (testing with or without context) x 2 (time of test) x 3 (item position) ANOVA with repeated measures on the last two variables and number of correctly remembered animal names as dependent measure was computed.

The ANOVA yielded two significant main effects for *item position* and *time of test*. For *item position*, we found $F(2, 72) = 18.48$, $p < .001$. The animal name, which directly followed the cue was remembered best (52 %) followed by the second animal name (39 %) and the third animal name (32 %). *Item position* did not interact with any of the other variables.

Time of test also had a significant influence on the results, $F(1, 36) = 89.42$, $p < .001$. In the second test phase participants remembered more than twice as much targets (56 %) compared to the first test phase (26 %). *Time of test* did not interact with any of the other variables.

There were no other significant effects, neither an effect of *learning condition* ($F < 1$) nor an effect of *testing condition*, $F(1, 36) = 1.21$, $p < .276$. The interaction between *learning* and *testing condition* was also not significant ($F < 1$). In Figure 2 the mean percentage of correctly remembered animal names in the four experimental conditions is shown. The results are summarized over the first and second time of test. Most surprisingly, memory performance was worst in the condition where learning and testing took place in the context of the landscape.

2.3   Discussion

Obviously, the manipulations of the surrounding context in the learning and the test phase did not lead to the expected effects. On a descriptive level, the results are even contrary to our expectations and to typical results about context effects reported in the literature (e.g. Smith, Glenberg & Bjork, 1978; Smith, 1988). Memory performance was worst in the condition, where learning and testing took place in the landscape-context, and best in the condition, where learning was with and testing without context. We interpret this result as an unfortunate consequence of the form of the stimulus presentation on the screen. Most of the objects, which constituted the surrounding landscape, had a larger size than the visible part of the road. Therefore, it was difficult to identify the objects while seeing only parts. Possibly, our participants in the context-condition paid more attention to the identification of the objects than to the animal names on the road. Therefore, in Experiment 2a the visible part of the screen was enlarged, so that two animal names were visible at the same time while riding along the road and the surrounding objects of the landscape could be identified easily. Also, in the test phase each cue was presented immediately without scrolling the window across the map as in Experiment 1.

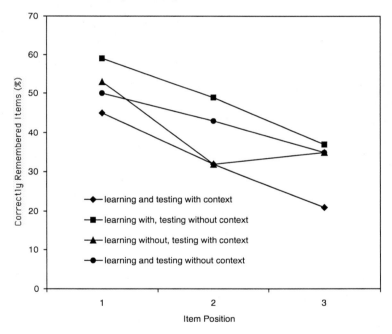

Fig. 2. Mean percentage of correctly remembered animal names as a function experimental condition and of item position in Experiment

3   Experiment 2a

The main goal of this experiment was similar to the one in Experiment 1. As a consequence of the results from Experiment 1 the procedure was changed as described below. Otherwise the experimental design remained identical.

3.1   Method

Participants. In this Experiment, 120 students participated in 40-minute sessions and were paid for their services. Most of them were psychology students at the University of Trier.

Material. The same route was used as in Experiment 1 but the visible part of the road was now made large enough to present two of the to-be-learned stimuli at the same time. By this change, the objects of the surrounding landscape could be identified more easily (see Figure 3).

Fig. 3. A part of the road with landscape (Experiment 2a)

A second consequence of the larger window was that now the context of the cue and the context of the second stimulus overlapped to some degree. Therefore, only the responses to the third stimulus can be analyzed for a mediate (or generalized) context effect in this experiment.

In the experimental conditions without context, only the rectangles with the animal names were visible, but not the street and the landscape. All other aspects of the presentation remained the same.

Design. The dependent variable was again the number of correctly remembered targets. The independent variables were presence of the surrounding landscape in the learning condition (context or no-context) and the testing condition (context or no-context). These factors were combined in a two by two design. Four groups of 30 participants each were assigned to the four conditions. Additional independent variables were *position* of remembered *item* (first or second or third after the cue) and *time of test* (first or second test). Animal names, the to-be-remembered stimuli, were randomly assigned to the positions on the street.

Procedure. Participants were instructed to imagine that they had to leave a freeway because of a traffic jam and to continue their way on a country road. Along this road signs with animal names are posted – the remains of a paper chase. Participants were further instructed to learn these animal names in their correct order and their positions in the landscape. Those participants, who saw the animal names in the context of the landscape, were also told to pay attention to the surrounding landscape. While riding along the route in the learning phase the acquisition of the stimuli was passive, i.e. active navigation along the route was not possible.

The experiment consisted of two learning phases with two consecutive test phases. The memory test was again a cued recall test. As a cue participants saw a part of the road with two rectangles. In the rectangle at the bottom of the screen an animal name was visible, the upper rectangle contained a question mark. Participants had to recall the three animal names that had followed the cue. First, that animal name had to be reproduced, which belonged to the rectangle with the question mark. Then, the two animal names had to be recalled, which in the learning phase had followed on the road. The animal names had to be typed in using the keyboard. The cues were either presented within the surrounding landscape as context or without the landscape.

The first learning phase consisted of four rides along the route and was followed by the first test phase. In contrast to Experiment 1, nine stimuli were used as cues in order to enlarge the number of test items. As a consequence, the test procedure was changed slightly. The cues were presented in the order in which they had appeared on the route, starting with the first animal name on the road. Participants then had to recall the following three animal names. The third to-be-recalled animal name then served as the next cue and so on. After the first test phase, a second learning phase started consisting of three further rides along the route. The following second test phase was identical with the first test phase, i.e. the same cues were presented in the same order as before.

3.2   Results

A 2 (learning conditions) x 2 (testing conditions) x 2 (time of test) x 3 (item position) ANOVA with repeated measures on the last two variables and number of correctly recalled items as dependent measure was computed. The answers to the first and the last cue were not included in this ANOVA to avoid primacy or recency effects.

The ANOVA yielded a significant effect for *item position*, $F(2, 232) = 158.23$, $p <$ .001. The target, which directly followed the cue was remembered best (38 %), followed by the second animal name (25 %), and the third animal name (18 %). This effect was further qualified by an interaction between *item position* and *testing condition*, $F(2, 232) = 6.03$, $p < .003$. The results are shown in Table 1.

The interaction between *item position* and *learning condition* ($F(2, 232) = 1.58$, $p <$ .209) and the triple interaction ($F(2, 232) = 1.80$, p < .168) were not significant.

Table 1. Mean percentage of remembered animal names as a function of testing condition and item position

	position 1	position 2	position 3
Context condition of test phase			
With context	40%	26%	17%
Without context	35%	25%	20%

Time of test also had a significant influence on the results, $F(1, 116) = 117.20$, $p < .000$. In the second test phase participants remembered twice as many items (36 %) compared to the first test phase (18 %). This effect was further qualified by the significant interaction between *item position* and *time of test*, $F(2, 232) = 7.20$, $p < .001$. The results are shown in Table 2.

Table 2. Mean percentage of remembered animal names as a function of time of test and item position

	position 1	position 2	position 3
Test phase			
1st phase	27%	17%	14%
2nd phase	49%	34%	26%

There was also a significant interaction between *time of test* and *learning condition*, $F(1, 116) = 4.21$, $p < .043$. The interactions between *time of test* and *testing condition* ($F(1, 116) = 2.63$, $p < .108$) and the triple interaction ($F(1, 116) = 3.28$, $p < .073$) were not statistically significant. There were no main effects for *learning condition* ($F(1, 116) = 2.00$, $p < .160$) and *testing condition* ($F < 1$). Although the interaction between *learning condition* and *testing condition* is also not statistically significant, $F(1, 116) = 2.40$, $p < .124$, the results are given in Table 3.

Table 3. Mean percentage of remembered animal names as a function of learning condition and testing condition

	Context condition of the test phase	
	With context	Without context
Context condition of the learning phase		
With context	34%	26%
Without context	23%	28%

On a descriptive level they show that performance in the condition where learning and testing took place in the context of the landscape is highest, followed by the condition, where learning and testing were conducted without the surrounding landscape. Note, that this is also a same context condition. In the two experimental conditions, which involve a change of context between the learning and the test phase, memory performance is poorest. Insofar, the results are indicative for an immediate context effect as discussed in the literature.

In Figure 4 the mean percentages of correctly remembered targets are presented. The results are summarized for the first and second test phase. These results will be discussed together with the results from Experiment 2b.

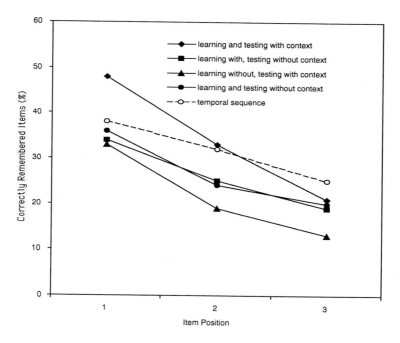

Fig. 4. Mean percentage of correctly remembered animal names as a function experimental condition and of item position in Experiment 2a and 2b

Two post-hoc analyses were performed. The first was a 2x2x3 analysis with the following factors, Factor A: *learning and testing with context* vs. *learning with context and testing without context*, Factor B: *time of test* (first or second), and Factor C *item position* (first, second, third). *Time of test* was significant ($F(1, 58) = 93.14, p < .001$), as was *item position* ($F(2, 116) = 85.77, p < .001$). More interesting, the interaction between Factor A and *item position* also reached significance: $F(2, 166) = 6.89$, $p < .001$. All other effects were not significant.

The second post-hoc analysis was also a 2x2x3 analysis with the following factors, Factor A: *learning and testing with context* vs. *learning without context and testing with context*, Factor B: *time of test* and Factor C: *item position*. Again *time of test* was significant ($F(1, 58) = 74.13$, $p < .001$) as was *item position* ($F(2, 116) = 91.94$, $p <$ .001). But this time the Factor A was also significant ($F(2, 58) = 5.54$, $p < .002$. This effect was modified by a significant interaction between Factor A and *time of test* ($F(1, 58) = 7.14$, $p < .010$).

4   Experiment 2b

This experiment is an addition to the previous one. Now a pure temporal sequence of the stimuli was used. This was motivated by findings of Schweizer and Janzen (1996) who found in priming studies different results for temporal presentation of items as compared to spatial orderings like a route. Furthermore, according to the assumptions by Ebbinghaus (1885/1985) immediate associations between neighboring items in a list should be stronger than indirect associations between items that are separated by other items in between. Merely temporal sequencing of the items might lead to a comparable memory performance and thus should be compared to possible contextual influences.

4.1   Method

Participants. 30 students at the University of Trier participated and were paid for their services.

Material. The same animal names as in Experiment 2a were used. The sequence of the animal names was varied randomly for each participant.

Procedure. The animal names were presented one at a time and each at the same position on the screen. Thus, the items were only separated by time. The number of learning phases and test phases was the same as in Experiment 2a. In a test phase, participants also had to recall the three animal names following a cue.

4.2   Results

A 2 (time of test) x 3 (item position) ANOVA with repeated measures was computed. Dependent measure was the number of correctly remembered animal names. Answers to the first and the last cue were not included in this ANOVA.

This ANOVA yielded a significant effect for *time of test*, $F(1, 29) = 27.53$, $p < .000$. In the second test phase more than twice as much animal names were correctly remembered (44%) compared to the first test phase (20%). *Item position* also had a significant influence on the results, $F(2, 58) = 23.08$, $p < .000$. Animal names following directly the cue being remembered best (item position 1: 38%, item position 2: 32% and item position 3: 25%). The interaction between *time of test* and *item position* was not significant, $F(2, 58) = 1.90$, $p < .158$. The mean percentage of correctly remembered animal names in this condition is also included in Figure 4.

5   Discussion

Influences of spatial context on memory for routes were investigated in a laboratory set-up. We were mostly interested in mediated context effects, i.e. better recall of not yet visible parts of a route, when the context of the cue word from the learning phase was present at recall. Higher recall rates especially for the third item when learning and testing took place in the same context would have been evidence for mediated context effects. Better recall rates for the first and second item when the animal names were learned and tested in the context of the landscape, would provide evidence for an immediate context effect. In these cases the cue and the recalled items had shared the same context at least partially. The third animal name after the cue had never been seen in the same spatial context with the cue-word.

However, our results provide only mixed evidence for mediated context effects. In the first post-hoc analysis we find no main effect between the condition where learning and testing took place in the context of the landscape and the condition where learning was conducted with and testing without context. However, the interaction of this factor with item position was significant. An inspection of Figure 4 shows that this is due to item positions one and two and that there is no difference anymore in position three. This would argue that there is an immediate context effect as shown by the first two positions but that we do not find a mediated effect because there is no difference at position three.

The second post-hoc analysis, however, revealed a main effect of context conditions *learning and testing with context* versus *learning without but testing with context*. As Figure 4 shows, this difference extends also to item position three. This result then speaks for a mediated context effect in this situation. The result may remind us of failures to find the correct way after having received a route description and then trying to find our way in the context of the real world. However, the result must be viewed with caution, because it is clearly a post-hoc result and was not expected before hand.

For the immediate context effect the recall rate is highest in the condition where in the learning and test phase the landscape is present (see Table 3). Although this result is not statistically significant, it provides a hint on immediate context effects. Congruence between learning and test phase, i.e. no change in context between the learning and the test phase, leads to better recall rates (see Grant, Bredahl, Clay, Ferrie, Groves, McDorman & Dark, 1998; Wippich & Mecklenbräuker, 1988). This is especially the case, when the environmental context provides richer memory traces.

It is interesting to note that the results of Experiment 2b, where the animal names were presented as single items and only separated by time are quite comparable with the results of Experiment 2a in the experimental condition where learning and testing occurred in the same context. The memory performance in the temporal condition (Experiment 2b) can be explained by the associative strengths between the items of the list. Immediate associations, i.e. between neighboring items, are stronger than between indirect associations. Thus, the mean percentage of correctly remembered stimuli decreases from item position 1 to item position 3. One has to keep in mind that in Experiment 2b the stimuli were presented separately, whereas in Experiment 2a always two animal names were presented at the same time. Thus, it was easier in Experiment 2b to concentrate on the items. Nonetheless, the memory performance is better in the context condition in Experiment 2a for item positions 1 and 2 compared to the temporal condition in Experiment 2b. This can be explained by contextual influences.

Not surprisingly, there is evidence for effects of practice. The recall rate is much higher in the second test phase compared to the first test phase – this effect is even stronger, when the animal names are learned in the presence of the surrounding landscape. All three stimuli are better remembered in the second test phase, especially the first item after the cue.

The results do provide only post-hoc evidence for a spatial generalization of context effects. But they are providing evidence for an immediate context effect. Richer memory traces facilitate retrieval. These context effects may occur as a result of incidental associations found between the general contextual stimuli and the list items. The contextual stimuli automatically activate and elicit the list responses when the same contextual stimuli are physically present during the acquisition phase and the retention test (Nixon & Kanak, 1985).

One aspect of our study that should not be neglected was that it was conducted in the laboratory and the route was presented on a computer screen. Our participants learned a two-dimensional map of a route in a fictitious landscape with animal names being placed on the route. It is quite possible that this context was not strong enough. In fact, the more general context of the laboratory room was present under all conditions. We did not transfer our subjects to a different room for recall. It is quite possible that a mediated context effect can be found more easily in real situations.

Events we experience and decisions we make are usually perceived as being embedded within a structure of other events. And, more important for the present discussion, the events that occur within one environment usually are related to the environment itself. While riding along a road in the real world, the environment may be perceived as causing specific events or requiring particular decisions, like making a specific turn when there are alternative routes. As Fernandez and Glenberg (1985, pp. 344) argue, the relationships between environmental context and events are integral to the representation of naturally occurring events.

The animal names, which were presented in our studies, possibly were not perceived as being closely related to the environment – the drawn road – in which they occurred nor were they perceived as causing or enabling each other (see Fernandez & Glenberg, 1985). The laboratory task our participants had to fulfill most probably is not characterized by strong links like those occurring between natural events. However, such links presumably underlie the anecdotal reports about generalized context effects that were mentioned at the outset.

Other experiments (Wender & Rothkegel; in preparation) found that in real environments mediated context effects may occur. If such context effects can be convincingly demonstrated this would have consequences for assumptions about spatial memory. Models of spatial memory like mental maps e.g., would then have to represent not only landmarks and the relationships between them but the context would also have to be included. In addition, different types of relations or associations between landmarks and context elements may be necessary.

Acknowledgments

This research was supported by a grant from the Deutsche Forschungsgemeinschaft (German Research Foundation) to Karl F. Wender (We 498/27-1). We would like to thank Claus C. Carbon, Frauke Faßbinder, Simone Knop, Oliver Lindemann and Pia Weigelt for their assistance with preparing and conducting the experiment and Tilman Knoll for assistance with data analysis.

References

Allen, G. L. (1987). Cognitive influences on the acquisition of route knowledge in children and adults. In P. Ellen & C. Thinus-Blanc (Eds.), *Cognitive processes and spatial orientation in animal and man*. Boston: Martinus Nijhoff.

Anooshian, L. (1996). Diversity within spatial cognition. Strategies underlying spatial knowledge. *Environment and Behavior, 28*, 471-493.

Bjork, R. A. & Richardson-Klavehn, A. (1989). On the puzzling relationship between environmental context and human memory. In C. Izawa (Ed.), *Current issues in cognitive processes: The Tulane Floweree Symposium on Cognition* (pp. 313-344). Hillsdale, NJ: Erlbaum.

Chown, E., Kaplan, S., & Kortenkamp, D. (1995). Prototypes, Location, and Associative Networks (PLAN): Towards a Unified Theory of Cognitive Mapping. *Cognitive Science, 19*, 1-51.

Dellu, F., Fauchey, V., Le Moal, M., & Simon, H. (1997). Extension of a new two-trial memory task in the rat: Influence of environmental context on recognition processes. *Neurobiology of Learning and Memory, 67*, 112-120.

Ebbinghaus, H. (1985). *Über das Gedächtnis. Untersuchungen zur experimentellen Psychologie* (1. Aufl., Leipzig 1885). Darmstadt: Wissenschaftliche Buchgesellschaft.

Fernandez, A. & Glenberg, A. M. (1985). Changing environmental context does not reliably affect memory. *Memory & Cognition, 13*, 333-345.

Gillund, G. & Shiffrin, R. M. (1984). A retrieval model for both recognition and recall. *Psychological Review, 91*, 1-67.

Godden, D. & Baddeley, A. D. (1975). Context-dependent memory in two natural environments: On land and under water. *British Journal of Psychology, 66*, 325-331.

Godden, D. & Baddeley, A. D. (1980). When does context influence recognition memory? *British Journal of Psychology, 71*, 99-104.

Grant, H. M., Bredahl, L. C., Clay, J., Ferrie, J., Groves, J. E., McDorman, T. A., & Dark, V. (1998). Context-dependent memory for meaningful material: Information for students. *Applied Cognitive Psychology, 12*, 617-623.

Jobe, J. B., Mellgren, R. L., Feinberg, R. A., Littlejohn, R. L., & Rigby, R. L. (1977). Patterning, partial reinforcements, and N-length effects of spaced trials as a function of reinforcement of retrieval cues. *Learning and Motivation, 8,* 77-97.

Mensink, G.-J. M. & Raaijmakers, J. G. W. (1988). A model for interference and forgetting. *Psychological Review, 95,* 434-455.

Nixon, S. J. & Kanak, J. (1981). The interactive effects of instructional set and environmental context changes on the serial position effect. *Bulletin of the Psychonomic Society, 18,* 237-240.

Nixon, S. J. & Kanak, J. (1985). A theoretical account of the effects of environmental context upon cognitive processes. *Bulletin of the Psychonomic Society, 23,* 139-142.

Searleman, A., & Herrmann, D. (1994). *Memory from a broader perspective.* McGraw-Hill, Inc.

Siegel, A. W. & White, S. H. (1975). The development of spatial representations of large-scale environments. In: H. W. Reese (Ed.), *Advances in child development and behavior.* Vol. 10. New York, Academic Press.

Smith, S. M. (1988). Environmental context-dependent memory. In: G. M. Davies, & D. M. Thomson (Eds.), *Memory in context: context in memory.* (pp. 13-34) New York, NY: Wiley.

Smith, S. M., Glenberg, A., & Bjork, R. A. (1978). Environmental context and human memory. *Memory & Cognition, 6 (4),* 342-353.

Schweizer, K. & Janzen, G. (1996). Zum Einfluß der Erwerbssituation auf die Raumkognition: Mentale Repräsentation der Blickpunktsequenz bei räumlichen Anordnungen. *Sprache & Kognition, 15, 4,* 217-233.

Stern, E. & Leiser, D. (1988). Levels of spatial knowledge and urban travel modeling. *Geographical Analysis, 20,* 140-155.

Wender, K. F. (1998). Kontexteffekte und Routenwissen. *Kognitionswissenschaft, 7,* 68-74.

Wender, K. F. & Rothkegel, R. (in preparation). On the structure of route knowledge.

Wippich, W. & Mecklenbräuker, S. (1988). Räumliche Kontexteffekte bei der Textverarbeitung: Eigenschafts- und Verhaltenserinnerungen. *Schweizerische Zeitschrift für Psychologie, 47 (1),* 37-46.

Zentall, T. R. (1970). Effects of context change on forgetting in rats. *Journal of Experimental Psychology, 86,* 440-448.

Preparing a Cup of Tea and Writing a Letter: Do Script-Based Actions Influence the Representation of a Real Environment?

Monika Wagener, Silvia Mecklenbräuker, Werner Wippich,
Jörg E. Saathoff, and André Melzer

Department of Psychology
University of Trier, D - 54286 Trier, Germany
{monika,mecklen,wippich,evert,melzer}@cogpsy.uni-trier.de

Abstract. Two experiments were conducted to examine the effects of having people carry out a sequence of actions in an environment on the spatial representation of the environment. The actions were linked by a common theme (e.g., writing a letter). In Experiment 1, the spatial memory test consisted of an implicit and an explicit distance estimation task. Participants who carried out a sequence of script-based actions inside a room showed poor spatial knowledge for this particular room (as compared to a control room or control participants) in disregarding actual distances in their estimations. This deficit could be due to a loss of or to a poorer encoding of spatial information. The results of Experiment 2, however, suggest that the effects observed in Experiment 1 seem to depend on the spatial task used. With a positioning task at testing, we could not find any evidence that could be attributed to an action-based change of a spatial mental representation. In sum, the general hypothesis of action-based influences on mental spatial representations was not corroborated by convincing data.

1 Introduction

Nearly all human activities require people being able to store information about spatial arrangements. Therefore, in cognitive psychology, mental representations (e.g., Herrmann, 1993; Tack, 1995) of spatial information are assumed. When people learn the locations of objects in an environment, they typically also acquire non-spatial (e.g., semantic) information about these objects. There is no doubt that spatial memories can be influenced by this non-spatial information (e.g., Hirtle & Mascolo, 1986; McNamara, Halpin, & Hardy, 1992; McNamara & LeSueur, 1989; Sadalla, Staplin, & Burroughs, 1979). For instance, McNamara, Halpin, and Hardy (1992) conducted a series of experiments investigating people's ability to integrate non-spatial information about an object with their knowledge of the object's location in space. Their results indicate that the spatial and non-spatial information were encoded in a common memory representation.

Ch. Freksa et al. (Eds.): Spatial Cognition II, LNAI 1849, pp. 363-386, 2000.

The current research was conducted as a means to examine interactions of spatial information and actions (i.e., simple activities). Spatial representations do not only play an important role in navigational decision making. One of their most important functions is to improve the potential to plan and to perform adequate actions in complex environments. And when people learn about the locations of objects, they typically also acquire information about activities, which have been performed at particular locations.

In a recent study (Mecklenbräuker, Wippich, Wagener, & Saathoff, 1998) we were able to demonstrate that previously acquired spatial information, that is, information about locations along a learned route, can be associated with imagined or with symbolically performed actions (i.e., simple activities like folding a handkerchief). The ability of remembering actions was influenced by spatial information, that is, locations proved to be effective cues for recalling actions. However, we have not found any evidence that previously acquired spatial representations can be altered by performing or imagining simple actions at certain locations.

Based on these findings, we tried to identify conditions under which the enactment of actions might influence the representation of a spatial setting. One important factor might be the *point in time* at which the enactment of actions can affect a spatial representation. In our former experiments participants first learned a spatial setting until a criterion was reached and then symbolically performed activities at different locations along a route. Perhaps performing activities can not alter a stable spatial representation, but it can influence the construction of a spatial representation. This possibility is indicated by results from developmental studies conducted by Cohen and colleagues (e.g., Cohen & Cohen, 1982; Cohen, Cohen, & Cohen, 1988) which show the utility of activities for the construction of spatial representations. For example, Cohen and Cohen (1982) and Cohen et al. (1988) assessed the effects of just walking among locations versus performing isolated activities at locations versus performing activities that were functionally linked across locations (through a letter writing and mailing theme). First and sixth graders had to walk through a novel environment constructed in a large, otherwise empty classroom. The results (i.e., distance estimates) from the Cohen and Cohen (1982) study suggest that when the environment can be viewed in its entirety, the theme serves to facilitate the acquisition of spatial knowledge for the entire space (i.e., distance estimates were more accurate).

Moreover, these studies point to a second important factor, that is the *type of activities performed*. Whereas we (Mecklenbräuker et al., 1998) used simple activities, which were neither related to each other nor to the locations at which they were performed, Cohen and colleagues used activities that were functionally linked across locations. Furthermore, they used a *real environment*.

In the present study, these three potentially relevant factors (point in time, type of activity performed, real environment) were taken into account. Our participants had to perform activities that were linked by a common theme and were related to the corresponding locations or objects in a real environment (a simulated environment, i.e., an apartment with three rooms). Furthermore, participants had to perform these activities in a *novel* environment, that is, participants simultaneously performed activities and acquired new spatial knowledge.

A fourth alteration to our former experiments concerned the measures of spatial knowledge. Studies of spatial cognition have used a diversity of tasks for assessing spatial knowledge (e.g., Evans, 1980). Some of the more common ones have been distance estimation (cf. Rothkegel, Wender, & Schumacher, 1998), location and orientation judgments (e.g., Wender, Wagener, & Rothkegel, 1997), and navigation. We (Mecklenbräuker et al., 1998) have mainly used distance estimations. Perhaps this spatial measure is not sensitive enough to indicate effects of performing actions on spatial representations. Perhaps distance estimations can be made without taking into account information about action. When asked for distance estimations, participants are given a so-called explicit task of spatial memory: All experimental or pre-experimental spatial experiences the participants have made are explicitly referred to before the participants are asked to express their spatial knowledge. In recent years implicit tests of memory have become very popular. The defining criterion of the distinction between explicit and implicit tests is the instruction given at the time of testing. During an explicit test participants are told to recollect events from an earlier time. During an implicit test they are told to simply perform a task as well as possible without mentioning the recollection of earlier experiences they have made while performing the task although the earlier experiences might affect performance on the tasks.

In the present study, we used two different kinds of tasks for assessing spatial knowledge (distance estimations and positioning of objects). Both tasks were given not only with traditional explicit, but also with implicit instructions. Implicit spatial measures have at least two advantages. First, they minimize performance demands. Siegel (1981) has suggested that tasks should be used that minimize performance demands so that the contents of spatial representations can be assessed more accurately. Second, findings from implicit memory research suggest that implicit tests primarily reflect automatic usage of specific priorly made experiences. Moreover, in our opinion, information which is assigned to spatial representations is normally used automatically rather than consciously. Only if problems occur, for example, if we loose our way, consciously controlled use of memory will be likely. Such considerations strongly suggest that we should use implicit measures in spatial cognition research. Anooshian and Seibert (1996) have recently emphasized the importance of automatic influences of priorly made experiences in place recognition. According to the model of way finding by place recognition proposed by Cornell, Heth, and Alberts (1994), people rely on the familiarity associated with particular places when making navigational decisions.

This paper is separated into two parts: First a report of a pilot study which was conducted to construct two comparable sets of activities that were both linked by a common theme will be presented (Part 2). Second, we will report two experiments that examined the influence of performing these activities on the representation of a real environment (Part 3). In the first experiment, participants were given a distance estimation task, at first with an implicit, then with an explicit instruction. In order to examine the dependency of results from the spatial measures employed, we used a different task (implicit and explicit) in the second experiment: A positioning task where participants had to indicate the location of objects in the previously seen environment. The main goal of this study was to assess the influence of the enactment of actions on the mental representation of the real environment in which the actions took place.

2   The Construction of Two Comparable Scripts

First, we sought activities that could be performed in a medium scale environment such as an apartment. These activities had to be linked by a common theme by belonging to a certain script. Two sets of activities were constructed to assure that possible effects were not specifically related to the script. These two sets had to be comparable with regard to important characteristics in the realization and use of scripts.

2.1   Important Characteristics of Scripts

Usually a script is defined as a predetermined, stereotyped sequence of actions that defines a well-known situation. In other words, a script is a mental representation of "what is supposed to happen" in a particular circumstance (Schank & Abelson, 1977).

As shown by Mandler and Murphy (1983) the structure of scripts is highly subjective and depends on the format of the given information. A relevant aspect is to determine what scenes a given script consists of (Hue & Erikson, 1991). For instance, scenes of a restaurant script are "Entering", "Ordering", "Eating", "Exiting" (Schank & Abelson, 1977). Single actions might belong to a more complex action or to a unit belonging to a certain theme. Another important factor that affects the ability of remembering script-based information is the typicality or relevance of specific events and actions to the theme. Many studies demonstrate that typical actions and events are recalled or recognized less well than atypical ones. For instance, in a study by Smith and Graesser (1981), typical events, those anticipated by the script, were recalled more poorly once the scores had been corrected for guessing. The atypical events had an advantage over the typical ones – they were specifically stored during comprehension, since they were details that could not have been anticipated by the script. Nakamura, Graesser, Zimmerman, and Riha (1985) could replicate this finding with a more naturalistic setting such as a classroom lecture following incidental learning conditions. Another relevant characteristic is the temporal order of the script-based activities. Furthermore, it is advisable to adjust the contents of a script to the population to be investigated. That means, we have to know whether important actions or objects are missing or whether mentioned actions or objects are obviously unnecessary or even disturbing. We constructed two scripts that should be comparable with regard to the previously described characteristics.

2.2   Pilot Study: The Construction of the Scripts

From several scripted action sequences, which can be performed in our simulated environment, we chose two equally complex scripts: Preparing a cup of tea in a kitchen and writing a letter on a computer in a study. Each script contained many actions, detailed according to the level of resolution. The level of resolution determines which objects are mentioned and how complex the movements are. Twenty actions were generated for each script. The 40 action phrases were each printed on a cardboard (6.7 cm x 9.4 cm). For each script, 24 students of the University of Trier were required to perform several tasks.

First, the participants were asked to arrange the action phrases in a temporal sequence that seemed to be most commonly used. Subsequently, they had to rate their typicality on a 4-point scale (with 1 being typical and 4 atypical). In order to assess the scenes and the hierarchical structure of the scripts, the participants were then asked to cluster the actions according to thematic categories in at least two and at most eight clusters. After performing interpolated activities, participants were required to freely recall the actions and all pieces of furniture and other objects that had been mentioned in the actions. Finally, they were asked to state any missing and disturbing actions.

First, we computed the mean positions for the activities from the temporal ordering task. As expected, the participants hardly differed from each other or from our predetermined temporal order. Therefore, it was not necessary to adjust the positions of the actions regarding the temporal order.

The data from the clustering task were analyzed by a cluster analysis. The results were clear for the letter writing script: three clusters emerged with 5 to 7 actions each. For the other script some problems emerged because the cluster analysis revealed a larger number of clusters with some consisting of one action only. Therefore, to make this script comparable to the first one, we changed some actions based on participants' answers about missing and disturbing actions. Participants recalled on average 16.2 actions alone and 13.2 actions together with the correct location with no difference between the two scripts.

Based on this pilot study, we constructed two scripts with eight script-based actions each. Because of the limited size of the apartment, the number of objects to be used in each script was restricted to six. To vary the intensity of acting, two of the objects were each involved in two different actions, and the other four objects were involved in only one action. The resulting two sets of script-based actions (see Table 1; note that the original phrases were in German) could be compared to each other regarding the temporal order, the typicality, and the recall probability of the actions. The mean typicality rating was 1.9 (i.e., rather typical), with a range from 1.2 to 2.4 for the letter writing script, and also 1.9 for the tea script (range: 1.4 to 2.2). The mean recall probability was .81 for the letter writing script and .83 for the tea script. Furthermore, the actions belonged to the same level of the hierarchical structure as indicated by the cluster analysis.

Table 1. The two scripts that are used for the two experiments (crucial objects are emphasized)

Script: Writing a Letter	Script: Preparing Tea
1. pick up paper from the **container** and put it into the printer	1. fill the kettle with water at the **sink**
2. sit down at the **computer** and press the Enter key	2. put the kettle on the **range**
3. get the letter from the **printer**	3. pick up a teabag from the **cupboard**
4. get the envelope from the **book-shelf**	4. at the **sideboard** put the teabag into the teapot
5. put the letter into the envelope at the **book-shelf**	5. place the teapot on the **kitchen table**
6. get the address from the **filing cabinet**	6. pour boiling water into the teapot at the **kitchen table**
7. sit down at the **desk** and write the address on the envelope	7. pick up the box of sugar from the **kitchen shelf** and open it
8. glue one of the stamps from the **desk** on the envelope	8. refill the sugar bowl at the **kitchen shelf**

3   Study: Actions and Spatial Knowledge

Two experiments were conducted to examine the influence of performing activities, linked through a common theme, on the construction of the mental representation of the real environment in which these activities took place. The two experiments only differed in the tasks used for assessing spatial knowledge: an implicit and explicit distance estimation task in Experiment 1, and an implicit and explicit positioning task in Experiment 2. The spatial setting (a simulated environment, i.e., an apartment with three rooms: kitchen, living room, and study) and the acquisition phase with incidental learning of the spatial setting were identical. In each experiment one group of participants had to perform the activities linked by the *letter writing script* in the study, whereas the other experimental group performed the *tea script* activities in the kitchen. One goal was to compare how performing linked activities influences the mental representation as compared to the enactment of unrelated activities in the other two rooms. The notion "unrelated" refers to activities that have no common temporal order and are not strongly related to the objects that were used to perform them. These conditions are met by a search task. Hence, each participant was also required to search for a lost key. Besides the experimental groups (the script groups), we had two control groups (one for each script). The control participants walked through the apartment, looked at the pieces of furniture and heard (read by the experimenter) the names of the same objects in the very same order as mentioned in the corresponding scripts. However, they did not perform any activities. They were also read the names of the objects, which were present at the locations where the respective script group searched for the key.

3.1   Experiment 1: Implicit and Explicit Distance Estimation

In general, we expected more action-related spatial information to be acquired when participants performed linked activities. Performing linked activities should lead to a stronger link between the locations where the activities were performed (i.e., the spatial information). In other words, for the script groups (i.e., the experimental groups), the locations in the corresponding script room should be closer together. Accordingly, these groups should estimate the distances in their script room as shorter in comparison to the control group and the distances in the comparable control room (e.g., the kitchen for the letter writing script). These expectations should hold for the implicit distance estimation task. For the explicit task, participants might be able to correct their estimations. Therefore, for this task, two alternative hypotheses were possible. The first hypothesis was that distances in the script room are estimated as shorter (as was expected for the implicit task). The second hypothesis was that the participants can correct their estimations and are more precise for the script room compared to the control room or the control group. The latter expectation is consistent with the results obtained with children in the developmental studies by Cohen and colleagues (1982, 1988).

3.1.1   Method

Participants. The participants were mainly students of the University of Trier from different majors. Twenty-seven women and eleven men (twelve in each script group, eight in the hearing-kitchen group, and six in the hearing-study group) completed this experiment in approximately 90 minutes. They were paid for their participation.

Material. As action phrases we used the two script lists described in Part 2 with eight short actions on each list. The search task also had a list of eight short actions, but the list was not organized hierarchically. The task consisted of finding a key in the apartment.

We used a large conference room as a spatial setting. Using poster boards we constructed an apartment (5 x 8 m) with three identical rooms and a hall, from which every room could be entered. Furthermore, each room was accessible from the neighboring rooms. An outline of the rooms including the furniture is shown in Figure 1. The rooms were identical in size and the two rooms at either end were symmetrical concerning the entrances and a column that could not be removed (indicated as a black circle in Figure 1). To establish a wall-like structure the poster boards were augmented with thick paper. Hence, it was not possible to see the other rooms through the open space underneath the boards. Every constructed room had a large window to ensure enough natural light throughout the time of experimenting.

We used mainly real furniture as objects such as tables, bookshelves, or computers with tables. The kitchen and the study were furnished with six experimental items and two filler items (see Figure 1) to accommodate the constraints given by the two scripts. They were placed so that the positions of the items in the kitchen formed a mirror image of the positions of items in the study and vice versa. The living room was needed for the search task and to separate both script rooms. Three experimental items and three additional filler items were placed in the living room.

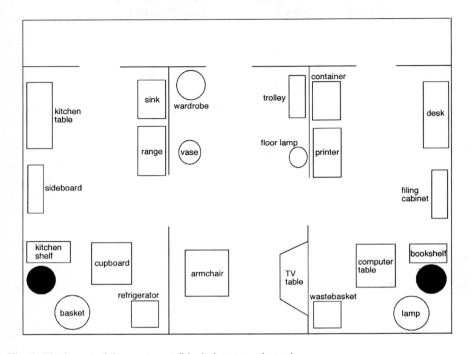

Fig. 1. The layout of the apartment (black dots are columns)

The positions of the objects were chosen such that these were a variety of distances for the distance estimation task and to balance the number of mentions of the objects in the script compared to the distances. Table 2 shows the relevant pairs of objects with their corresponding actual distances, the number of mentions, and the number of intervening objects in the script. The objects of the pair *bookshelf - desk*, for instance, have an actual distance of 2.26 m from each other, one of the objects is mentioned twice during the script-phase, and the number of objects that are mentioned in the script between these two items is one. Table 2 shows that there were 15 pairs of objects in each script room. Furthermore, pairs of diverse items from the other rooms were added (see Table 2 for description of pairs). Altogether there were 76 pairs of objects for the explicit distance estimation task.

For the implicit distance estimation task, we used additional rooms such as a bathroom and a bedroom for testing in order to conceal the relevance of the prior experience for the task at hand. More furniture was also needed for the new rooms as well as for the old rooms. In this way, we had a total of 92 pairs of objects for the implicit distance estimation.

Table 2. Pairs of objects from the script rooms used for distance estimation separated by distance categories (first object pair from the study, second object pair from the kitchen)

short distance	actual distance (cm)	mentions	interv. obj.
bookshelf - filing cabinet / kitchen shelf - sideboard	91	2 - 1	0/2
container - printer / sink - range	96	1 - 1	1/0
computer table - bookshelf / cupboard - kitchen shelf	116	1 - 2	1/3
filing cabinet - desk / sideboard - kitchen table	135	1 - 2	0/0
computer table - filing cabinet / cupboard - sideboard	168	1 - 1	3/0
medium distance			
printer - computer table / range - cupboard	208	1 - 1	0/0
container - desk / sink - kitchen table	209	1 - 2	5/3
printer - desk / range - kitchen table	221	1 - 2	3/2
printer - filing cabinet / range - sideboard	224	1 - 1	2/1
bookshelf - desk / kitchen shelf - kitchen table	226	2 - 2	1/0
long distance			
printer - bookshelf / range - kitchen shelf	260	1 - 2	1/4
container - filing cabinet / sink - sideboard	265	1 - 1	4/2
computer table - desk / cupboard - kitchen table	278	1 - 2	4/1
container - computer table / sink - cupboard	296	1 - 1	0/1
container - bookshelf / sink - kitchen shelf	326	1 - 2	2/5

Note: Number of mentions are identical for objects from both rooms, but number of intervening objects are mostly different.

Since the distance estimation was a ratio-estimation, a reference distance was necessary from which the respective ratio could be computed. This was the long and narrow hall of the apartment where either end was marked by a picture (Anna Freud and Sigmund Freud, respectively). The difference between both pictures was 543 cm.

Procedure. The experiment was divided in two main phases: The acquisition phase and the test phase. The test phase consisted of three memory tests: (1) implicit distance estimation, (2) explicit distance estimation, and (3) a description of the apartment while walking through the apartment. We will not report the data of the third test in this paper.

The *acquisition phase* started with the entrance of the participants through the door of the living room. The procedure was slightly different for the four groups of participants. First, the procedure for the experimental groups and the control groups differed in whether they heard a complete script or simply heard the names of the objects mentioned in the scripts. The control participants (*hearing* groups) were told that they had to walk through the apartment looking at the furniture. The experimenter named the same fixed objects in the same order as mentioned in the scripts. The experimental participants (*acting* groups) were told one action phrase at a time and subsequently carried out the described action as accurately as possible. Objects such as stamps, a teapot etc. were provided. The participants who heard or acted with the objects in the study (writing a letter) were called the *study* groups. The participants who heard or acted with the objects in the kitchen were called the *kitchen* groups. The study groups went immediately from the living room to the study to perform the letter writing script or to hear the names of the objects in the script. The kitchen groups went immediately to the kitchen to perform the tea preparing script or to hear the names of the objects of that script.

After the visit to either script room, all participants were guided into the living room to either start with the search task or to hear the names of the objects of the search task. The experimenter read every necessary action to them as it was read during the script-performing task. At the last location the search task was successful: the key was indeed found and was later replaced by the experimenter for the next participant. The separate procedures for each group were the following: The *acting-study* group first entered the study to perform the actions given by the script *writing a letter*. Then they proceeded to the living room in order to search for the key in the living room and in the kitchen. The *acting-kitchen* group first entered the kitchen to perform the actions given by the script *preparing tea*. Afterwards they walked into the living room to start the search task in the living room and in the study. The *hearing-study* group first visited the study to hear the names of the fixed objects in the same order that was given by the script *writing a letter*. Then they entered the living room to proceed with the names of the fixed objects of the search task for the living room and the kitchen. The *hearing-kitchen* group first visited the kitchen to hear the names of the fixed objects in the same order that was given by the script *preparing tea*. Afterwards they walked into the living room to proceed with the names of the fixed objects of the search task for the living room and the study. Note, that participants were not required to learn the spatial setting or the actions (incidental learning conditions).

Having completed the search task, the participants immediately went into the hall to walk the reference path between Anna and Sigmund Freud back and forth. Thereafter they were asked to enter the test room to estimate distances at a Macintosh G3 computer. The instructions for the distance estimation task were displayed on a 21" screen. The test phase started with a fixation cross for each pair. The reference path was displayed simultaneously with each pair whose inter-object distance had to be judged. The procedure for collecting distance estimates was similar to that used by Mecklenbräuker et. al. (1998): The participants had to indicate the appropriate distance between the two objects in relation to the reference distance at a horizontal line with a slash. Differences between the above mentioned procedure and ours were: (a) the task was presented on a computer, (b) the slash had to be indicated with a mouse click, and (c) the scale on which the distance had to be indicated was a long and thin rectangle. In addition to the scale, the current estimation was given as a percentage value of the reference distance. Each *explicit* estimation was followed by a judgement of certainty of the previously performed estimation on a five-point scale, with increasing numbers meaning increasing certainty.

This certainty measure was omitted during the *implicit* estimation task. The implicit task always preceded the explicit estimation. The participants were told to estimate distances between pieces of furniture that would usually happen to be in a common five-room apartment, including kitchen and bath. The participants were motivated for this task by the desire of the authors to learn about a normative representation of an apartment such as this one for future research. When the participants finished the implicit estimation task, the explicit estimation task followed immediately with a short instruction telling the participants to now remember the correct distances of the visited apartment.

In conclusion we had the following independent variables. Between subjects variables were: *type of acquisition* (acting, hearing) and *type of script* (writing a letter, preparing tea). Within subjects variables were: (1) *activities* performed in a certain room (script, search), (2) *room* (study, kitchen), (3) *distance* between pairs of objects, (4) *number of mentions* of the objects in the script (once or twice), and (5) *number of intervening objects* with respect to the script list. Distance between pairs of objects (3) was separated into two variables. Because we had 15 actual distances for each script room we categorized these distances into short, medium, and long distances as can be seen in Table 2. Thus the factor *distance* in ANOVAs refers to this categorization and has a control function. Regression analyses were computed with all *actual distances* in one room. The number of mentions (4) in a script is a variable with which the intensity of the participants' engagement is varied. If an object is mentioned twice it means that participants had to perform two actions there or heard the name twice. An object can be mentioned never, once, or twice. The maximum number of intervening objects that we mentioned in the script between two objects constituting one pair in the distance estimation task is five. That means, the hearing groups heard zero to five names of objects between any two objects for which they estimated the distances later on. The acting group performed actions at zero to five objects between any two objects for which they estimated distances. In a way, this resembles a kind of route distance, because participants had to walk from one object to the other. When the number of intervening objects increases the path they have to walk increases also. As dependent measures we had the distance estimates from the implicit estimation task together with the decision times, the estimates from the explicit estimation task together with the decision times and the certainty judgements.

3.1.2   Results

We will report the results of the different dependent measures separately. The results of the explicit distance estimation will be described first, followed by the implicit distance estimates. The significance level was set at $\alpha = .05$ for all statistical tests. As estimates of effect size we calculated partial R^2 values. They are based on total between-subjects variances for a between-subjects factor and on total within-subjects variances for a within-subjects factor.

Explicit Estimates. The computation of the decision times revealed no relevant results mainly because the standard deviation was usually high. The reason for this probably is that we recorded the mouse click on the *continue* button and not the last click for the estimate on the scale. The only interesting result was that the decision time was obviously slower for pairs of objects where at least one object was never mentioned. There was no difference between acting and hearing groups.

The next step was the computation of the estimates from the values in pixel to corresponding metrical values. We used the reference distance of 543 cm as 100% and converted the estimates according to the pixel value of the latter metrical value. This procedure resulted in distance values in cm for each pair of objects.

In a first analysis of variance (ANOVA), we submitted only the estimates for the critical pairs from the script rooms. To control the differences in actual distances, we computed the ratio of estimated distance to actual distance for each pair and participant. The ANOVA included as between factors the type of acquisition (acting vs. hearing) and the type of script (letter writing vs. preparing tea). As within factors the activities (script vs. search) and the distance (short, medium, long) were included. The only statistically reliable effect was the type of distance, $F(2,68) = 86.65$, $R^2 = .72$, with mean values of .84, .48 and .43 for short, medium and long distances, respectively. That means, only short distances were estimated close to the length of the actual distances. Medium and long distances were drastically underestimated. On further consideration, we noticed that this was especially true for the script groups in their corresponding script room. Nevertheless, none of these differences proved to be reliable.

Table 3. Correlations between real and estimated distances for the explicit distance estimation task, separated by group and type of room (s = significant, ns = not significant)

group	all rooms	pairs from study	pairs from kitchen
acting study	r = 0.67 s	r = 0.02 ns	r = 0.70 s
acting kitchen	r = 0.70 s	r = 0.58 s	r = 0.15 ns
hearing study	r = 0.73 s	r = 0.57 s	r = 0.61 s
hearing kitchen	r = 0.73 s	r = 0.33 ns	r = 0.40 ns

The next step was to compare the distance estimates with the real distances. The correlation coefficients separated by the four groups and the different rooms are shown in Table 3. As can be seen from this table, there was no correlation between real and estimated distances for the acting-study group for object pairs from the study. In contrast, we found a significant correlation for this group for objects from the kitchen, which served as a search room for this group. The hearing-study group (i.e., the control group) showed significant correlations between real and estimated distances for object pairs from the study and the kitchen. This result suggests that the difference between the script room and the search room observed in the acting-study group is due to the performance of the script-based actions. Moreover, a similar pattern of results was obtained for the other script (although the data were less clear for the control group, i.e., the hearing-kitchen group). For the acting-kitchen group we found a significant correlation between real and estimated distances for object pairs from the search room (i.e., the study), but not for object pairs from the script room (i.e., the kitchen).

The low correlation coefficients indicate an uncertainty of the estimation process. We computed the corresponding judgements of certainty for each of the four groups separately. First we converted the values from 1 to 5 to comparable values between 0 and 1, with 0 as completely uncertain and 1 as completely certain. Surprisingly, all participants judged their estimates to be rather certain. The means were always above .5 (see Figure 2). More surprisingly, there was a reliable interaction between the type of script and the corresponding script room, $F(1,34) = 26.06$, $R^2 = .43$. That means the study group had an increase in certainty for the pairs of objects from the study as compared to the kitchen and vice versa for the kitchen group. For the acting groups this was definitely not in accordance with their actual performance in estimating the distances of their corresponding script room.

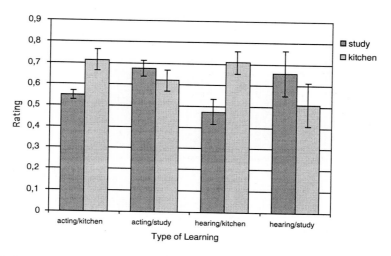

Fig. 2. Certainty ratings as a function of type of acquisition separated by rooms (error bars indicate the standard error of the mean)

A multiple regression analysis with the dependent measure being the estimates for the study pairs and the independent measures being the actual distance, the number of mentions and the number of intervening objects revealed an improvement of the correlation coefficient due to the number of mentions. This improvement could be observed both for the acting groups and for the hearing groups. Therefore, we computed an ANOVA only for the factor *number of mentions*. Figure 3 shows the mean estimates in cm for the object pairs from the study as a function of the number of mentions separated by groups of participants. Note, that the mean actual distance for object pairs with both objects being mentioned once and the mean actual distance for pairs with at least one object being mentioned twice did not differ. The influence of the number of mentions was statistically reliable, $F(1,34) = 14.39$, $R^2 = .30$, as was the interaction with the type of script, $F(1,34) = 4.69$, $R^2 = .12$. That means, the distance between objects with at least one object mentioned twice was estimated longer than the distance between objects mentioned only once. This was only true for both study groups. However, there was no difference between the acting and the hearing group. In fact, the difference between one and two mentions was even larger (not reliably) for the hearing group. The distance estimates for the object pairs from the kitchen yielded comparable results.

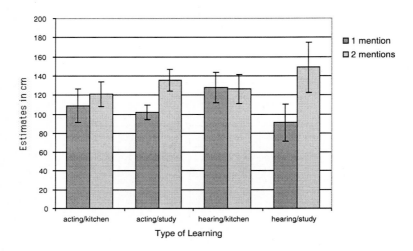

Fig. 3. Mean explicit estimates for object pairs from the study as a function of number of mentions in the script separated by type of acquisition and type of script (error bars indicate the standard error of the mean)

Implicit Estimates. The implicit distance estimates were analyzed in exactly the same way as the explicit estimates. It was expected that the actual distances have even less influence on the implicit estimates than they had on the explicit estimates. This was indeed the case with correlations between real and estimated distances from .008 to .24. However, a reliable influence could be shown in an ANOVA with the number of intervening objects (0-5) as the within factor and the type of acquisition (acting vs. hearing) and the type of script (letter writing vs. preparing tea) as the between factors. The type of acquisition revealed a statistically reliable influence on the implicit measures: The acting groups showed lower estimates as compared to the hearing groups.

Furthermore, the number of mentions in the script exerted a similar influence on the implicit distance estimates as on the explicit ones. Again the estimates increased when objects were mentioned twice during the script session.

3.1.3   Discussion

The most important result is the independence of distance estimates from the actual distances for the acting groups in their corresponding script rooms. Obviously, participants who had performed script-based activities judged all distances in the script room as almost identical. This result does not correspond to the findings from developmental studies conducted by Cohen and colleagues (see Part 1). They could show that performing activities that were functionally linked across locations facilitated the acquisition of spatial knowledge, that is, led to estimates closer related to actual distances. However, the Cohen et al. experiments and the present one differed in several aspects. For example, the participants in the Cohen et al. study were school-age children, whereas our participants were young adults (university students) with a stable kitchen and study schema. These schemata might have influenced the performance in the distance estimation task. Furthermore, whereas Cohen et al. used an empty classroom, our spatial setting – a simulated environment – was more realistic, and thus easily evoked the above-mentioned schemata. Perhaps more importantly, we employed an incidental learning procedure, that is, participants were not informed about the later spatial memory task and were not told to focus their attention on the spatial relations between the objects (i.e., the pieces of furniture). Presumably, most participants in the acting groups focused their attention on the performance of the activities, whereas at least some of the control participants might have suspected that they would later be tested on their spatial knowledge and focused their attention on the spatial relations between the objects. If this was the case, it is not surprising that the control participants were not less but in fact more precise in their distance estimates for the script rooms than the participants in the acting groups. In contrast, the children of all three groups (walk only, perform isolated activities, perform linked activities) in the Cohen et al. study were encouraged to pay attention to the distances among environmental objects. Furthermore, our script-based activities were more strongly linked and structured than the activities used in the Cohen et al. experiments. These strong links possibly influenced the subjective perception of closeness of the objects on which the activities were performed. Further differences concerned the concrete procedures used in the acquisition phase and in the distance estimation task. In the acquisition phase, the children in the linked activities group (and in the other two groups) had three walks (not one as in our experiment) and performed the activities during the second walk. That means, the children could focus their full attention on the spatial relations among the objects during the first and third walk. In the test phase, the children had to reconstruct the distances, whereas a ratio estimation procedure was used in the present experiment. In sum, there are numerous differences between the Cohen et al. study and our experiment that might be responsible for the differing results.

An important question of the present experiment was whether an implicit distance estimation task would lead to different results as compared to an explicit version of the same task. The implicit and the explicit tasks led to similar results. This finding suggests that spatial information is used automatically rather than consciously controlled.

At least three alternative interpretations might account for our main result, that is, the lack of a correlation between estimated and actual distances for those participants who acted in the script room. *First*, objects were linked together by performing activities and these links led to perceiving subjectively closer spatial connections and to a loss of spatial information. *Second*, performing activities did not lead to a loss of spatial information. Rather, when performing activities, participants focused their attention on acting and not on gathering spatial information. In other words, less spatial information was encoded. *Third*, the observed effects had nothing to do with the spatial representation of the real environment (as alternatives 1 and 2 suggest), but were an artifact of the spatial testing situation. When asked for distance estimates for pairs of objects at whose locations script-based activities had been performed, participants might have based their judgments – at least partially – on activated script information. That means, they reflected what activities they had performed with the objects of a given pair whose distance they had to judge.

If alternatives 1 or 2 turn out to be true, then a similar pattern of results should be demonstrated with a different task for assessing spatial knowledge. Therefore, in Experiment 2, we used a positioning task. Because this spatial task requires participants to locate single objects in a free order, it is unlikely that script information activated by the activities is relevant to the testing situation. In other words, if our results can be attributed to the specific spatial representation constructed in the acquisition phase, then Experiment 2 should lead to a similar pattern of results as Experiment 1. On the contrary, if our results are specific for the spatial task used in Experiment 1 (alternative 3), a different pattern of results should be observed in Experiment 2.

3.2   Experiment 2: Reconstructing Configurations from Positioning Data

The second experiment is a replication of the first one, with the exception of the dependent measure or task for the participants. As mentioned above the estimation of distances is an abstract measure that singles out one aspect of a spatial configuration. As a more holistic approach the complete reconstruction of the learned spatial lay-out seems to be more appropriate. Two problems have to be taken into consideration here.

First, it has to be ensured that personal skills such as the ability to draw two-dimensional maps would not distort the results. As Evans (1980) stated, in spatial cognition research, the possibility of receiving biased results is quite high. A drawing test gives the participants more freedom, because objects can be designed differently. Nevertheless, most people differ very much in drawing objects correctly on a map. Therefore, we decided to use a computer-aided positioning task. That is, the objects and their size were given and displayed on the computer screen. They had to be positioned in a two-dimensional map of the learned spatial layout through a drag and drop procedure via the mouse.

Second, the process of complete reconstruction always involves more stimuli than single measures such as distance estimates can provide. In displaying the map with the increasing number of objects that were already positioned by the participants more and more spatial information is gained. Hence, using holistic configurational measures is usually much easier and yields more correct results than using single measures. The participants of our first experiment underestimated distances drastically and showed poor performance, especially when acting in the script room before estimating distances. Therefore the holistic approach seems to be more appropriate for answering the question whether the poor performance of the acting groups was due to an impoverished spatial representation caused by a loss of or by lesser encoding of external information (see the Discussion in 3.1.3).

3.2.1   Method

Participants. As in Experiment 1 the participants were mainly students of the University of Trier. None of the participants from Experiment 1 took part in Experiment 2. The 23 women and 18 men completed this experiment in about the same time as in the first experiment. There were 12 participants in each of the two acting groups. In the hearing groups there were nine participants in the kitchen group, and eight participants in the study group. They were paid for their participation.

Material. The same spatial setting was used for this experiment. This was also true for the script actions and the search task. For the dependent measure we constructed a positioning task at the same Macintosh Computer as we used in Experiment 1. For this task a two-dimensional map of the apartment was drawn with the locations missing but with windows and doors included. The same objects that were used for the distance estimation task in the first experiment were then displayed as abstract two-dimensional sketches in a comparable size to the scale of the map of the apartment. For the implicit positioning task two rooms – the bathroom and the bedroom – were added to the apartment. The furniture that had to be positioned in these additional rooms was the same as the additional furniture, which was used in the implicit distance estimation task in the first experiment.

Procedure. As in our first experiment the participants attended two different phases of the experiment. The first phase, the incidental acquisition of the environment, was identical with Experiment 1. The participants performed the same script-based activities in the same sequence (acting groups), or heard the same object names (hearing groups) as in Experiment 1. The second phase, the testing phase, differed in as much as the task was more complex and involved every important object at the same time. After walking the same reference path, participants immediately entered the test room to complete the positioning task on the computer with a 21" monitor. First, the implicit positioning task was employed. Participants were asked to arrange objects on a floor plan of an apartment with five rooms. They were told to make use of their knowledge of common arrangements of furniture in a comparable apartment. The floor plan was divided into two units: First, the two additional rooms were displayed one after the other. After completion of the positioning task for each of these two rooms the three experimental rooms were shown simultaneously. Participants were instructed to use at least eight objects per room from the set of ten objects, which were provided at the left and right hand side of the map on the screen. For the three experimental rooms all old objects were shown simultaneously with six new objects. These new objects were the same as those used in the implicit distance estimation task.

Basically, the subsequent explicit positioning task was the same as the implicit task. The instruction, however, stated clearly that this time participants should remember the exact position of the objects in the apartment they had acted in or where they had heard the names of the objects. The additional two rooms were omitted and only the second unit of the floor plans and the old objects were displayed. As in the first experiment the explicit task always followed the implicit task.

For both tasks participants were able to move the objects freely via the mouse (drag and drop). They had the possibility of turning objects clockwise in 90° steps when pushing the mouse button together with the control key. Hence, participants were able to adjust the position of elongated objects to the walls if necessary. They were allowed to adjust the positions as often as they found it necessary. When they were satisfied with the configuration they pressed the space key. The movements of the mouse, the time for each movement, and each final position were recorded. Certainty measures were not collected.

3.2.2   Results

The data were analyzed according to three different aspects of how well participants reconstructed the original configuration of the apartment: (1) percentage of objects that were placed in the correct room; (2) absolute distance between studied and reconstructed object location; (3) relative accuracy of the reconstruction as a whole. Finally, we will present the results of an analysis that matches the analysis of the distance estimates in Experiment 1. That is, the distances between critical pairs of the reconstructed apartment were compared and analyzed. Data from the implicit positioning task will not be reported in this paper. The resulting configurations were highly inhomogeneous and needed a special strategy of analysis that would not match with our line of analysis reported here. The level of significance for the results was again set at $\alpha = .05$.

Placement of Objects in the Correct Room. The percentage of objects that were placed in the correct room was computed. Before analyzing the data, we calculated two error terms for each reconstruction, because some participants had rotated the apartment about 180 degrees. First, distances (in cm) between target positions of the original and the reconstructed configuration were computed for each object. Second, reconstructed positions were compared to a rotated version of the original configuration, that is, target positions were recoded relative to maximum values of the coordinates of the apartment. A participant's reconstruction was considered rotated if the mean of the second error term was smaller than the mean of the first error term. This was true for five participants (11.9%). Because only an entire 180-degree rotation of the configuration as a whole would have preserved inter-object relations, we decided not to correct other forms of distortions.

Most of the objects were placed in the correct rooms (M = .89). Means were virtually identical in the different conditions. This was confirmed in a 2 (type of acquisition: acting vs. hearing) x 2 (activities: script vs. search) ANOVA, with type of acquisition serving as a between-subjects variable and activities as a within-subjects variable that was performed on the percentage of correctly placed objects. The ANOVA yielded no statistically reliable effects. Thus, neither was there an influence due to the type of acquisition (with means of .88 for the *acting-group* and .89 for the *hearing-group*), nor was there a reliable interaction.

With respect to the high level of performance, we cannot fully neglect the possibility of the lack of significant effects being due to ceiling effects. Moreover, the dependent measure used in the analysis described here may be insensitive to specific spatial knowledge.

Distance between Studied and Reconstructed Object Location. To determine absolute relocation errors, distances (in cm) between target positions of the original (or rotated) and the reconstructed configuration were computed for each object. In an initial analysis, positioning data from all critical objects (i.e., a maximum of six objects each in the kitchen and in the study) were included in the computation of relocation errors, irrespective of whether or not objects had been placed in the correct room. To minimize distortions of relocation errors that would result from drastically misplaced objects, group means were calculated on the basis of each participant's median for each room. An ANOVA that included the type of script (preparing tea vs. writing a letter), the room participants acted or heard the script in (kitchen vs. study), and number of mentions of objects in the encoding list (once vs. twice) showed no significant effects concerning the type of script. Hence, we conducted an ANOVA where data from both script groups were collapsed. This analysis with one between-subjects variable (type of acquisition: acting vs. hearing) and one within-subjects variable (activities: script vs. search) yielded only a marginally significant result: Objects that were placed in the script room (M = 112.66 cm) were found to be estimated slightly closer to their original position than the objects in the search room (M = 136.94 cm). There was no reliable difference between types of acquisition, and the interaction did not reach significance.

Next, we computed relative relocation errors, that is, only positioning data from critical objects that had been placed in the correct room were included in this analysis. Figure 4 shows the mean distances between target positions and reconstructed positions that were based on each participant's median for each activity. Again, data were submitted to a 2 x 2 ANOVA. The interaction, including both variables (type of acquisition and activities) was the only marginally significant result, $F(1,37) = 3.25$, $R^2 = .08$, $p < .10$. Further analyses indicated that, for the *hearing-group*, the performance for objects they had heard in the search rooms was poorer than the performance for objects they had previously heard in the script room, $F(1,37) = 5.37$, $R^2 = .13$. However, there was no effect with regard to the *acting-group* and no significant difference between the two groups, either for the room with the script activity, or for the room with the search activity.

Relative Accuracy of the Reconstruction. It is possible that participants' reconstructed configurations contain systematic transformations, in which the absolute positions differ from the target locations but the relative positions are preserved. A translation, a shift of all objects along the axis of a room, is an example for such a transformation.

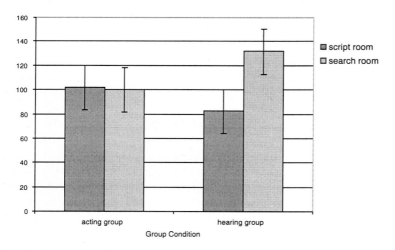

Fig. 4. Relative relocation errors in cm as a function of type of acquisition separated by activity (script vs. search) (error bars indicate the standard error of the mean)

To test whether the reconstructed configurations preserved relative relations regardless of absolute positions, we analyzed the angular relations in the configurations. We invented straight lines between all 15 pairs of the six critical objects in one of both script rooms (kitchen and study). Then we compared the directions of these lines in the reconstructed configuration for each participant to the direction of the corresponding lines in the target configuration and calculated the angle between these two lines. The direction of the deviation was not considered. This angle measures how much the angular relation in the reconstructed configuration deviates from the corresponding relation in the target configuration. The median of these 15 angles was used as the overall measure of relative accuracy.

A three-way ANOVA with two between-subjects variables (type of acquisition: acting vs. hearing; type of script: preparing tea vs. writing a letter) and one within-subject variable (room: kitchen vs. study) yielded the following significant results: There was an interaction between type of script and room, $F(1,33) = 15.68$, $R^2 = .32$. The reconstructed configurations of the room in which the participants performed actions or heard names of objects given by a script were less accurate than the reconstructed configurations of the room in which they searched for an object or heard names of objects given by the search task. Both study groups reconstructed the kitchen more accurately than the study. Both kitchen groups reconstructed the study better than the kitchen. Figure 5 illustrates this interaction.

Furthermore, there was a main effect of the type of script, $F(1,33) = 4.83$, $R^2 = .13$, indicating that the reconstruction of the kitchen groups ($M = 32.2$) was more accurate than the reconstruction of the study groups ($M = 59.2$).

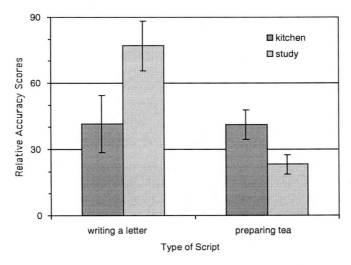

Fig. 5. Relative accuracy scores as a function of type of script and room

Distance between Object Pairs. In a separate analysis we addressed the important question whether distances between specific pairs of objects that are inferred from the reconstructed configuration from the positioning task in Experiment 2, differ from distance estimates in the estimation task in Experiment 1. Note that direct comparisons of separate experiments are problematic. Hence, we analyzed the inferred distances in Experiment 2 separately, but in the same way as in Experiment 1. Distances between the 30 critical pairs of objects (15 pairs of objects from each of the two script rooms) were computed. Subsequently, we calculated the ratio between actual distance and computed distance for each pair as the dependent measure. These ratio values were then submitted to an ANOVA with the between factors "type of acquisition" (acting, hearing) and "type of script" (preparing tea, letter writing) and the within factors "activities" (script, search) and "distances" (short, medium, long).

As in Experiment 1 the factor distance revealed a statistically reliable effect with mean values of 1.91, 1.07, and .98 for short, medium, and long distances, respectively. An interaction between type of script and activities emerged, showing that the mean values for the study groups (writing a letter) were higher in their corresponding script room, that is, the study (1.45) than in the search room (1.18), whereas the kitchen groups (preparing tea) showed lower values in their script room (1.27) compared to the search room (1.38).

Furthermore, no effect of the type of acquisition nor any interaction with this factor proved to be significant. That is, there was no reliable difference, regardless of whether participants performed script activities (1.41) or heard the names of the objects (1.29).

3.2.3   Discussion

In general, the results of this experiment did not help to clarify those of the first experiment. The positioning task did not show any evidence of an action-based influence on the representation of spatial information. For the percentage of objects that were placed in the correct room, the results showed a high level of accuracy and the non-significant effects could be due to ceiling effects. For the relocation error scores, however, effects of ceiling were not relevant. Nevertheless, no clear trend emerged showing a clear-cut difference between the acting and the hearing groups restricted to the critical room (i.e., the script room). The analysis of the relative accuracy of the reconstruction (the configurational measure) showed that all participants demonstrated better spatial performance in those cases where less linked activities had to be performed (e.g., the kitchen for the acting-study group) or where fewer objects had been named in the acquisition phase (e.g., the kitchen for the hearing study group). This pattern of findings indicate that being engaged in an activity in the environment may distract participants from a systematic encoding of relevant configurational spatial information. However, this encoding deficit was not specific to the performance of activities in the environment and may reflect a general distraction concerning configurational spatial information. Participants may have focused their attention on single locations and have paid less attention to relational spatial information.

Moreover, the positioning task was used to calculate distances between specific pairs of objects that were inferred from the reconstructed configuration. As compared with the results of the first experiment, we could not detect any specific deficit in the acting groups in the reconstructed distance estimates. Thus, the independence of distance estimates from the actual distances as found in the first experiment for the acting groups in their corresponding script rooms was not replicated. In general, the observed effects within the distance estimation task do not seem to reflect consequences of a general spatial representation but may be due to the test situation as suggested in the discussion of the first experiment.

4   Concluding Remarks

The present study showed little evidence that performing activities can affect the use of spatial information in a real environment. In a distance estimation task (Experiment 1), participants who performed activities showed poor spatial knowledge of the room in which they acted in, disregarding actual distances in their estimates. This deficit could be due to a loss of spatial information or to a poorer encoding of spatial information. But results of Experiment 2 do not corroborate these interpretations. With a positioning task at testing, we reconstructed the inter-object distances and did not find any influence that could be attributed to an action-based change of a spatial mental representation. Therefore, the effects observed in Experiment 1 seem to be a result of the spatial task used. When given pairs of objects in a distance estimation task, participants possibly base their judgments – at least in part – on the activities performed at these objects. Contrary to the first experiment, the second experiment showed that each kind of activity – performing linked activities or only walking through the room and hearing the names of the corresponding objects – had a detrimental effect on spatial performance when a configurational measure was used. This general distraction effect is presumably due to the fact that only few relevant spatial relational information was encoded. Moreover, it might indicate that the encoding of at least some characteristics of the spatial environment require controlled, resource-dependent processing operations.

In sum, the overall pattern of results suggests that information about actions is not fully integrated in the spatial representation. Possibly, performing activities only once does not suffice for a full integration. Our results indicate that not fully integrated and even separately represented spatial and action information are stored. If the action information is reactivated in the spatial testing phase – as it is presumably the case when distance estimates are required – this reactivation may overshadow the use of stored spatial information and may lead to a leveling in the estimation of actual distances. In contrast, if the action information is not reactivated in a positioning task (or is activated to a lesser extent), specific action effects cannot be observed. These conclusions are preliminary and should be tested more strongly. For example, an explicit reactivation of the action information in the testing phase should have no effects on a distance estimation task, whereas for a positioning task, effects as in Experiment 1 should be expected. So far, however, the general hypothesis of action-based influences on mental spatial *representations* is still found to be wanting and has not been corroborated by convincing data.

Acknowledgments

We would like to thank Martina Lück, Johannes Rakoczy, Volker Schmidt, Stefanie Ahlke, and Kai Lotze for their assistance with data collection, and an anonymous reviewer for very helpful comments on an earlier version of this paper.

References

Anooshian, L.J., & Seibert, P.S. (1996). Diversity with spatial cognition: Memory processes underlying place recognition. *Applied Cognitive Psychology, 10*, 281-299.

Cohen, R., Cohen, S.L., & Cohen B. (1988). The role of functional activity for children's spatial representations of large-scale environments with barriers. *Merrill-Palmer Quarterly, 34*, 115-129.

Cohen, S.L., & Cohen, R. (1982). Distance estimates of children as a function of type of activity in the environment. *Child Development, 53*, 834-837.

Cornell, E.H., Heth, C.D., & Alberts, D.M. (1994). Place recognition and way finding by children and adults. *Memory & Cognition, 22,* 633-643.

Evans, G.W. (1980). Environmental Cognition. *Psychological Bulletin, 88,* 259-287.

Herrmann, T. (1993). Mentale Repräsentation - ein erläuterungsbedürftiger Begriff. In J. Engelkamp & T. Pechmann (Hrsg.), *Mentale Repräsentation* (S. 17-30). Bern: Huber.

Hirtle, S.C., & Mascolo, M.F. (1986). Effect of semantic clustering on the memory of spatial locations. *Journal of Experimental Psychology: Learning, Memory, and Cognition, 12,* 182-189.

Hue, C.-W., & Erikson, J.R., (1991). Normative studies of sequence strength and scene structure of 30 scripts. *American Journal of Psychology, 104,* 229-240.

Mandler, J. M. & Murphy, C. M. (1983), Subjective estimates of script structure. *Journal of Experimental Psychology: Learning, Memory, and Cognition, 9,* 534-543.

McNamara, T.P., & LeSueur, L.L. (1989). Mental representations of spatial and nonspatial relations. *Quarterly Journal of Experimental Psychology, 41A,* 215-233.

McNamara, T.P., Halpin, J.A., & Hardy, J.K. (1992). The representation and integration in memory of spatial and nonspatial information. *Memory & Cognition, 20,* 519-532.

Mecklenbräuker, S., Wippich, W., Wagener, M., & Saathoff, J.E. (1998). Spatial information and actions. In D. Freksa, C. Habel, & K.F. Wender (Eds.), *Spatial cognition: An interdisciplinary approach to representing and processing spatial knowledge* (pp. 39-61). Berlin: Springer.

Nakamura, G.V., Graesser, A.C., Zimmerman, J.A., & Riha, J. (1985). Script processing in a natural situation. *Memory & Cognition, 13,* 140-144.

Rothkegel, R., Wender, K.F., & Schumacher, S. (1998). Judging spatial relations from memory. In D. Freksa, C. Habel, & K.F. Wender (Eds.), *Spatial cognition: An interdisciplinary approach to representing and processing spatial knowledge* (pp. 79-105). Berlin: Springer.

Sadalla, E.K., Staplin, L.J., & Burroughs, W.J. (1979). Retrieval processes in distance cognition. *Memory & Cognition, 7,* 291-296.

Schank, R.C., & Abelson, R.P. (1977). *Scripts, plans, goals and understanding.* Hillsdale: Erlbaum.

Siegel, A.W. (1981). The externalization of cognitive maps by children and adults: In search of ways to ask better questions. In L.S. Liben, A.H. Patterson, & N. Newcombe (Eds.), *Spatial representation and behavior across the life span* (pp. 167-194). New York: Academic Press.

Smith, D.A., & Graesser, A.C. (1981). Memory for actions in scripted activities as a function of typicality, retention interval, and retrieval task. *Memory & Cognition, 9,* 550-559.

Tack, W.H. (1995). Repräsentation menschlichen Wissens. In D. Dörner & E. van der Meer (Hrsg.), *Das Gedächtnis. Probleme - Tends - Perspektiven* (S. 53-74). Göttingen: Hogrefe.

Wender, K.F., Wagener, M., & Rothkegel, R. (1997). Measures of spatial memory and routes of learning. *Psychological Research, 59,* 269-278.

Action Related Determinants of Spatial Coding in Perception and Memory

Bernhard Hommel[1,2] & Lothar Knuf[2]

[1]Unit of Experimental and Theoretical Psychology University of Leiden, P.O. Box 9555, 2300 RB LEIDEN, The Netherlands
[2]Max Planck Institute for Psychological Research, Amalienstrasse 33, D-80799 Munich, Germany
`hommel@fsw.leidenuniv.nl; knuf@mpipf-muenchen.mpg.de`

Abstract. Cognitive representations of spatial object relations are known to be affected by spatial as well as nonspatial characteristics of the stimulus configuration. We review findings from our lab suggesting that at least part of the effects of nonspatial factors originate already in perception and, hence, reflect principles of perceptual rather than memory organization. Moreover, we present evidence that action related factors can also affect the organization of spatial information in perception and memory. A theoretical account of these effects is proposed, which assumes that cognitive object representations integrate spatial and nonspatial stimulus information as well as information about object related actions.

1   Introduction

Our environment consists of objects and relations between objects. However, the term 'object' is relative in multiple respect. Most objects consist of more than one level and each level might have its own object character. For example, cities consist of houses and streets, houses consist of rooms, rooms of ceiling, floor and walls, etc. What might reasonably be termed 'object' from a certain perspective might just as reasonably be termed element of an object or group of objects from another. Accordingly, spatial relations can be of global or local nature, depending on the context and point of view: A given room can be to the left or right of another room in the same house, to a house in the same street, to a street in the same city, and so forth. Human perception, cognition, and action can be directed to objects as defined on any of these different levels, which raises the question of how spatial layouts are represented in the human cognitive system. To approach this issue, several authors have proposed that complex spatial layouts are represented in a hierarchical fashion (see, e.g., McNamara, Hardy, & Hirtle, 1989; Navon, 1977; Palmer, 1977). The main assumption is that object representations, which themselves might already be the result of a cognitive clustering of subordinate objects and features, are combined into object clusters with a certain spatial intra-object structure on a superordinate level (Palmer, 1977; Watt, 1988).

The idea of hierarchical representation or, more liberally, of cognitive object clusters raises the question of the criteria according to which perceptual and memory

Ch. Freksa et al. (Eds.): Spatial Cognition II, LNAI 1849, pp. 387-398, 2000.

information about spatial layouts is structured and organized. In this article, we focus on two types of criteria. On the one hand, exogenous factors, such as similarities between the color and shape of objects, may suggest the integration of similar objects into a common cognitive group or category. On the other hand, we will argue that endogenous factors, such as knowledge about nonperceivable characteristics of objects or the actions they afford, may also contribute to the cognitive organization of object information. Indeed, as we will discuss in turn, there is evidence for the impact of both exogenous and endogenous factors on the cognitive clustering of spatial layouts.

2   The Impact of Exogenous Factors on Spatial Coding: Color and Shape

In the literature there are several indications that spatial memories are influenced by nonspatial attributes of, or relations between, the respective objects, such as linguistic (Bower, Karlin, & Dueck, 1975; Daniel, 1972), semantic or episodic (Hirtle, & Mascolo, 1986; McNamara & LeSueur, 1989; Sadalla, Staplin, & Burroughs, 1979), and functional (McNamara, Halpin, & Hardy, 1992) information. For example, in experiments reported by Hirtle and Mascolo (1986), participants memorized maps in which place names fell into two semantic cluster: names of recreational facilities (e.g., Golf Course or Dock) and names of city buildings (e.g., Post Office or Bank). Locations were arranged in such a way that, although places belonging to the same semantic cluster were spatially grouped on the map, the Euclidean distance of one recreational facility was shorter to the cluster of the city buildings than to any other recreational facility, and vice versa. However, when subjects were asked to estimate inter-object distances on the basis of memory information, they showed a clear tendency to (mis)locate these critical places closer to their fellow category members then to members of the other cluster. These findings suggest that nonspatial information is automatically encoded and integrated into, or together with, spatial representations to a degree that sometimes even alters or distorts the original information in systematic ways (for an overview see McNamara, 1991; Tversky, 1981). Such findings have been used to question the possibility that an adequate model of a spatial representation can be built around a strictly Euclidean conception (e.g., Hirtle & Jonides, 1985; McNamara et al., 1984). As an alternative model, several authors have therefore proposed that spatial areas are arranged hierarchically, so that judgements across clusters require knowledge of spatial arrangements within each cluster plus knowledge of the spatial arrangement of the superordinate structures (e.g., McNamara, 1986; Stevens & Coupe, 1978).

What kind of processes might be responsible for such kind of cognitive organization? Obvious candidates are memory processes which may aim at reducing the perceptual input to minimize storage costs and optimize later retrieval – an idea that is implicit or explicit in most studies on distortions of spatial memories (e.g., Tversky, 1981). However, hierarchical coding could also be a result of perceptual processes, which may not passively register sensory evidence, but actively integrate the available information into a structured whole or *Gestalt* (Baylis & Driver, 1993; Navon, 1977; for the difference between perception and memory, see Werner &

Schmidt, this volume). If so, the distortions and clustering effects observed so far in memory tasks may not so much reflect organizational principles of memory processes, but rather be a more or less direct consequence of distortions and clustering tendencies in perception.

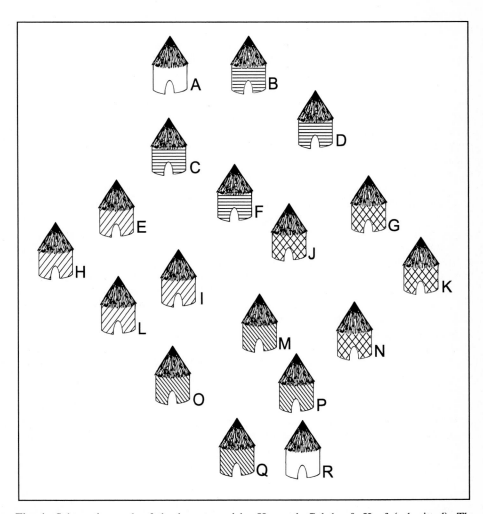

Fig. 1: Schematic graph of the layout used by Hommel, Gehrke & Knuf (submitted). The letters indicating the locations were not shown to the subjects; instead each house was identified by a nonsense "name" (i.e., a meaningless syllable like "DUS", omitted here) appearing in its center. Color and shape (indicated here by texture) were systematically varied (see text).

In a recent series of experiments we found support for a perceptual interpretation of clustering effects in spatial memory (Gehrke & Hommel, 1998; Hommel, Gehrke & Knuf, submitted). We presented our subjects with visual map-like configurations of up to 18 objects, which looked like houses of an imaginary city (see Figure 1). Color or shape served as exogenous factors that was assumed to induce cognitive clustering. That is, either the coloring or the shape of the objects was chosen in such a way that the configuration could be subdivided into three or four perceptual clusters (e.g., houses at locations B, C, D, and F were red, houses at locations E, H, I, and L green, etc.). To test if color and shape affected the coding of the spatial information between the objects, participants were asked to perform two spatial tasks, an unspeeded estimation of Euclidean distances — a common measure in memory experiments — and the speeded verification of sentences describing spatial relations (e.g., "is house A above house B?") — a task often used in perceptual experiments. Participants performed under three conditions in three consecutive sessions: In a *perceptual* session, the configuration was constantly visible; in a *memory* session, participants first memorized the configuration and then performed the task without seeing it. The third *perceptual/memory* session was identical to the first one, hence the configuration was again visible while the two tasks were performed. We expected all tasks to reveal the same pattern of results: objectively identical Euclidean distances between two given objects should be estimated smaller (the verification of spatial relations should proceed more quickly) when both objects were elements of the same (identical color or shape) than of different visual groups.

As expected, the time it took to verify the spatial relation between the members of a pair was shorter when both members shared a perceptual feature and, hence, belonged to the same "perceptual group" (e.g., relation E-L was judged faster than I-O). Interestingly, this was true for all three sessions and the size of clustering effects did not increase if the judgments were made from memory. This suggests that memory processes do not determine, and perhaps do not even contribute to the coding of spatial information. Apparently, the spatial layout was coded in a way that led to the integration of information about perceptually similar objects into common cognitive clusters, so that later access to information about objects belonging to the same cluster was easier.

Surprisingly, distance estimations were not affected at all by the feature-overlap manipulation. On the one hand, this may have been due to a lower sensitivity of estimation measures as compared to reaction time measures – a theoretically uninteresting possibility. On the other hand, it is conceivable that these two measures tap into different processes. For instance, it may be that reaction time measures are sensitive to the cognitive organization of spatial (and nonspatial) information, whereas estimation measures are sensitive to the quality of spatial information (veridical versus biased spatial memories; see McNamara, 1991, for a broader discussion of this issue). If so, our findings might indicate that nonspatial factors do not really alter the quality of spatial memories, although they can affect the way knowledge about spatial relations is organized and the likelihood that these relations are encoded (see McNamara & LeSueur, 1989). At this point, the evidence available does not allow us to distinguish between these alternative interpretations; yet, it is interesting to note that evidence of cognitive clustering can be obtained in the absence of qualitative distortions and, hence the two measures (if not the processes they measure) can be dissociated. At any rate, we consistently obtained strong effects of exogenous variables on reaction times, which is the measure we will focus on in the following.

3   The Coding of Perceptual Events and Action Plans

The findings reported so far suggest that objects forming a spatial array are cognitively represented by clusters of object codes that facilitate intra-cluster as compared to inter-cluster processing. In our previous studies, we used exogenous factors to induce cognitive clustering, that is, manipulations of perceptual object features. Before we continue discuss evidence of the impact of endogenous, action related factors on spatial coding, it is instructive to consider how cognitive clustering might work in detail.

According to the *Theory of Event Coding* (TEC) proposed by Hommel, Müsseler, Aschersleben, and Prinz (1999), perceived objects are cognitively coded in two steps. First, the features of the respective objects are registered and the corresponding feature codes activated. Second, the distributed feature codes are integrated into a coherent event representation. Figure 2A sketches how this might work in tasks like those employed by Hommel and colleagues (Gehrke & Hommel, 1998; Hommel et al., submitted)[1]. Assume an array consisting of three equally-spaced objects is presented, say, three huts named DUS, FAY, and MOB, from left to right, with DUS and FAY appearing in red and MOB in green. Let us further assume that the task is to verify statements regarding the relative horizontal location, such as "is DUS left from FAY?" (yes/no). To solve such a task, especially when the display is no longer visible, requires the integration of at least two features or attributes: the name and the (absolute or relative) location of each object. However, if these two features would be the only ones involved, it would be hard to understand why the verification performance of our subjects was affected by color and shape manipulations. Therefore, it seems reasonable to assume that color and shape, as well as other features of the objects, also became a part of the object representations, just as depicted in Figure 2A.

If so, the representations of DUS and FAY would be associated through the use of the same color code (red), and all three representations would be associated by a common shape code (hut). Consider, for instance, what happens when the retrieval cue "DUS" is presented (as in the question "is DUS left from FAY?"). This cue would access the object representation of DUS, so that the corresponding spatial location can be determined. However, given that the codes "red" and "hut" are part of this representation, activation is spread to the other object representations with which those codes are shared. This means that accessing the representation of DUS also

---

[1] At this point, our findings do not necessarily require the assumption of different representational levels or hierarchical coding, which is why we chose to present a nonhierarchical account. This is not to deny that more complex spatial arrays or maps can be or actually are coded in a hierarchical fashion, and it would be easy to incooperate hierarchies into a TEC-type model. However, our main point here is that nonspatial information is integrated into object representations, and that this effectively links the representations sharing some of this information—an issue that we feel is independent from the hierarchy issue.

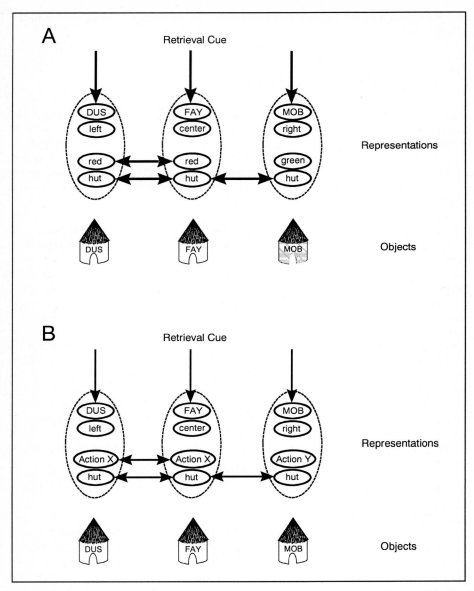

Fig. 2: A simplified model of the impact of nonspatial information on the verification of spatial propositions. *A*. The red stimulus huts DUS and FAY, and the green hut MOB are cognitively represented by integrated object features referring to the object's name, location, color, and shape. Horizontal arrows indicate the use of common feature codes and the mutual priming produced by that. *B*. The three huts are represented by codes referring to the name, location, and shape of the stimuli, as well as information about associated actions.

leads to a strong activation of the representation of FAY (via both color and shape links) and to a weaker activation of the representation of MOB (via the shape link only). Accordingly, if then the representation of FAY is accessed, it is already primed, which facilitates the retrieval of (e.g., spatial) information from this representation as compared to retrieval from the less strongly primed representation of MOB. Hence the better performance with greater feature overlap.

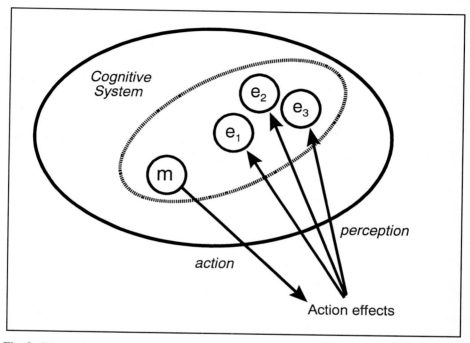

Fig. 3: Schematic graph of the cognitive representation of action through action concepts, i.e., integrated structures of motor patterns, represented by m, and action-effect codes, represented by e_1, e_2, and e_3.

A further interesting assumption of TEC and the related *Action-Concept Model* (ACM) of Hommel (1997, 1998) is that perceived and produced events (i.e., "perceptions" and "actions") are cognitively coded in the same way, namely in terms of the perceivable features they produce. The key idea is that actions are performed to produce perceivable outcomes—action effects. However, this means that the actor has to have acquired some knowledge about which action is likely to produce which outcome. How it is acquired is sketched in Figure 3. First, a motor pattern m is activated (intentionally or unintentionally), which leads to some kind of movement, that again produces perceivable action effects. For instance, pressing a particular key of the left side of one's body will be represented by codes referring to the spatial and nonspatial features of the key being pressed (including possible further consequences of keypressing, such as the consequent launch of a rocket) and the finger pressing it, as delivered through the available sensory channels. These effects are perceived and internally coded through the activation of codes representing the features of the action effects, say, codes e_1, e_2, and e_3. As the activation of m and e_1, e_2, and e_3 will overlap in time—a situation that is known to support learning—they will become associated, thus forming what Hommel (1997) called an *action concept*. Action concepts can be assumed to represent the building blocks of intentional action, because they provide the necessary precondition for goal-directed action planning: Once motor patterns and outcome representations are associated, activating the intended action effect is sufficient to select and implement the required motor pattern.

An important implication of this view on the cognitive representation of action is that there might be no qualitative difference between feature codes representing perceived events and feature codes representing actions or action plans. If so, there is no reason why codes of object related actions should not become part of object representations, which in the present context offers an interesting hypothesis. Assume that our three huts DUS, FAY, and MOB all have the same shape and color, as sketched in Figure 2B. However, whereas DUS and FAY are associated with some action X, MOB is associated with another action Y. If actions are represented in terms of their features, this situation should be very similar to that with the green and the two red huts. The representations of DUS and FAY are more strongly connected with each other through the common use of Action X features (whatever these may be) than with the representation of MOB. Accordingly, spatial relations between DUS and FAY should be easier to judge than, say, between FAY and MOB. Whether this is so, we attempted to find out in another series of experiments, which we will now review.

4   The Impact of Endogenous Factors on Spatial Coding: Simple and Complex Action

The TEC framework offers a number of arguments why endogenous, action related factors might play a similarly important role in spatial coding as has been shown for exogenous factors. Obviously, we should be able to test this hypothesis with the same experimental technique and under comparable conditions as for the exogenous studies, which is what we did (Hommel, Knuf, & Gehrke, submitted). The spatial arrangement of the stimulus configuration was the same as before (see Figure 1), the only difference was that a homogenous stimulus set was used (black on white drawings of 18 identically shaped houses).

In a first session, subjects were to perform simple keypressing actions to form different cognitive clusters. In each trial, one of the houses would flash and the subject would press one of three or four response keys. The mapping of keys upon house locations followed the same logic as in our exogenous studies. For instance, one key was to be pressed in response to locations B, C, D, and F, another key in response to locations E, H, I, and L, and so forth. The mapping was fixed but not communicated to the participants, so that they had to find out the correct mapping by trial and error. Once they produced correct consecutive responses to all locations, this first mapping-acquisition phase ended. The remaining sessions were set up as in the exogenous studies, that is, the first was a *perceptual* session with the configuration being visible, a *memory* session with judgments based on the memorized configuration, and a final *perceptual/memory* session, where the configuration was again visible. Although we sometimes also collected distance estimations (which again produced only null effects), our main measure was reaction time for the verification of spatial relationships.

In a first experiment, the location-response mapping had no effect, that is, pairs mapped onto the same response were not judged faster than other pairs. However, there was some indication of an effect in the first (perceptual) session, but not in the following sessions. One reason for that may be the decay of mapping information. As the perceptual session immediately followed the mapping-acquisition phase, the location-response associations must have been more activated than in the following sessions, which took place one or more days later. Consequently, we (i.e., Hommel et al., submitted) conducted a replication, where the subjects were to perform the keypressing task in the beginning of both the perceptual and the following memory session, so that the location-response associations were to be reactivated. As expected, this produced a substantial mapping effect: Verification times were faster for pairs mapped onto the same response (e.g., E-L) than for pairs mapped onto different responses (e.g., I-O). Obviously, response related information was integrated into the representations of our visual objects. Consequently, accessing those representations in order to extract spatial information for the verification task reactivated codes related to the associated response key, which, in turn, must have primed representations associated with the same response.

It is conceivable that this effect is restricted to the very simple, spatial responses we employed. In the cognitive system, the keypressing responses might have been coded as just another set of locations used in the wider context of the verification task. That is, house locations and response locations might have been coded in a very similar way and on the same feature dimensions, which might have facilitated the integration of stimulus and response information. Although this would not make the demonstration of such an integration less interesting, it would certainly restrict the degree to which our findings can be generalized. Another constraint, theoretically perhaps even more interesting, may have to do with the semantic relationship between our stimuli and responses. Given that the responses were spatially defined, they can in some sense be understood as "directed toward" the associated houses or house locations (although the spatial stimulus-response mapping was balanced across subjects). It may well be that it was this kind of "semantic" relationship that induced the integration of response information into object representations.

To investigate the possible roles of dimensional overlap and of the semantic relationship between stimuli and responses, Hommel et al. (submitted) conducted a further experiment with more complex everyday actions. In one group of subjects, particular houses were associated with house related activities that the subjects had to perform, such as coloring the picture of a house, constructing a small house with wooden building blocks, opening a door of a toy house, and operating a door knocker. Again, three to four houses were mapped onto a common activity. In another group of subjects, the activities were all unrelated to houses, such as reading aloud a weather report, tying a shoe, telling the time, and taking a sip of water.

Of course, all actions were extended in space (as any action is), but their main goal was clearly not spatially defined as was the case in our first action experiments. Thus, if actions would be integrated into object representations only if they are defined on the same (spatial) dimension as the objects, no integration should take place in this experiment, hence no mapping effect should be observed. However, if the crucial factor would be the semantic relationship between objects and actions, evidence for integration should be obtained with house related activities but not with house unrelated activities. Indeed, this is what the findings show. If the actions were house related, verification was faster with pairs mapped onto the same activity rather than to different activities. This was true in all sessions. In contrast, there was no effect of mapping in any session if the actions were unrelated to houses. This shows that the integration of action related information into object representations is by no means restricted to simple, strictly spatially defined actions. However, integration is limited by some kind of relevance factor, that is, actions are integrated only if they have something to do with the object at hand.

4   Conclusion

The findings reported here add to the available evidence showing that the spatial coding of object arrays is affected by the nonspatial attributes of these objects. Apparently, information about the spatial characteristics of objects, or relations between objects, are not stored in a way that allows for a selective retrieval of this information. Instead, spatial and nonspatial object information is automatically integrated into a common object representation, so that later retrieval of one piece of information also reactivates the information associated with it (McNamara, 1991).

However, there are two novel points we wanted to make. First, in all our studies we consistently found the same effects under perceptual conditions, i.e., judgments about visible displays, and memory conditions, where the relevant information had to be retrieved from memory. In fact, even the sizes of these effects were the same, leaving no room for any moderating factor. As pointed out by Gehrke and Hommel (1998) and Hommel et al. (submitted), this does not seem to fit with the often implicit but sometimes explicit (e.g., Tversky, 1981) idea that interactions between spatial and nonspatial codes reflect the organizational principles of memory processes. Rather, it seems that effects in memory tasks reflect the way the stored information has been organized in the process of perceptual encoding, and, therefore, if anything, reflect organizational principles of perceptual processes. Yet, although this is an interesting, parsimonious, and provocative hypothesis that nicely accounts for the findings reported here, there are considerable methodological differences between the available studies on interactions between spatial and nonspatial information. Therefore, it would seem premature to generalize our own failure to find differences between perceptual and memory conditions before the possible differences between the experimental tasks used hitherto are explored somewhat more deeply.

Our second point is that spatial coding is not only affected by exogenous factors, such as stimulus features and contextual properties, but by internal, action related factors as well. Apparently, people do not only group things that look the same, but also those things that afford similar actions. Apart from our action studies, such a conclusion is further suggested by observations of Merril and Baird (1987) and Carlson-Radvansky, Covey, and Lattanzi (in press). In the Merril and Baird study, students were asked to sort names of familiar local campus buildings. On the basis of a cluster analysis, two sorting criteria were identified: the spatial proximity between, and the functions of the buildings. As an instance of the latter criterion, all the dormitories, fraternities, and classrooms were sorted together. If one assumes that sorting behavior reflects the way the sorted items are organized in memory, the function of objects seems to provide at least one principle for this organization.

Carlson-Radvansky et al. (in press) instructed their subjects to place the picture of one object "above" or "below" a reference object. Among other things, the functional relatedness between the two objects was manipulated. Functional relatedness was high if, for example, a toothpaste tube was to be placed above or below a toothbrush, and it was low if a tube of oil paint was to be placed above or below a toothbrush. When the reference objects were presented in an asymmetrical fashion, such as with a lateral depiction of a toothbrush (i.e., with the bristles at one end of the brush), the horizontal placement was found to be biased towards the "functional parts" of the reference object. For instance, the toothpaste tube was not placed above or below the center of the toothbrush but closer to the bristles. Moreover, this bias was more pronounced with functionally related than with unrelated object pairs, hence, for the toothbrush example, stronger for the toothpaste tube than for tube of oil paint. Given that the functions of objects were not relevant for Carlson-Radvansky et al.'s task at all, this finding strongly suggests that functional, action related information is an integrated ingredient of object representations.

Taken altogether, we feel that the implication of TEC or ACM that the cognitive representation of stimulus layouts might be enriched by action related information or action affordances in the sense of Gibson (1979), is worthwhile to pursue. Indeed, if people are able to acquire and store spatial and nonspatial information about their environment, this ability should stand first and foremost in the service of action planning and action control. After all, representing information without knowing what it is good for does not seem to be an overly useful strategy.

References

Baylis, G.C., & Driver, J. (1993). Visual attention and objects: Evidence for hierarchical coding of location. *Journal of Experimental Psychology: Human Perception and Performance, 19*, 451-470.

Bower, G. H., Karlin, M. B., & Dueck, A. (1975). Comprehension and memory for pictures. *Memory & Cognition, 3*(2), 216-220.

Carlson-Radvansky, L.A., Covey, E.S., & Lattanzi, K.M. (in press). "What" effects on "where": Functional influences on spatial relation. *Psychological Science.*

Daniel, T. C., & Ellis, H. C. (1972). Stimulus codability and long-term recognition memory for visual form. *Journal of Experimental Psychology, . 93*(1), 83-89.

Gehrke, J., & Hommel, B. (1998). The impact of exogenous factors on spatial coding in perception and memory. In C. Freksa, C. Habel, & K.F. Wender (Eds.), *Spatial cognition: An interdisciplinary approach to representing and processing spatial knowledge* (pp. 64-77). Berlin: Springer.

Gibson, J.J. (1979). *The ecological approach to visual perception.* Boston, MA: Houghton Mifflin.

Hirtle, S.C., & Jonides, J. (1985). Evidence for hierarchies in cognitive maps. *Memory and Cognition, 13*, 208-217.

Hirtle, S.C., & Mascolo, M.F. (1986). Effect of semantic clustering on the memory of spatial locations. *Journal of Experimental Psychology: Learning, Memory, and Cognition, 12*, 182-189.

Hommel, B. (1997). Toward an action-concept model of stimulus-response compatibility. In B. Hommel & W. Prinz (Eds.), *Theoretical issues in stimulus-response compatibility* (pp. 281-320). Amsterdam: North-Holland.

Hommel, B. (1998). Observing one 's own action—and what it leads to. In J. S. Jordan (Ed.), *Systems theory and apriori aspects of perception* (pp. 143-179). Amsterdam: North-Holland.

Hommel, B., Gehrke, J., & Knuf, L. (submitted). Hierarchical coding in the perception and memory of spatial layouts. Manuscript submitted for publication.

Hommel, B., Knuf, L., & Gehrke, J. (submitted). The impact of action related knowledge on the organization of spatial information. Manuscript submitted for publication.

Hommel, B., Müsseler, J., Aschersleben, G., & Prinz, W. (1999). The theory of event coding (TEC): A framework for perception and action. Manuscript submitted for publication.

McNamara, T.P. (1986). Mental representation of spatial relations. *Cognitive Psychology, 18,* 87-121.

McNamara, T.P. (1991). Memory's view of space. *The Psychology of Learning and Motivation, 27,* 147-186.

McNamara, T.P., Halpin, J.A., Hardy, J.K. (1992). Spatial and temporal contributions to the structure of spatial memory. *Journal of Experimental Psychology: Learning, Memory and Cognition, 18,* 555-564.

McNamara, T.P., Hardy, J.K., & Hirtle, S.C. (1989). Subjective hierarchies in spatial memory. *Journal of Experimental Psychology: Learning, Memory, and Cognition, 15,* 211-227.

McNamara, T.P., & LeSueur, L.L. (1989). Mental representations of spatial and nonspatial relations. *Quarterly Journal of Experimental Psychology, 41,* 215-233.

McNamara, T.P., Ratcliff, R., & McKoon, G. (1984). The mental representation of knowledge acquired from maps. *Journal of Experimental Psychology: Learning, Memory and Cognition, 10,* 723-732.

Merrill, A.A., & Baird, J.C. (1987). Semantic and spatial factors in environmental memory. *Memory & Cognition, 15,* 101-108.

Navon, D. (1977). Forest before trees: The precedence of global features in visual perception. *Cognitive Psychology, 9,* 353-383.

Palmer, S.E. (1977). Hierarchical structure in perceptual representation. *Cognitive Psychology, 9,* 441-474.

Sadalla, E.K., Burroughs, W.J., & Staplin, L.J. (1980). Reference points in spatial cognition. *Journal of Experimental Psychology: Human Learning and Memory, 6,* 516-528.

Stevens, A., & Coupe, P. (1978). Distortions in judged spatial relations. *Cognitive Psychology, 10,* 422-427.

Tversky, B. (1991). Spatial mental models. *The Psychology of Learning and Motivation, 27,* 109-145.

Watt, R.J. (1988). *Visual processing: Computational, psychophysical, and cognitive research.* Hillsdale, NJ: Erlbaum.

Investigation of Age and Sex Effects in Spatial Cognitions as Assessed in a Locomotor Maze and in a 2-D Computer Maze[1]

Bernd Leplow[1], Doris Höll[2], Lingju Zeng[2] and Maximilian Mehdorn[3]

[1]Dept. of Psychology, Martin-Luther-University, Halle, Brandbergweg 23
D-06099 Halle, Germany
[2]Dept. of Psychology, Christian-Albrechts-University of Kiel, Olshausenstr. 62,
D-24098 Kiel, Germany
[3]Clinic for Neurosurgery, Christian-Albrechts-University of Kiel, Weimarer Str. 8, D-24106
Kiel, Germany

Abstract. Spatial behavior was assessed by a locomotor maze and a comparable 2-D computer version of this maze. Probable age and sex effects were studied before and after puberty, and in later adulthood. No sex differences were found in either task. In the locomotor task, age groups didn't differ with respect to exploration and orientation. The challenge provided by the PC task was greater for children and middle-aged adults. During acquisition of the locomotor maze children were inferior compared to the two adult groups, students tended to be slightly better than middle-aged adults. In the PC task differences in acquisition were mostly found between the children's group and the two adult groups. Spatial memory errors showed a developmental course. In both tasks spatial performance in adults followed a "one-trial" course, whereas in children a gradual decline of errors across learning trials was seen. It is concluded that apparent similarities between the motor and non-motor task may account for some fundamental strategic processes underlying these different spatial tasks.

Keywords: spatial cognition, spatial orientation, spatial memory, memory, orientation

1 Introduction

Spatial functions are most frequently assessed by means of 2-D computer setups or, more recently, by virtual reality (VR) technology. Since computer paradigms allow extensive variation of stimulus conditions and response requirements according to the experimenter's needs, these methods are clearly advantageous if compared to more traditional modes of assessment (e.g., pointing tasks) and are thus widely used in the field of experimental spatial cognition research. But despite the large number of

---

[1] This research was supported by the DFG governmental program "Spatial Cognition" (Le 846/2-2).

Ch. Freksa et al. (Eds.): Spatial Cognition II, LNAI 1849, pp. 399-418, 2000.

benefits there is a number of shortcomings if machine-based approaches are used exclusively. First, spatial functions have developed evolutionary by means of gross motor behavior. This component of spatial functioning is excluded or restricted to quite artificial joy stick or computer mouse movements. Second, prominent theories about spatial behavior underscore extensive exploration behavior at least for the formation of a sophisticated spatial representation which allows the organisms to take detours, returns, and short-cuts [29]. But it is still unclear whether or not this exploration behavior requires active locomotion or to which extent visual scanning is sufficient. Third, there is no consistent evidence whether or not spatial knowledge acquired on a computer transfers to the real world. Fourth, nearly nothing is known about different learning strategies activated in PC and real environments and about different abilities to use this knowledge if environmental conditions or behavioral challenges have changed. Fifth, PC and real world environments may attach spatial functions in different ways within development and aging, but this topic has not been studied yet.

Evolutionary aspects have recently been discussed especially with respect to the time of the emergence of spatial differences between sexes [37]. These differences have been attributed to hippocampal size e.g.[38], to male and female foraging [4], [10], to male and female dispersal [39], [23], and to the effects of hormonal changes in puberty [16], [17], [40]. With respect to the latter, accumulating evidence shows that estrogens in females suppress special types of spatial functioning. This effect accounts for the onset of most of the sex differences around puberty, as well as for hormonal and seasonal variations in spatial competence [44], [22].

It has been shown that positive effects of gross motor behavior on the acquisition of a spatial layout is beneficial especially in very young infants [11], [3], [21]. This effect can be explained by the visual tracking of the target location while moving around [1]. But in preschoolers and school children it has also been shown that spatial memory improves after having had the opportunity to explore a new environment actively. This effect was especially pronounced in preschoolers [11], [18]. An own investigation has shown that seven year olds learned a spatial layout equally well irrespectively if they had acquired a complex spatial layout within a non-motor/active choice condition or within an active locomotor/active choice condition. But the locomotor/active choice group was advantageous if a probe trail had to be performed in which the frames of references were dissociated [6]. Thus, locomotor behavior may be especially valuable if more complex spatial tasks have to be performed.

In order to disentangle active choice and active locomotion Foreman, Foreman, Cummings, and Owen [12] subjected preschoolers to a radial, maze analogue either by means of active locomotion or by means of a wheelchair. In both groups either the participants or the experimenter decided where to go. In sum, the results of two experiments showed that spatial learning was worse in the wheelchair/no-decision group. The three other groups did hardly differ. The experiments seem to suggest that active choice and locomotor behavior may be beneficial especially in younger ages. Peruch, Vercher, and Gauthier [33] found a comparable effect in adults using a wayfinding paradigm in a VR paradigm. But since no differences between groups were found in a very similar task performed by Wilson [45], the role of active locomotor behavior has to be a subject of further investigation, especially with respect to free versus restricted choice conditions.

Transfer from VR to real world has quite frequently been investigated within the very recent years [45]. This has been studied in the fields of architecture e.g. [46], [41], [36], [43], neurological applications e.g. [34], [35], [7], and emergency training [8]. An own investigation has clearly shown, that acquiring locations within a locomotor maze is dramatically improved if a VR-based pretraining was performed [15]. But if the spatial layout was learned once, VR pretraining did not affect later use of spatial information if the frames of reference were dissociated. In the same study we were able to show that even misleading VR pretraining had beneficial effects for the acquisition phase of the locomotor maze training procedure even though the exploration phase was significantly prolonged. These results showed that even restricted or misleading VR pretraining may improve the formation of a spatial representation of a real world environment. Taken these reports together, VR-based spatial knowledge seems to transfer to the real world, but it is still unclear to which degree and under which circumstances this transfer is possible. Research is especially required in order to understand the underlying learning strategies which may be elicited differentially by PC- and real-world setups.

Developmental and aging effects have been studied in rodents showing that an egocentric type of spatial behavior develops first, followed by landmark orientation. Development of spatial behavior is completed if place orientation is possible. In aged rats the decline of spatial competence seems to follow a reversed order [2]. In humans there are only a very few studies in which the relationships between aging, spatial abilities, and sex differences are investigated e.g. [19], [9]. Spatial experiments in rats are frequently performed by means of Radial Maze setups which date back to the work of Olton and coworkers [30], [31]. With this method research is especially addressed to the investigation of different types of spatial memory errors. If, for example, only subsets of the arms are baited, two types of errors can be differentiated: Reentering an arm which was visited within the same trial accounts for a spatial "working memory error", whereas entering an arm which has never been baited accounts for a spatial "reference memory error". In this latter case the subject did not learn the locations of the arms relative to the spatial layout outside the maze. Contrary to the reference memory failure working memory errors indicate an inability of updating the spatial positions of an ongoing behavior.

Another, widely used method is the Morris Water Maze [28], in which rodents have to swim towards a hidden platform. This procedure is especially suitable for the investigation of different orientation strategies which have been termed "position responses", "cue responses" and "place responses" [24]. In humans, there are only few attempts to study spatial functions within paradigms which are functionally equivalent to the setups used in animal research. Thus, Foreman, Warry, and Murray [13] used a radial maze analogue and found in children from eighteen months to five years of age that reference memory develops earlier than working memory and that the differences between groups of six year olds either actively and passively moving around or with and without freedom of choice were best reflected by the reference memory component of the spatial task [14].

Overman, Pate, Moore, and Peuster [32] constructed a radial maze which was roughly similar to that of Foreman et al. [12]. In the absence of any sex differences it was shown that - depending on the testing procedure used - place representations could be established from age five to seven onwards and that proximal cues ameliorated performance. Working memory seemed to be fully developed in children above the age of five. Moreover, Overman et al. [32] adopted the water maze paradigm by using a large cardboard "pool" which was filled with plastic packing chips and in which a hidden "treasure chest" had to be found. It was shown that place learning was not fully developed until the children had reached approximately the age of seven. Only a very few other setups were constructed which included locomotor behavior in humans: The "sandbox" task used for children followed a very similar principle [20] and the "invisible sensor task" is a hidden platform paradigm used for adults [5].

In our own research we attempted to combine the radial maze and water maze paradigms for humans [26],[27]. First results obtained in children showed that five and twelve years olds differed significantly in that the former were bound to a cue strategy whereas the latter were all able to display place orientation [25]. Furthermore, we could clearly relate this development of place orientation to the utilization of the distal cue configuration [42]. Sex differences were never found but we were not sure if these sex differences actually do not exist when spatial functions are assessed within a locomotor paradigm or if the absence of sex differences was due to the prepubertal ages of most of our participants under investigation. Thus, the present paper is addressed to the questions if spatial competence as assessed with the Kiel Locomotor Maze improves further beyond puberty, if sex differences will be seen, especially after puberty, and if the developmental course of acquiring spatial information is comparable to the Kiel Locomotor Maze if a 2-D PC-based layout of the maze is used.

2   Method

Apparatus. The apparatus has been described in detail in the preceding volume of this book so only a very brief description will be given here. Participants had to explore a wooden platform of about 3.60 m in diameter. In each of the four corners of the experimental chamber one out of four distinct extramaze cues was attached (cross, square, circle, and triangle of 30 x 30 cm each). No other cues were visible except for objects of about 3x3x3 cm, which were located at predefined positions on the experimental platform (Figure 1a).

The participant's task was to identify and remember the location of five out of twenty "hidden platforms". These hidden locations were made of twenty capacity detectors which were semi-irregularly mounted below the wooden platform. If a participant thought a location was correct he or she stepped on it. When stepping on a correct location a 160 Hertz-tone was elicited, signaling the first successful visit of one of the five locations. If all of these five locations had been visited the first acquisition trial was terminated. Stepping on to an incorrect location did not yield a feedback tone and a spatial "reference memory" error was recorded. Feedback was also not provided if a correct location was visited more than once within the same trial. In this case a spatial "working memory" error was recorded. The acquisition phase was continued unless two subsequent errorless trials were performed.

a)

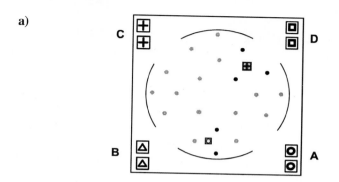

b)

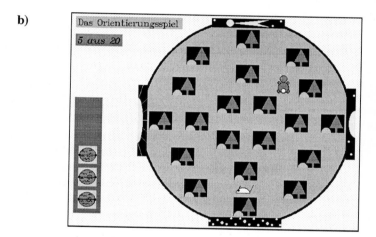

Fig. 1. Spatial layout of the locomotor setup (a) and the 2-D PC setup (b)

The 2-D PC-version of the maze was equivalent except that the twenty boxes were simultaneously seen (Figure 1b). Like in the locomotor maze the participant had to choose five out of the twenty boxes presented on the screen. In the case of a correct choice the box changed its color for about two seconds and a token was sampled in a column at the left side of the screen. As in the real maze a second visit of this box as well as a visit to an incorrect box would not change its original appearance, thus informing the participant about the occurrence of either a spatial working memory error or a spatial reference memory error.

3   Experiment 1: Spatial Learning within a Real and a 2-D Environment

3.1   Participants

For the locomotor experiment forty-eight healthy participants (24 females) were recruited from three age groups of 16 participants (8 females) each. Eleven year olds were chosen because this age lies in the prepubertal part of the lifespan and is well within the developmental phase when the formation of a place representation is possible [25]. This age group was compared with a group of young (mean age: 29,4) and middle aged adults (mean age: 55,9).

For the 2-D PC task an independent sample of forty-eight healthy participants (24 females) from three age groups of 16 participants (8 females) each was recruited. The children's group was slightly younger than in the locomotor experiment (mean age: 9.3) but at this age, place orientation is already well established. Young adults were also slightly younger than in the locomotor experiments (mean age: 22.1) but the middle-aged adults showed the same mean age (56.1).

3.2   Results

One middle-aged female had to be excluded because she was completely disorientated and the abortion criterion of 30 minutes of acquisition time was elapsed. No sex differences emerged between sexes in terms of trials to criterion nor of spatial memory errors and IRI. Therefore data were combined across sexes for all analyses. The same result was seen in the 2-D PC task: Since not even a tendency for a significant sex difference was seen in any one of the variables in any one of the age groups under investigation, both sexes were combined for further analysis.

Exploration Behavior. During the exploration phase the average number of locations visited in the locomotor maze did not differ significantly between the three age groups (Chi-Squared (2) = 0.77, p = 0.681). The youngest group visited 19,7 locations, the student group did so with 17,4 locations, and the oldest group stepped onto 16,7 detectors. Groups also didn't differ in the IRI measure (Chi-Squared (2) = 3.60, p = 0.165). Both children and middle-aged adults needed a mean of 2,9 seconds for two subsequent visits whereas students yielded an IRI of 3,7 seconds. Thus, since both IRI and number of locations visited were comparable between the three groups it can be concluded that the challenge provided by the experimental setup was equivalent for each group.

In the PC task the number of positions visited during exploration differed significantly between the three age groups (Chi-Squared (2) = 7.16, p = 0.028). The youngest group visited 22.3 positions, the student group did so with 14.2 positions and the oldest group visited 16.2 positions. Post hoc analysis showed that children visited more locations than students (z = -2.34, p = 0.02), but not than middle-aged adults (z = - 0.873, p = 0.383). The student visited fewer locations than middle-aged adults (z = -2.146, p = 0.032). Mean time spent between each of two locations also differed significantly between the three age groups (Chi-Squared (2) = 13.402, p < 0.000). Children needed a mean IRI of 4.3 seconds, young adults needed 2.9 seconds for two subsequent visits whereas middle-aged participants yielded an IRI of 5.0 seconds. Middle-aged participants were slower than the student's group (z = -3.37, p < 0.001) but not slower than the children's group. Children were also slower than students (z = -2.75, p = 0.006). Thus, contrary to the locomotor task the challenge provided by the PC-task affected the three age groups differentially.

Acquisition Rates. Despite comparable exploration behavior in the locomotor task, groups differed markedly with respect to the number of trials needed to achieve criterion (chi-square (2) = 24.46, p < 0.001). Figure 2a shows the average number of trials required to reach the criterion of two successive error-free trials.

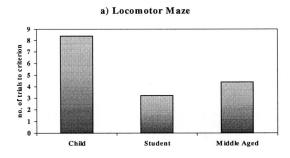

Fig. 2a. Number of trials to achieve criterion in the locomotor setup and b), in the 2-D PC setup

Groups needed 8.4, 3.2, and 4.4 trials to criterion, from the youngest to the oldest group, respectively. The youngest group was inferior to both the student and the middle-aged group (z = -4.32, and -3.72, p's < 0.001), and the difference between the latter groups also reached significance (z = -2.16, p = 0.031).

Though the PC task seemed to be slightly easier than the locomotor task, groups also differed significantly with respect to the number of trials needed to achieve criterion (chi-square (2) = 7.502, p = 0.023; Figure 2b). Groups needed 5.6, 3.8, and 3.9 trials to criterion, from the youngest of the oldest group, respectively. The youngest group was inferior to both the student and the middle-aged group (z = -2.32, and -2.38, p's < 0.020 and p < 0.017), but the difference between the latter two groups didn't reach significance (z = -0.138, p = 0.89).

In the locomotor experiment different mean numbers of errors were obtained between groups (chi-square (2) = 26.88, p < 0.001; Figure 3a). While children displayed a mean of 6.1 errors, the two other age groups performed nearly perfect. Mean error rates were .30 and 1.2, respectively. Again, the youngest group was inferior with respect to the student's group and the middle-aged group (z = -4.80, and -4.19, p's < 0.000), and students outperformed middle-aged adults (z = -2.55, p = 0.020).

b) PC Task

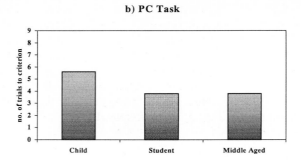

Fig. 2b. Number of trials to achieve criterion in the 2-D PC setup

a) Locomotor Maze

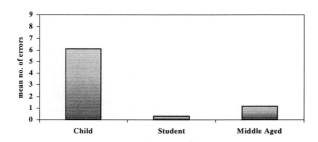

Fig. 3a. Mean number of errors within acquisition a) in the locomotor setup and b) in the 2-D PC setup

The differences found with respect to the number of trials to reach criterion and to the mean error rates were not reflected by differences in IRIs (chi-square (2) = 1.61, p = 0.446; Figure 4a).

Performance in the PC experiment was widely comparable to the locomotor experiment, at least for children (Figure 3b). Mean number of errors differed significantly between groups (chi-square (2) = 8.38, p = 0.015). While children displayed a mean of 6.6 errors, the two other age groups performed much better. Mean error rates were 2.8 for the student's group and 2.9 for the middle-aged group. Again, the youngest group was inferior with respect to the two other groups (z = -2.30, p < 0.022 and z = -2.65, p < 0.007, for the student and the middle-aged group, respectively), whereas the difference between the latter groups did not reach significance (z = -0.059, p = 0.956).

b) PC Task

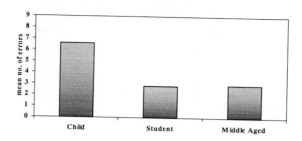

Fig. 3b. Mean number of errors within acquisition in the 2-D PC setup

a) **Locomotor Maze**   b) **PC Task**

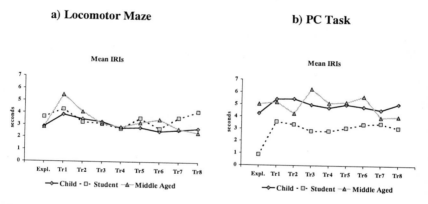

Fig. 4. Mean inter-response intervals (IRI) across training trials a), in the locomotor setup and b), in the 2-D PC setup

The differences found were also reflected by differences in the IRI measure (chi-square (2) = 14.616, p = 0.001). Young adults were faster than both the youngest group (z = -3.58, p < 0.000) and the oldest group (z = -2.977, p < 0.002), whereas the youngest and the oldest groups didn't differ (Figures 4b). Mean IRI for children was 5.1 seconds, for students 3.6 seconds and the middle-aged group needed a mean time of 4.9 seconds.

Types of Spatial Errors. For the locomotor task figure 5a shows the course of the three types of spatial memory errors for each age group, respectively. It can be seen that the learning curves of the two adult groups roughly followed the "one-trial"-principle [29], whereas in children the courses of the memory error types decline gradually across trials. For all three types of spatial memory errors groups differed significantly (chi-squares (2) = 30.2, 29.6, 32.0, p < 0.001; for reference memory errors, working memory errors, and reference-working memory errors, respectively).

a) Locomotor Maze

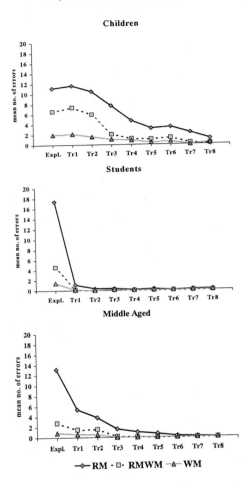

Fig. 5a. Course of the three types of spatial memory errors in the locomotor setup

Post hoc comparisons showed that children displayed significant higher rates of all three types of spatial memory errors if compared both to students and to middle-aged adults (z < -4.09, p < 0.001, respectively), and that students were better than middle-aged adults in both types of reference memory components (z < -2.40; p < 0.013, respectively). Working memory errors were very rare in each of the investigated groups, followed by the combined working-reference memory, and reference memory error. Table 1 shows that during development and adulthood the relative amount of reference memory errors increases, whereas the relative proportion of the two other types of spatial memory errors decreases.

b) PC Task

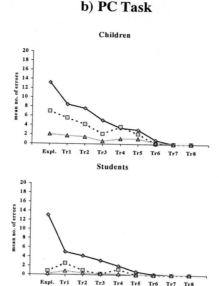

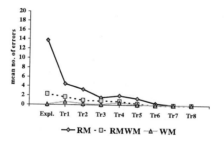

Fig. 5b. Course of the three types of spatial memory errors in the 2-D PC setup

Figure 5b shows the course of the three types of spatial memory errors for each age group in the PC task. The courses of the error types are more or less identical to the courses obtained from the locomotor experiment (Figure 5a). Type of memory errors differed significantly in the reference memory component of the task (chi-square (2) = 7.15, p = 0.028), its working memory component (chi-square (2) = 10.66, p = 0.005), and its reference-working memory component (chi-square (2) = 7.972, p = 0.005). Post hoc comparisons showed that children displayed significant higher rates of all three types of spatial memory errors if compared to students (z < -2.00, p < 0.050) and to middle-aged adults (z < -2.969, p < 0.026). Contrary to the results obtained with the locomotor maze, students and middle-aged adults didn't differ in the 2-D PC task. Again, working memory errors were very rare in each group, followed by the combined working-reference memory error, and again, most of the errors displayed were of the reference memory error type.

Table 1.

A. ACQUISITION	locomotor maze			2-D PC maze		
	children	students	midd. aged	children	students	midd. aged
mean error scores						
reference memory	3.8	0.2	0.9	3.6	1.9	1.6
reference-working memory	1.7	0.03	0.2	2.3	0.6	0.59
working memory	0.6	0.04	0.1	0.8	0.2	0.21
percentage scores						
reference memory	63.5	88.5	75.7	65.7	80.0	81.68
reference-working memory	26.9	6.7	11.7	23.8	14.4	13.99
working memory	9.6	4.8	12.6	10.5	5.6	4.32

B. ORIENTATION						
	children	students	midd. aged	children	students	midd. aged
mean error scores						
reference memory	0.8	2.3	2.7	4.4	3.3	3.13
reference-working memory	0.3	0.6	0.5	1.1	0.6	0.25
working memory	0.6	0.0	0.0	0.4	0.0	0.0
percentage scores						
reference memory	62.9	78.7	83.3	66.0	89.1	93.75
reference-working memory	27.3	15.1	12.3	25.4	8.3	6.25
working memory	9.8	6.2	4.4	8.9	2.7	0

note: children = children's group; students = student's group, midd. aged = middle-aged adults.

These differences are also reflected by the proportions of the respective error types (Table 1). Thus, as in the locomotor experiment, the relative amount of reference memory errors increases during development and adulthood and the relative proportion of the two other types of spatial memory errors decreases. But contrary to the locomotor task the main dissociation of error types between age groups seems to occur between childhood and young adults.

4   Experiment 2: Orientation within a Real and a 2-D Environment

After having learnt the spatial layout of either the locomotor maze or the 2-D PC-version the starting position was rotated 180° with respect to the initial starting point. Simultaneously the proximal cues are rotated 180° too so that the viewer's perspective of the new starting position is identical to that of the initial starting point - as long as distal cues are not taken into account. If the participant relies his or her orientation behavior on the proximal cues or if he or she is oriented with respect to egocentric responses this probe trial will not be successfully mastered. But if the participant is oriented with respect to the distal cues the "places" of the hidden locations are correctly identified.

4.1   Results

Again, none of the comparisons became significant with respect to sex differences, neither for the locomotor maze nor for the 2-D PC tasks, respectively. Thus, data were again combined for the two sexes.

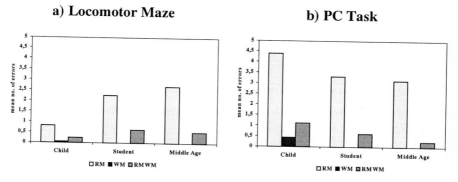

Fig. 6. Mean number of errors for each group within the orientation test trial a), in the locomotor setup and b), in the 2-D PC setup

Global Performance. In the locomotor task orientation performance did not differ between groups with respect to spatial memory errors (Chi-Squared (2) = 2.60, p = 0.201; Figure 6a) Children visited a mean of 1.1 locations, whereas students visited 2.9, and middle-aged visited 3.2 locations. But the groups were differentially affected with respect to the IRI measure (Chi-Squared (2) = 11.30, p = 0.004; Figure 7). Children displayed an IRI of 2.3 seconds, students needed 6.8 seconds for two subsequent visits whereas middle-aged participants yielded an IRI of 7.0 seconds. Thus, children were significantly faster than both the student's group (z = -3.0, p = 0.003) and the middle-aged group (z = -2.79, p < 0.005), whereas students and middle-aged adults didn't differ from each other.

In the PC task the number of positions visited also didn't differ with respect to spatial memory errors (Chi-Squared (2) = 0.141, p = 0.932; Figure 6b). The youngest group visited 5.9 locations, the student group did so with 3.9 locations and the oldest group visited 3.38 locations. But the three groups again differed with respect to IRI (Chi-Squared (2) = 19.17, p < 0.000; Figure 7). Children needed an IRI of 4.5, young adults needed 3.7 seconds for two subsequent visits whereas middle-aged participants yielded an IRI of 7.1 seconds. Thus, the middle-aged group was significantly slower than both the children's group (z = -2.997, p = 0.002) and the student's group (z = -3.958, p < 0.001), and children and students also differed from each other (z = -2.75, p = 0.040).

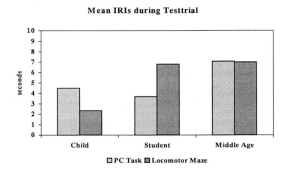

Fig.7. Mean inter-response intervals (IRI) within the orientation test trial in the locomotor setup and in the 2-D PC setup

Types of Spatial Errors. For the locomotor task figure 6a shows the amounts of the three types of spatial memory errors for each age group, respectively. Groups didn't differ with respect to one of the memory error types but most of the errors displayed still were of the reference memory error type. In general, mean memory error rates were extremely low, especially the working memory components of the two tasks. Contrary to the acquisition phase figure 6b and table 1 show that the error type scores were slightly higher in the 2-D PC task if compared to the locomotor experiment. Moreover, in the PC task groups differed with respect to the working memory component (chi-square (2) = 6.261, p < 0.044). But like in the locomotor experiment groups neither differed with respect to reference memory errors (chi-square (2) = 0.97, p = 0.953) nor to reference-working memory errors (chi-square (2) = 0.796, p = 0.672), and no post hoc comparison for the working memory component reached significance. Despite the difference found in the working memory error the pattern of error distributions is comparable for both the locomotor and the 2-D PC task (Table 1).

5   Discussion

The present investigation yielded three main results: First, we never found sex differences, second, apparent similarities between the motor and 2-D task were seen, and third, despite these similarities there are important differences elicited by the motor and 2-D version of our spatial task. Especially, during exploration, locomotor behavior did not differ between age groups but the challenge provided by the PC task was greater for children and, to a less degree, for middle-aged adults. During acquisition of the locomotor maze the children's performance was clearly below that of the two adult groups, and students tended to be slightly better than middle-aged adults. In the PC task differences in acquisition were mostly found between the children's group and the two adult groups. In both experiments and in each of the age groups reference memory errors were far most frequent, though the proportion of reference memory errors increased and that of working memory errors decreased with age. In both experiments spatial performance in adults followed a "one-trial" course, whereas in children a gradual decline across learning trials of the various spatial error types was seen. Following acquisition, orientation differences between groups were only seen with respect to reaction times in the PC task. Though training was performed up to two successive errorless trials reference memory error rates of the PC task were slightly above the error rates in the locomotor experiment.

The complete absence of sex effects was not expected, especially not in the PC version of the maze and not after puberty. According to hormonal theories about sex differences post-pubertal estrogen expression should have suppressed spatial abilities in females. Since the PC task puts high demands onto mental rotation capacity, especially in the orientation phase of the experiment, suppression should have been apparent at least in this orientation test. But our experimental results do not support this hypothesis. Since mental rotation is the most reliable sex effect [44], the lack of differences has to be explained. In a PC version of a locomotor experiment the participant's starting position cannot be rotated physically. For this reason the rotation effect was induced by rotating the distal cues. Moreover, proximal cues were also rotated in order to produce the same local view (with respect to the proximal cues) like in the acquisition part of the experiment. Thus, there are two ways to solve the search problem: the participant could rotate both his/her starting point and the proximal cues back to their respective initial position. This would put high demands on the mental rotation factor. This may be reflected by the slightly enhanced error rates in the orientation phase of the PC task if compared to the locomotor experiment. Or, the second way to get out of the conflict situation of dissociated frames of reference, is to orient one's search with respect to the distal cue configuration, i.e., to define the "places" of the target locations. Only if such a non-rotational strategy had been used no sex differences can be expected.

Developmental studies undertaken in our group [25] and by Overman et al. [32] suggest that such non-rotational place strategies develop after age seven. In this case our samples have been far above this critical age. And since place strategies should not necessarily been affected by estrogen expression the absence of sex effects may be understandable. But it has to be mentioned that our sample sizes have been quite small. Within each cell we only included eight participant of each sex so that the effect had to be rather strong to yield a statistically significant difference. Thus, the absence of sex effects is not a proof that these effects actually do not exist. But since we have investigated a large number of different samples with about several hundreds of clinical and non-clinical participants of nearly every age without ever finding a sex effect we tend to believe that these effects actually do not exist in the type of task we are using. Moreover the failure to identify sex effects in tasks like this is in accordance with the literature e.g. [32].

The results obtained within the exploration and acquisition phases of the respective motor and the 2-D experiments were strikingly similar. This is not only surprising because only gross motor behavior and far distance movements elicit an exhaustive variety of egocentric information. Moreover, in the locomotor paradigm the participant was provided with a step-by-step view which successively changes with each step, whereas in the PC maze all 20 locations were always simultaneously visible. Thus, locomotor behavior is more prone to elicit "route knowledge" whereas in the PC task the development of "survey knowledge" is stressed. But how could these psychologically distinct challenges lead to nearly the same maze performance? Before answering this question it has to be mentioned that equivalence of performance was not complete. First, the exploration phase of the locomotor task was obviously easier than that of the 2-D PC task. Second, children needed less trials to achieve the criterion in the PC task than in the locomotor experiment. Third, IRIs differed between age groups only in the PC task. These differences were seen in the exploration phase, within acquisition and in the orientation test. Contrary to these age effects, differences of this type were never seen in the locomotor maze experiment. And fourth, many more memory errors were displayed in the PC orientation test. This effect was especially seen in children. Thus, procedural differences between the two spatial tasks mentioned in the above paragraph may account for these results. But these effects were obviously not strong enough to counterbalance the strong influence which was elicited by visual exploration. This exploration was driven by active choice and may have led to comparable solution strategies. Perhaps the special features of our maze with its two clusters of correct locations enhances this effect.

The participant's own choices and not the locomotor behavior per se may explain the considerably good performance in the two experimental setups. Thus, exploration would rely more on a variety of local views and different spatial foci of attention than on the magnitude of vestibular, proprioceptive an kinesthetic cues the organism is provided with when navigating actively within a spatial layout. Therefore, from the equivalent performance in the two spatial tasks it may be concluded that the information obtained from *active* visual search is crucial for strategy formation which in turn guides spatial behavior. "Going right", "move towards the proximal cue x", "obtain the response sequence xyz", "always take the same starting position", or "take into account the distal cue configuration" are examples for strategies based on visual exploration.

The large proportion of reference memory errors also fits with the literature [14]. Interestingly, this type of spatial memory error reflects development in that its proportion increases with adulthood whereas the proportion of working memory errors decreases. Thus, though working memory errors are comparably rare they tend to be more frequent during childhood. Moreover, considerable rates of working memory errors are seen up to acquisition trial two and children also show high proportions of reference memory errors up to acquisition trial four. Contrary to this pattern of results, working memory errors are nearly absent in adults whereas reference memory errors showed rates from 76% to 91%. Furthermore, reference memory errors in adults sharply drop following exploration. This strongly resembles "one-trial" learning predicted by O'Keefe & Nadel [29] as being one essential characteristic for the formation of a "cognitive map", the most sophisticated mental representation of space.

In sum, the experiments presented in this paper show that striking similarities of two completely different versions the same basic task may lead to a better understanding of the underlying cognitive functions necessary to solve spatial tasks of this type. But the differences between these tasks, which are only seen if detailed analyses are performed, highlight the special contribution of motor performance in spatial behavior of different age groups.

Acknowledgments

The authors are indebted to Dipl.-Ing. Arne Herzog, an engineer who intensively supported us by working on our hardware and data recording techniques. In addition, we wish to thank, cand. phil. Franka Weber, can. phil. Angelika Langer and cand. phil. Roy Murphy who worked in this project as student research assistants.

References

1. Acredolo, L.P., Adams, A. & Goodwyn, S.W.(1984).The role of self-produced movement and visual tracking in infant spatial orientation. Journal of Experimental Child Psychology, 38,312-327.
2. Barnes, C.A., Nadel, L. & Honig, W.K. (1980). Spatial memory deficits in senescent rats. Canadian Journal of Psychology, 34, 29-29.
3. Bertenthal, B.I., Campos, J.J. & Barrett, K.C. (1984). Self-produced locomotion: An organizer of emotional, cognitive, and social development in infancy. In R. Emde & R. Harmon (Eds.), Continuities and discontinuities in development (pp. 175-210). New York: Plenum.
4. Blumenshine, R.J. & Cavallo, J.A. (1992). Scavenging and human evolution. Scientific American, 267 (4), 90-96.
5. Bohbot, V. D., Kalinka, M., Stepankova, K., Spackova, N., Petrides, M. & Nadel, L. (1998). Spatial memory deficits in patients with lesions to the right hippocampus and the right parahippocampal cortex. Neuropsychologia, 36, 1217 - 1238.

6. Böller, M. (1999). Die Effekte der aktiven und passiven Lokomotion auf den Erwerb räumlicher Anordnungen bei gesunden 7-jährigen Kindern. Diploma-Thesis at the Dept. of Psychology, University of Kiel.

7. Brooks, B.M., McNeil, J.E., Rose, F.D., Greenwood, J.R., Attree, E.A. & Leadbetter, A.G. (1999). Route learning in a case of amnesia: A preliminary investigation into the efficacy of training in a virtual environment. Journal of Neuropsychological Rehabilitation, in press.

8. Chi, D.M., Kokkevis, E., Ogunyemi, O., Bindiganavale, R., Hollick, M.J., Clarke, J.R., Webber, B.L. & Badler, N.I. (1997). Simulated casualities and medics for emergency training. Studies of Health Technology and Informatics, 39, 486-494.

9. Dollinger, S.M.C. (1995). Mental rotation performance: age, sex, and visual field differences. Developmental Neuropsychology, 11, 215 - 222.

10. Eals, M. & Silverman, I. (1994). The hunter-gatherer theory of spatial sex differences: proximate factors mediating the female advantage in recall of object arrays. Ethological Sociobiology, 15, 95 - 105.

11. Feldman, A. & Acredolo, L. (1979). The Effect of Active versus Passive Exploration on Memory for Spatial Location in Children. Child Development 50, 698-704.

12. Foreman, N., Foreman, D., Cummings, A. & Owens, S. (1990). Locomotion, active choice, and spatial memory in children. The Journal of General Psychology, 117, 215-232.

13. Foreman, N., Warry, R. & Murray, P. (1990). Development of reference and working spatial memory in preschool children. The Journal of General Psychology, 117 (3), 267-276.

14. Foreman, N., Gillet, R. & Jones, S. (1994). Choice autonomy and memory for spatial locations in six-year-old children. British Journal of Psychology, 85, 17-27.

15. Foreman, N., Stirk, J., Pohl, J., Mandelkow; L., Lehnung, M., Herzog, A. & Leplow, B. (2000). Spatial information transfer from virtual to real versions of the Kiel locomotor maze. Behavioural Brain Research in press.

16. Hampson, E. (1990). Estrogen-related variations in human spatial and articulatory-motor skills. Psychoneuroendocrinology, 15, 97 -111.

17. Hampson, E. & Kimura, D. (1988). Reciprocal effects of hormonal fluctuations on human motor and perceptual spatial skills. Behavioral Neuroscience, 102, 456 - 459.

18. Herman, J.F., Kolker; R.G. & Shaw, M.L. (1982). Effects of motor activity on children's intentional and incidental memory for spatial locations. Child Development, 53, 239-244.

19. Herman, J.F. & Bruce, P.R. (1983). Adult's mental rotation of spatial information: effects of age, sex and cerebral laterality. Experimental Aging Research, 9, 83 - 85.

20. Huttenlocher, J., Newcombe, N., & Sandberg, E.H. (1994). The coding of spatial location in children. Cognitive Psychology, 27, 115-147.

21. Kermoin, R. & Campos, J.J. (1988). Locomotor Experience: A Facilitator of Spatial Cognitive Development. Child Development, 59, 908-917.

22. Kimura, D. & Hampson, E. (1994). Cognitive pattern in men and women is influenced by fluctuations of sex hormones. Current Directions of Psychological Sciences, 3, 57 - 61.

23. Koenig, W.D (1989). Sex-biased dispersal in the contemporary United States. Ethological Sociobiology, 10, 29 -54.

24. Kolb, D. & Wishaw, I.Q. (1996). Neuropsychologie, 2. Auflage, Heidelberg; Berlin; Oxford: Spektrum Akad. Verlag.

25. Lehnung, M. Leplow, B., Friege, L., Herzog, A. & Ferstl, R. (1998). Development of spatial memory and spatial orientation in preschoolers and primary school children. British Journal of Psychology, 89, 463-480.

26. Leplow, B. (1994). Diesseits von Zeit und Raum: Zur Neuropsychologie der räumlichen Orientierung (Neuropsychology of spatial orientation). Habilitation thesis University of Kiel.

27. Leplow, B., Höll, D., Zeng, L. & Mehdorn, M. (1998). Spatial Orientation and Spatial Memory Within a 'Locomotor Maze' for Humans. In Chr. Freksa, Chr Habel and K. F. Wender (Eds.) Lecture Notes of Artificial Intelligence 1404/ Computer Sciences/ Spatial Cognition, pp 429-446, Springer: Berlin.

28. Morris, R.G.M. (1981). Spatial localization does not require the presence of local cues. Learning and Motivation, 12, 239-260.

29. O'Keefe, J. & Nadel, L. (1978). The hippocampus as a cognitive map. Oxford University Press.

30. Olton, D.S. & Samuelson, R.J. (1976). Remembrance of places passed: Spatial memory in rats. Journal of Experimental Psychology: Animal Behavior Processes, 2, 97-116.

31. Olton, D.S., Becker, J.T. & Handelmann, G.E. (1979). Hippocampus, space, and memory. The Behavioral and Brain Sciences, 2, 313-365.

32. Overman, W.H., Pate, B.J., Moore, K. & Peuster, A. (1996). Ontogenecy of place learning in children as measured in the radial arm maze, Morris search task, and open field task. Behavioral Neuroscience, 110, 1205-1228.

33. Peruch, P., Vercher, J. & Gauthier, G.M. (1995). Acquisition of spatial knowledge through visual exploration of simulated environments. Ecological Psychology, 7, 1-20.

34. Rose, F.D., Attree, E.A. & Johnson, D.A. (1996). Virtual Reality: An assistive technology in neurological rehabilitation. Current Opinion in Neurology, 9, 461 - 467.

35. Rose, F.D., Johnson, D.A., Attree, E.A., Leadbetter, A.G. & Andrews, T.K. (1996). Virtual reality in neurological rehabilitation. British Journal of Therapy and Rehabilitation, 3, 223 - 228.

36. Ruddle, R.A., Payne, S.J. & Jones, D.M. (1996) Navigating buildings in "desktop" virtual environments: Experimental investigations using extended navigational experience. Journal of Experimental Psychology: Applied, 3, 143-159.

37. Sherry D.F. & Hampson, E. (1997). Evolution and the hormonal control of sexually-dimorphic spatial abilities in humans. Trends in Cognitive Sciences, 1, 50-55.
38. Sherry, D.F., Jacobs, L.F. & Gaulin, S.C. (1992). Spatial memory and adaptive specialization of the hippocampus. Trends in Neuroscience, 15, 298 - 303.
39. Silverman, I. & Eals, M. (1992). In: J.H. Barkow, L. Cosmides & J. Tooby (Eds.), The adaptive Mind: Evolutionary Psychology and the Generation of Culture, (pp. 533 - 549). Oxford: University Press.
40. Silverman, I. & Phillips, K. (1993). Effects of estrogen changes in during the menstrual cycle on spatial performance. Ethological Sociobiology, 14, 257 - 270.
41. Stanton, D., Wilson, P. & Foreman, N. (1996). Using virtual reality to aid spatial awareness in disabled children. In: P. Sharkey (Ed.), Proceedings of the First International Conference on Disability, Virtual Reality, and Associated Technologies. Reading, Berkshire; pp. 93-101.
42. Taeger, B. (1999). Die Entwicklung von Orientierungsstrategien im Raum anhand von distalen Hinweisreizen. Diploma-Thesis at the Dept. of Psychology, University of Kiel.
43. Thorndyke, P.W. & Hayes-Roth, B. (1982) Differences in spatial knowledge acquired from maps and navigation. Cognitive Psychology, 14, 560-589.
44. Voyer, D., Voyer, S. & Bryden, M.P. (1995). Magnitude of sex differences in spatial abilities: A meta-analysis and consideration of critical variables. Psychological Bulletin, 117, 250-270.
45. Wilson, P.N. (1997). Use of virtual reality computing in spatial learning research. In: N. Foreman, R. Gillet et al. (Eds.), A handbook of spatial research paradigms and methodologies, Vol. 1: Spatial cognition in the child and adult (pp.181-206). Hove, England, UK: Psychology Press: Erlbaum (UK).
46. Wilson, P., Foreman, N. & Tlauka, M. (1997) Transfer of spatial information from a virtual to a real environment. Human Factors, 39, 526-531.

Author Index

Baier, Volker 145
Barkowsky, Thomas 41, 54, 100, 225
Belingard, Loïc 253
Blok, Connie 16
Brauer, Wilfried 115, 145
Bülthoff, Heinrich H. 317

Cabedo, Lledó Museros 225
Casakin, Hernan 54
Christou, Chris 317

Eisenkolb, Andreas . . . 115, 145, 265
Eschenbach, Carola 127
Eyferth, Klaus 157

Frank, Andrew U. 80
Freksa, Christian 54, 100

Gärtner, Holger 157

Habel, Christopher 127
Hagen, Cornelius 198
Herrmann, Theo 277, 295
Hirtle, Stephen C. 31
Höll, Doris 399
Hommel, Bernhard 387
Hörnig, Robin 157

Isli, Amar 225

Janzen, Gabriele 277

Katz, Steffi 277
Klippel, Alexander 54
Knauff, Markus 184
Knuf, Lothar 387
Krieg-Brückner, Bernd 295

Kulik, Lars 127, 239

Lankenau, Axel 265
Latecki, Longin Jan 41
Leplow, Bernd 399

Mecklenbräuker, Silvia 363
Mehdorn, Maximilian 399
Melzer, André 363
Müller, Rolf 265
Moratz, Reinhard 100, 225
Musto, Alexandra 115, 145, 265

Péruch, Patrick 253

Rauh, Reinhold 184, 239
Renz, Jochen 184
Richter, Kai-Florian 41
Röfer, Thomas 115, 265
Röhrbein, Florian 145
Rothkegel, Rainer 348
Ruddle, Roy A. 333

Saathoff, Jörg E. 363
Schill, Kerstin 115, 145
Schlieder, Christoph 198
Schmid, Ute 212
Schmidt, Thomas 169
Schumacher, Sabine 348
Schweizer, Karin 277
Stein, Klaus 115, 265
Strohecker, Carol 1

Thinus-Blanc, Catherine 253
Tschander, Ladina 127
Tversky, Barbara 72

Wagener, Monika 363

Wender, Karl Friedrich 348

Werner, Steffen 169, 295

Wiebrock, Sylvia 212

Wippich, Werner 363

Wittenburg, Lars 212

Wysotzki, Fritz 212

Zeng, Lingju 399

Lecture Notes in Artificial Intelligence (LNAI)

Vol. 1674: D. Floreano, J.-D. Nicoud, F. Mondada (Eds.), Advances in Artificial Life. Proceedings, 1999. XVI, 737 pages. 1999.

Vol. 1688: P. Bouquet, L. Serafini, P. Brézillon, M. Benerecetti, F. Castellani (Eds.), Modeling and Using Context. Proceedings, 1999. XII, 528 pages. 1999.

Vol. 1692: V. Matoušek, P. Mautner, J. Ocelíková, P. Sojka (Eds.), Text, Speech, and Dialogue. Proceedings, 1999. XI, 396 pages. 1999.

Vol. 1695: P. Barahona, J.J. Alferes (Eds.), Progress in Artificial Intelligence. Proceedings, 1999. XI, 385 pages. 1999.

Vol. 1699: S. Albayrak (Ed.), Intelligent Agents for Telecommunication Applications. Proceedings, 1999. IX, 191 pages. 1999.

Vol. 1701: W. Burgard, T. Christaller, A.B. Cremers (Eds.), KI-99: Advances in Artificial Intelligence. Proceedings, 1999. XI, 311 pages. 1999.

Vol. 1704: Jan M. Żytkow, J. Rauch (Eds.), Principles of Data Mining and Knowledge Discovery. Proceedings, 1999. XIV, 593 pages. 1999.

Vol. 1705: H. Ganzinger, D. McAllester, A. Voronkov (Eds.), Logic for Programming and Automated Reasoning. Proceedings, 1999. XII, 397 pages. 1999.

Vol. 1711: N. Zhong, A. Skowron, S. Ohsuga (Eds.), New Directions in Rough Sets, Data Mining, and Granular-Soft Computing. Proceedings, 1999. XIV, 558 pages. 1999.

Vol. 1712: H. Boley, A Tight, Practical Integration of Relations and Functions. XI, 169 pages. 1999.

Vol. 1714: M.T. Pazienza (Eds.), Information Extraction. IX, 165 pages. 1999.

Vol. 1715: P. Perner, M. Petrou (Eds.), Machine Learning and Data Mining in Pattern Recognition. Proceedings, 1999. VIII, 217 pages. 1999.

Vol. 1720: O. Watanabe, T. Yokomori (Eds.), Algorithmic Learning Theory. Proceedings, 1999. XI, 365 pages. 1999.

Vol. 1721: S. Arikawa, K. Furukawa (Eds.), Discovery Science. Proceedings, 1999. XI, 374 pages. 1999.

Vol. 1724: H.I. Christensen, H. Bunke, H. Noltemeier (Eds.), Sensor Based Intelligent Robots. Proceedings, 1998. VIII, 327 pages. 1999.

Vol. 1730: M. Gelfond, N. Leone, G. Pfeifer (Eds.), Logic Programming and Nonmonotonic Reasoning. Proceedings, 1999. XI, 391 pages. 1999.

Vol. 1733: H. Nakashima, C. Zhang (Eds.), Approaches to Intelligent Agents. Proceedings, 1999. XII, 241 pages. 1999.

Vol. 1735: J.W. Amtrup, Incremental Speech Translation. XV, 200 pages. 1999.

Vol. 1739: A. Braffort, R. Gherbi, S. Gibet, J. Richardson, D. Teil (Eds.), Gesture-Based Communication in Human-Computer Interaction. Proceedings, 1999. XI, 333 pages. 1999.

Vol. 1744: S. Staab, Grading Knowledge: Extracting Degree Information from Texts. X, 187 pages. 1999.

Vol. 1747: N. Foo (Ed.), Adavanced Topics in Artificial Intelligence. Proceedings, 1999. XV, 500 pages. 1999.

Vol. 1757: N.R. Jennings, Y. Lespérance (Eds.), Intelligent Agents VI. Proceedings, 1999. XII, 380 pages. 2000.

Vol. 1759: M.J. Zaki, C.-T. Ho (Eds.), Large-Scale Parallel Data Mining. VIII, 261 pages. 2000.

Vol. 1760: J.-J. Ch. Meyer, P.-Y. Schobbens (Eds.), Formal Models of Agents. Poceedings. VIII, 253 pages. 1999.

Vol. 1761: R. Caferra, G. Salzer (Eds.), Automated Deduction in Classical and Non-Classical Logics. Proceedings. VIII, 299 pages. 2000.

Vol. 1771: P. Lambrix, Part-Whole Reasoning in an Object-Centered Framework. XII, 195 pages. 2000.

Vol. 1772: M. Beetz, Concurrent Reactive Plans. XVI, 213 pages. 2000.

Vol. 1775: M. Thielscher, Challenges for Action Theories. XIII, 138 pages. 2000.

Vol. 1778: S. Wermter, R. Sun (Eds.), Hybrid Neural Systems. IX, 403 pages. 2000.

Vol. 1792: E. Lamma, P. Mello (Eds.), AI*IA 99: Advances in Artificial Intelligence. Proceedings, 1999. XI, 392 pages. 2000.

Vol. 1793: O. Cairo, L.E. Sucar, F.J. Cantu (Eds.), MICAI 2000: Advances in Artificial Intelligence. Proceedings, 2000. XIV, 750 pages. 2000.

Vol. 1794: H. Kirchner, C. Ringeissen (Eds.), Frontiers of Combining Systems. Proceedings, 2000. X, 291 pages. 2000.

Vol. 1805: T. Terano, H. Liu, A.L.P. Chen (Eds.), Knowledge Discovery and Data Mining. Proceedings, 2000. XIV, 460 pages. 2000.

Vol. 1810: R. López de Mántaras, E. Plaza (Eds.), Machine Learning: ECML 2000. Proceedings, 2000. XII, 460 pages. 2000.

Vol. 1822: H.H. Hamilton, Advances in Artificial Intelligence. Proceedings, 2000. XII, 450 pages. 2000.

Vol. 1831: D. McAllester (Ed.), Automated Deduction – CADE-17. Proceedings, 2000. XIII, 520 pages. 2000.

Vol. 1835: D. N. Christodoulakis (Ed.), Natural Language Processing – NLP 2000. Proceedings, 2000. XII, 438 pages. 2000.

Vol. 1849: C. Freksa, W. Brauer, C. Habel, K.F. Wender (Eds.), Spatial Cognition II. XI, 420 pages. 2000.

Lecture Notes in Computer Science

Vol. 1782: G. Smolka (Ed.), Programming Languages and Systems. Proceedings, 2000. XIII, 429 pages. 2000.

Vol. 1783: T. Maibaum (Ed.), Fundamental Approaches to Software Engineering. Proceedings, 2000. XIII, 375 pages. 2000.

Vol. 1784: J. Tiuryn (Eds.), Foundations of Software Science and Computation Structures. Proceedings, 2000. X, 391 pages. 2000.

Vol. 1785: S. Graf, M. Schwartzbach (Eds.), Tools and Algorithms for the Construction and Analysis of Systems. Proceedings, 2000. XIV, 552 pages. 2000.

Vol. 1786: B.H. Haverkort, H.C. Bohnenkamp, C.U. Smith (Eds.), Computer Performance Evaluation. Proceedings, 2000. XIV, 383 pages. 2000.

Vol. 1787: J. Song (Ed.), Indormation Security and Cryptology – ICISC'99. Proceedings, 1999. XI, 279 pages. 2000.

Vol. 1789: B. Wangler, L. Bergman (Eds.), Advanced Information Systems Engineering. Proceedings, 2000. XII, 524 pages. 2000.

Vol. 1790: N. Lynch, B.H. Krogh (Eds.), Hybrid Systems: Computation and Control. Proceedings, 2000. XII, 465 pages. 2000.

Vol. 1792: E. Lamma, P. Mello (Eds.), AI*IA 99: Advances in Artificial Intelligence. Proceedings, 1999. XI, 392 pages. 2000. (Subseries LNAI).

Vol. 1793: O. Cairo, L.E. Sucar, F.J. Cantu (Eds.), MICAI 2000: Advances in Artificial Intelligence. Proceedings, 2000. XIV, 750 pages. 2000. (Subseries LNAI).

Vol. 1794: H. Kirchner, C. Ringeissen (Eds.), Frontiers of Combining Systems. Proceedings, 2000. X, 291 pages. 2000. (Subseries LNAI).

Vol. 1795: J. Sventek, G. Coulson (Eds.), Middleware 2000. Proceedings, 2000. XI, 436 pages. 2000.

Vol. 1796: B. Christianson, B. Crispo, J.A. Malcolm, M. Roe (Eds.), Security Protocols. Proceedings, 1999. XII, 229 pages. 2000.

Vol. 1800: J. Rolim et al. (Eds.), Parallel and Distributed Processing. Proceedings, 2000. XXIII, 1311 pages. 2000.

Vol. 1801: J. Miller, A. Thompson, P. Thomson, T.C. Fogarty (Eds.), Evolvable Systems: From Biology to Hardware. Proceedings, 2000. X, 286 pages. 2000.

Vol. 1802: R. Poli, W. Banzhaf, W.B. Langdon, J. Miller, P. Nordin, T.C. Fogarty (Eds.), Genetic Programming. Proceedings, 2000. X, 361 pages. 2000.

Vol. 1803: S. Cagnoni et al. (Eds.), Real-World Applications of Evolutionary Computing. Proceedings, 2000. XII, 396 pages. 2000.

Vol. 1805: T. Terano, H. Liu, A.L.P. Chen (Eds.), Knowledge Discovery and Data Mining. Proceedings, 2000. XIV, 460 pages. 2000. (Subseries LNAI).

Vol. 1806: W. van der Aalst, J. Desel, A. Oberweis (Eds.), Business Process Management. VIII, 391 pages. 2000.

Vol. 1807: B. Preneel (Ed.), Advances in Cryptology – EUROCRYPT 2000. Proceedings, 2000. XVIII, 608 pages. 2000.

Vol. 1811: S.W. Lee, H.. Bülthoff, T. Poggio (Eds.), Biologically Motivated Computer Vision. Proceedings, 2000. XIV, 656 pages. 2000.

Vol. 1815: G. Pujolle, H. Perros, S. Fdida, U. Körner, I. Stavrakakis (Eds.), Networking 2000 – Broadband Communications, High Performance Networking, and Performance of Communication Networks. Proceedings, 2000. XX, 981 pages. 2000.

Vol. 1816: T. Rus (Ed.), Algebraic Methodology and Software Technology. Proceedings, 2000. XI, 545 pages. 2000.

Vol. 1817: A. Bossi (Ed.), Logic-Based Program Synthesis and Transformation. Proceedings, 1999. VIII, 313 pages. 2000.

Vol. 1818: C.G. Omidyar (Ed.), Mobile and Wireless Communications Networks. Proceedings, 2000. VIII, 187 pages. 2000.

Vol. 1822: H.H. Hamilton, Advances in Artificial Intelligence. Proceedings, 2000. XII, 450 pages. 2000. (Subseries LNAI).

Vol. 1823: M. Bubak, H. Afsarmanesh, R. Williams, B. Hertzberger (Eds.), High Performance Computing and Networking. Proceedings, 2000. XVIII, 719 pages. 2000.

Vol. 1824: J. Palsberg (Ed.), Static Analysis. Proceedings, 2000. VIII, 433 pages. 2000.

Vol. 1830: P. Kropf, G. Babin, J. Plaice, H. Unger (Eds.), Distributed Communities on the Web. Proceedings, 2000. X, 203 pages. 2000.

Vol. 1831: D. McAllester (Ed.), Automated Deduction – CADE-17. Proceedings, 2000. XIII, 520 pages. 2000. (Subseries LNAI).

Vol. 1835: D. N. Christodoulakis (Ed.), Natural Language Processing – NLP 2000. Proceedings, 2000. XII, 438 pages. 2000. (Subseries LNAI).

Vol. 1839: G. Gauthier, C. Frasson, K. VanLehn (Eds.), Intelligent Tutoring Systems. Proceedings, 2000. XIX, 675 pages. 2000.

Vol. 1846: H. Lu, A. Zhou (Eds.), Web-Age Information Management. Proceedings, 2000. XIII, 462 pages. 2000.

Vol. 1848: R. Giancarlo, D. Sankoff (Eds.), Combinatorial Pattern Matching. Proceedings, 2000. XI, 423 pages. 2000.

Vol. 1849: C. Freksa, W. Brauer, C. Habel, K.F. Wender (Eds.), Spatial Cognition II. XI, 420 pages. 2000. (Subseries LNAI).

Vol. 1850: E. Bertino (Ed.), ECOOP 2000 – Object-Oriented Programming. Proceedings, 2000. XIII, 493 pages. 2000.